T0310119

ESD TESTING

ESD Series

ESD: Circuits and Devices, 2nd Edition
June 2015

ESD: Analog Circuits and Design
October 2014

Electrical Overstress (EOS): Devices, Circuits and Systems
October 2013

ESD Basics: From Semiconductor Manufacturing to Product Use
September 2012

ESD: Design and Synthesis
March 2011

ESD: Failure Mechanisms and Models
July 2009

Latchup
December 2007

ESD: RF Technology and Circuits
September 2006

ESD: Circuits and Devices
November 2005

ESD Physics and Devices
September 2004

ESD TESTING

FROM COMPONENTS TO SYSTEMS

Steven H. Voldman

IEEE Fellow, New York, USA

This edition first published 2017
© 2017 Wiley

Registered office
John Wiley & Sons Ltd, The Atrium, Southern Gate, Chichester, West Sussex, PO19 8SQ, United Kingdom

For details of our global editorial offices, for customer services and for information about how to apply for permission to reuse the copyright material in this book please see our website at www.wiley.com.

The right of the author to be identified as the author of this work has been asserted in accordance with the Copyright, Designs and Patents Act 1988.

All rights reserved. No part of this publication may be reproduced, stored in a retrieval system, or transmitted, in any form or by any means, electronic, mechanical, photocopying, recording or otherwise, except as permitted by the UK Copyright, Designs and Patents Act 1988, without the prior permission of the publisher.

Wiley also publishes its books in a variety of electronic formats. Some content that appears in print may not be available in electronic books.

Designations used by companies to distinguish their products are often claimed as trademarks. All brand names and product names used in this book are trade names, service marks, trademarks or registered trademarks of their respective owners. The publisher is not associated with any product or vendor mentioned in this book.

Limit of Liability/Disclaimer of Warranty: While the publisher and author have used their best efforts in preparing this book, they make no representations or warranties with respect to the accuracy or completeness of the contents of this book and specifically disclaim any implied warranties of merchantability or fitness for a particular purpose. It is sold on the understanding that the publisher is not engaged in rendering professional services and neither the publisher nor the author shall be liable for damages arising herefrom. If professional advice or other expert assistance is required, the services of a competent professional should be sought.

Library of Congress Cataloging-in-Publication Data

Names: Voldman, Steven H., author.
Title: ESD testing : from components to systems / Steven H. Voldman.
Other titles: Electrostatic discharge testing | ESD series.
Description: Chichester, UK ; Hoboken, NJ : John Wiley & Sons, 2016. | Series: ESD series | Includes bibliographical references and index.
Identifiers: LCCN 2016023736 (print) | LCCN 2016033086 (ebook) | ISBN 9780470511916 (cloth) | ISBN 9781118707142 (pdf) | ISBN 9781118707159 (epub)
Subjects: LCSH: Electronic circuits–Effect of radiation on. | Electronic apparatus and appliances–Testing. | Electric discharges–Detection. | Electric discharges–Measurement. | Electrostatics.
Classification: LCC TK7870.285 .V65 2016 (print) | LCC TK7870.285 (ebook) | DDC 621.3815/4–dc23
LC record available at https://lccn.loc.gov/2016023736

A catalogue record for this book is available from the British Library.

Set in 10/12pt, TimesLTStd by SPi Global, Chennai, India.
Printed and bound in Malaysia by Vivar Printing Sdn Bhd

1 2017

To My Parents
Carl and Blossom Voldman

Contents

About the Author

Dr Steven H. Voldman is the first IEEE Fellow in the field of electrostatic discharge (ESD) for "Contributions in ESD protection in CMOS, Silicon on Insulator and Silicon Germanium Technology." He received his BS in Engineering Science from the University of Buffalo (1979); a first MS EE (1981) from Massachusetts Institute of Technology (MIT); a second degree EE Degree (Engineer Degree) from MIT; an MS Engineering Physics (1986); and a PhD in electrical engineering (EE) (1991) from University of Vermont under IBM's Resident Study Fellow program.

Voldman was a member of the semiconductor development of IBM for 25 years. He was a member of the IBM's Bipolar SRAM, CMOS DRAM, CMOS logic, Silicon on Insulator (SOI), 3D memory team, BiCMOS and Silicon Germanium, RF CMOS, RF SOI, smart power technology development, and image processing technology teams. In 2007, Voldman joined the Qimonda Corporation as a member of the DRAM development team, working on 70, 58, 48, and 32 nm CMOS DRAM technology. In 2008, Voldman worked as a full-time ESD consultant for Taiwan Semiconductor Manufacturing Corporation (TSMC) supporting ESD and latchup development for 45 nm CMOS technology and a member of the TSMC Standard Cell Development team in Hsinchu, Taiwan. In 2009–2011, Steve became a Senior Principal Engineer working for the Intersil Corporation working on analog, power, and RF applications in RF CMOS, RF Silicon Germanium, and SOI. In 2013–2014, Dr Voldman was a consultant for the Samsung Electronics Corporation in Dongtan, South Korea.

Dr Voldman was chairman of the SEMATECH ESD Working Group from 1995 to 2000. In his SEMATECH Working Group, the effort focused on ESD technology benchmarking, the first transmission line pulse (TLP) standard development team, strategic planning, and JEDEC-ESD Association standards harmonization of the human body model (HBM) Standard. From 2000 to 2013, as Chairman of the ESD Association Work Group on TLP and very-fast TLP (VF-TLP), his team was responsible for initiating the first standard practice and standards for TLP and VF-TLP. Steven Voldman has been a member of the ESD Association Board of Directors and Education Committee. He initiated the "ESD on Campus" program that was established to bring ESD lectures and interaction to university faculty and students internationally; the ESD on Campus program has reached over 40 universities in the United States, Korea, Singapore, Taiwan, Senegal, Malaysia, Philippines, Thailand, India, and China. Dr Voldman teaches short courses and tutorials on ESD, latchup, patenting, and invention.

He is a recipient of 258 issued US patents and has written over 150 technical papers in the area of ESD and CMOS latchup. Since 2007, he has served as an expert witness in patent litigation and has also founded a limited liability corporation (LLC) consulting business supporting patents, patent writing, and patent litigation. In his LLC, Voldman served as an expert witness for cases on DRAM development, semiconductor development, integrated circuits, and ESD. He is presently writing patents for law firms. Steven Voldman provides tutorials and lectures on inventions, innovations, and patents in Malaysia, Sri Lanka, and the United States.

Dr Voldman also has written an article for *Scientific American* and is an author of the first book series on ESD, latchup, and EOS (nine books): *ESD: Physics and Devices*; *ESD: Circuits and Devices*; *ESD: RF Technology and Circuits*; *Latchup*; *ESD: Failure Mechanisms and Models*; *ESD: Design and Synthesis*; *ESD Basics: From Semiconductor Manufacturing to Product Use*; *Electrical Overstress (EOS): Devices, Circuits and Systems*; and *ESD: Analog Circuits and Design,* as well as a contributor to the book *Silicon Germanium: Technology, Modeling, and Design* and a chapter contributor to *Nanoelectronics: Nanowires, Molecular Electronics, and Nanodevices.* In addition, the International Chinese editions of book *ESD: Circuits and Devices*; *ESD: RF Technology and Circuits*; *ESD: Design and Synthesis*; and *ESD Basics: From Semiconductor Manufacturing to Product Use* are also released.

Preface

The book *ESD Testing: From Components to Systems* was targeted for the semiconductor process and device engineer, the circuit designer, the ESD/latchup test engineer, and the ESD engineer. In this book, a balance is established between the technology and testing.

The first goal of this book is to teach the ESD models used today. There are many ESD test models, and more types are being developed today and in the future.

The second goal is to show recent test systems and test standards. Significant change in both the test methodologies and issues are leading to proposal of new ESD models, introduction of new standards, and an impact on product diversity and product variety.

The third goal is to expose the reader to the growing number of new testing methodologies, concepts, and equipment. In this book, commercial test equipment is shown as an example to demonstrate the "state-of-the-art" of ESD testing. Significant progress has been made in recent years in ESD, EOS, and EMC.

The fourth goal, as previously done in the ESD book series, is to teach testing as an ESD design practice. ESD testing can be used as a design methodology or an ESD tool. ESD testing can lead to understanding of the fundamental practices of ESD design and the ESD design discipline. This practice uses ESD testing for "de-bugging" and diagnosis.

The fifth goal is to provide a book that can view the different test methods independently. Each chapter is independent so that the reader can study or read about a test model independent of the other test models.

The sixth goal is to provide a text where one can compare the interrelationship between one ESD model and another ESD model. In many cases, there is commonality between the test waveform, the test procedure, and even failure mechanisms.

The seventh goal is to provide a text structure similar to a standard or standard test method, but read easier than reading a standard document. The goal was also to reduce the level of details of the standard to simplify the understanding.

The book *ESD Testing: From Components to Systems* consists of the following:

Chapter 1 introduces the reader to fundamentals and concepts of the electrostatic discharge (ESD) models and issues.

Chapter 2 discusses the human body model (HBM). It discusses the purpose, scope, waveforms, test procedures, and test systems. In this chapter, both the wafer-level and product-level test methodologies are discussed. This chapter includes HBM failure mechanisms to circuit solutions. Alternative test methodologies such as sampling and split fixture methods are reviewed.

Chapter 3 discusses the machine model (MM). It discusses the purpose, scope, waveforms, test procedures, and test systems. In this chapter, both the wafer-level and product-level test methodologies are discussed. This chapter includes MM failure mechanisms to circuit solutions. Alternative test methodologies such as the small charge model (SCM) are discussed. In addition, correlation relations of HBM to MM ratio are analyzed and reviewed.

Chapter 4 discusses the charged device model (CDM). It discusses the purpose, scope, waveforms, CDM test procedures, and CDM test systems. This chapter includes CDM failure mechanisms to circuit solutions to avoid CDM failures. Alternative test methodologies such as the socketed device model (SDM) and charged board model (CBM) are discussed.

Chapter 5 discusses the transmission line pulse (TLP) methodology and its importance in the semiconductor industry and ESD development. It discusses the purpose, scope, waveforms, TLP pulsed *I–V* characteristics, TLP test procedures, and TLP test system configurations. TLP current source, time domain reflection (TDR), time domain transmission (TDT), and time domain reflection and transmission (TDRT) configurations is explained

Chapter 6 discusses the very fast transmission line pulse (VF-TLP) methodology. It discusses the purpose, scope, waveforms, VF-TLP pulsed *I–V* characteristics, VF-TLP test procedures, and VF-TLP test system configurations. Alternative test methods such as ultra fast transmission line pulse (UF-TLP) are discussed.

Chapter 7 discusses the system-level method, known as IEC 61000-4-2. It discusses the purpose, scope, IEC 61000-4-2 waveforms, IEC 61000-4-2 table configurations, and requirements. Failure mechanisms and circuit solutions to avoid failures are explained.

Chapter 8 discusses the human metal model (HMM) method. The HMM model has many similarities to the system-level method, known as IEC 61000-4-2. It discusses the purpose, scope, waveforms, HMM table configurations, and requirements as well as the distinctions and commonality to the IEC 61000-4-2 test method.

Chapter 9 discusses the system-level transient surge method, known as IEC 61000-4-5. It discusses the purpose, scope, IEC 61000-4-5 waveforms, IEC 61000-4-5 table configurations, and requirements. Failure mechanisms and circuit solutions to avoid failures are explained. The distinction from the IEC 61000-4-2 is highlighted.

Chapter 10 discusses the cable discharge event (CDE) method. It discusses the purpose, scope, waveforms, cable configurations, and impact on the pulse event. Examples of cable-induced failures are given, as well as circuit- and system-level solutions to avoid chip and system failures.

Chapter 11 discusses latchup. It addresses latchup testing, characterization, and design. It also addresses latchup test techniques for product-level testing. Technology benchmarking to ground rule development is also briefly discussed.

Chapter 12 discusses electrical overstress (EOS). It focuses on electrical and thermal safe operating area (SOA) and how EOS occurs. It also focuses on how to distinguish latchup from EOS events.

Chapter 13 discusses electromagnetic compatibility (EMC). It addresses ESD and EMC testing and characterization methods. It also serves as a brief introduction to this large subject matter.

Hopefully, the book covers the trends and directions of ESD testing discipline.
Enjoy the text, and enjoy the subject of ESD testing.

B"H
Steven H. Voldman
IEEE Fellow

Acknowledgments

I would like to thank the individuals who have helped me learn about experimental work, high current testing, high voltage testing, electrostatic discharge (ESD) testing, electrical overstress (EOS), and standards development. In the area of ESD, EOS, and latchup testing, I would like to thank for all the support received from SEMATECH, the ESD Association, and the JEDEC organizations.

I would like to thank the SEMATECH organization for allowing me to establish the SEMATECH ESD Work Group: this work group initiated the ESD technology benchmarking test structures, the JEDEC-ESD Association collaboration on ESD standard development, alternate test methods, and most important, the initiation of the transmission line pulse (TLP) standard development.

I thank the ESD Association ESD Work Group (WG) standard committees for many years of discussion on standard developments and on human body model (HBM), machine model (MM), charged device model (CDM), cable discharge event (CDE), human metal model (HMM), TLP testing, and very fast transmission line pulse (VF-TLP) testing. I also thank the ESD Association Standards Development Work Group 5.5 TLP testing committee. We were very fortunate to have a highly talented and motivated team to rapidly initiate the TLP and VF-TLP documents for the semiconductor industry; this included for the development of the TLP and VF-TLP standards, which was a significant accomplishment that has influenced the direction of ESD testing. I am thankful to my colleagues Robert Ashton, Jon Barth, David Bennett, Mike Chaine, Horst Gieser, Evan Grund, Leo G. Henry, Mike Hopkins, Hugh Hyatt, Mark Kelly, Tom Meuse, Doug Miller, Scott Ward, Kathy Muhonen, Nathaniel Peachey, Jeff Dunihoo, Keichi Hasegawa, Jin Min, Yoon Huh, and Wei Huang. I am also thankful to Tze Wee Chen of Stanford University for discussions on the ultra-fast transmission line pulse (UF-TLP) testing.

I am grateful to the Oryx Instrument ESD test development team for years of ESD test support and the Thermo Fisher Scientific team of David Bennett, Mike Hopkins, Tom Meuse, Tricia Rakey, and Kim Baltier. My sincere thanks goes to Jon Barth of Barth Electronics for usage of the images of the Barth test equipment for this text; Keichi Hasegawa of Hanwa Electronics for the images of the Hanwa test equipment; Yoon Huh and Jin Min of Amber Precision Instruments for the scanning images and the test equipment; Wei Huang for the images of the ESDEMC test equipment; Jeff Dunnihoo of Pragma Design Inc for the current reconstruction method images; the HPPI corporation for images of its TLP test equipment; and Chris O'Connor of UTI Inc. for transient latchup analysis.

I would like to thank the JEDEC organization's ESD committee.

This work was supported by the institutions that allowed me to teach and lecture at conferences, symposiums, industry, and universities; this gave me the motivation to develop the texts. I would like to thank for the years of support and the opportunity to provide lectures, invited talks, and tutorials at the International Physical and Failure Analysis (IPFA) in Singapore, the Electrical Overstress/Electrostatic Discharge (EOS/ESD) Symposium, the International Reliability Physics Symposium (IRPS), and the Taiwan Electrostatic Discharge Conference (T-ESDC), International Conference on Solid State and Integrated Circuit Technology (ICSICT), and ASICON.

Finally, I am immensely thankful to the ESD Association office for the support in the area of publications, standards developments, and conference activities – Lisa, Christine, and Terry. I also thank the publisher and staff of John Wiley and Sons, for including the text *ESD Testing: From Components to Systems* as part of the ESD book series.

To my children, Aaron Samuel Voldman and Rachel Pesha Voldman, good luck to both of you in the future.

And Betsy H. Brown, for her support on this text…

And of course, my parents, Carl and Blossom Voldman.

B"H
Dr Steven H. Voldman
IEEE Fellow

1

Introduction

1.1 Testing for ESD, EMI, EOS, EMC, and Latchup

In the electronics industry, testing of components and systems is a part of the process of qualifying and releasing products. Standards are established to provide methodology, process, and guidance to quantify the technology issue [1–14]. Testing is performed to evaluate the sensitivity and susceptibility of products to electric, magnetic, and electromagnetic events. These can be categorized into electrostatic discharge (ESD) [1–12], electrical overstress (EOS), electromagnetic interference (EMI), and electromagnetic compatibility (EMC) events, and latchup (Figure 1.1) [13]. In the electronic industry, tests and procedures have been established to quantify the influence of these events on components and systems associated with ESD, EOS, EMC, and latchup [15–24].

1.2 Component and System Level Testing

In the testing of electronics, different tests and procedures were established that tested components, and other tests for testing of systems. These tests have been established based on the environment that the components and systems experience in processing, assembly, shipping, to product use [1–24].

Figure 1.2 shows examples of component tests that are applied to wafer level, packaged and unpackaged products. Today, it is common to test semiconductor components for the following standards. These include the human body model (HBM) [1], machine model (MM) [2, 3], charged device model (CDM) [4, 5], to transmission line pulse (TLP) [6, 7], and very fast transmission line pulse (VF-TLP) [8, 9]. In the future chapters, these tests are discussed in depth.

Figure 1.3 shows examples of system level tests that are applied to systems to address the robustness to environments that the systems may experience in product use. For system level tests, it is now common to test for the IEC 61000-4-2 [10], human metal model (HMM) [11], IEC 61000-4-5 [12], and cable discharge events (CDE).

ESD Testing: From Components to Systems, First Edition. Steven H. Voldman.
© 2017 John Wiley & Sons, Ltd. Published 2017 by John Wiley & Sons, Ltd.

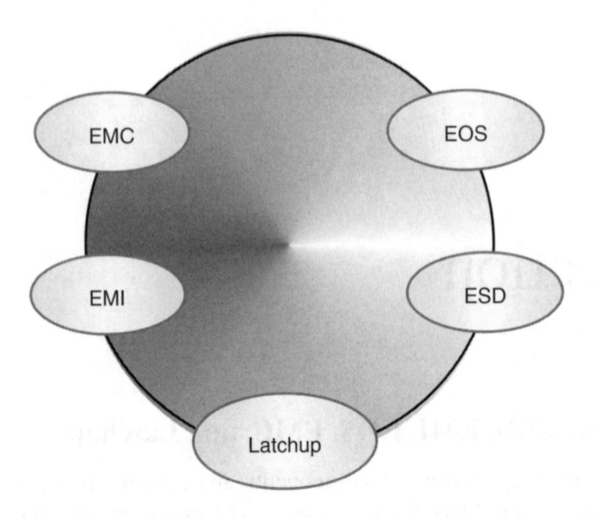

Figure 1.1 ESD, EMI, EOS, and EMC

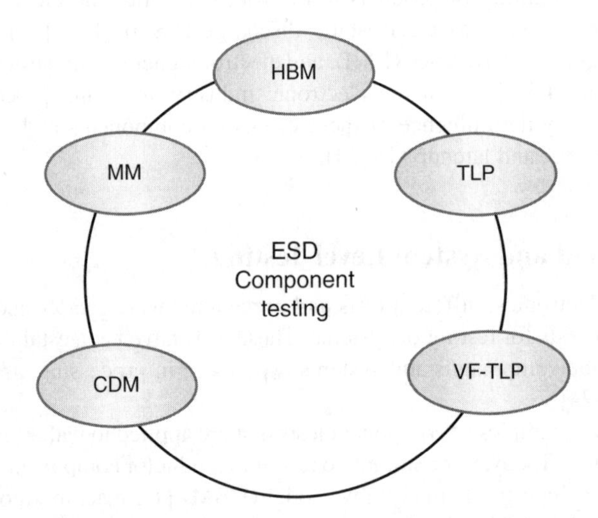

Figure 1.2 Component tests

1.3 Qualification Testing

Many of the tests are used for different purposes. Some electrical tests are established for characterization, whereas other tests have been established for qualification of components or systems. Qualification tests are performed to guarantee or insure quality and reliability in the system, or in the field. Figure 1.4 shows examples of qualification tests that are performed in the electronic industry. These qualification tests include standard practice (SP) documents, to standard test method (STM).

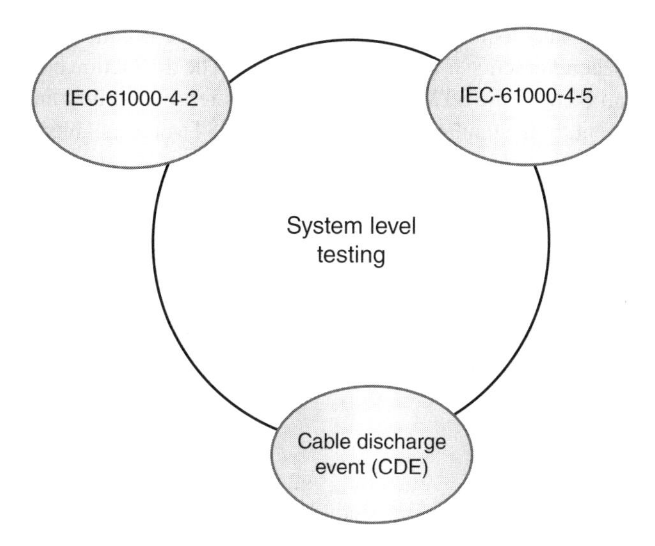

Figure 1.3 System level tests

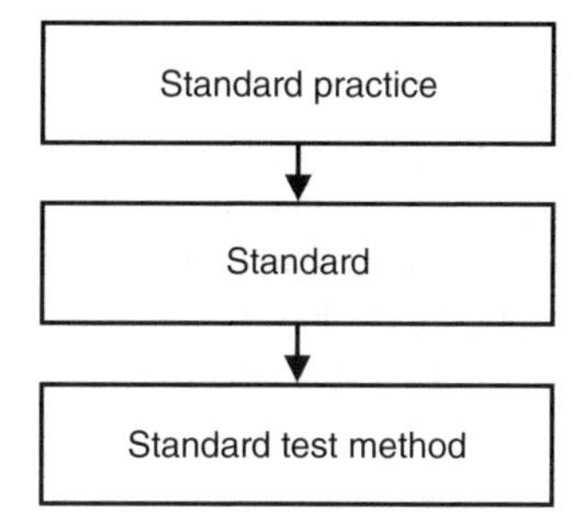

Figure 1.4 Qualification testing

1.4 ESD Standards

In the development of these qualification processes, different types of documents and processes are established. In standards development, practices and processes are established for the quality, reliability, and release of products to customers.

1.4.1 Standard Development – Standard Practice (SP) and Standard Test Methods (STMs)

In the development of these qualification processes, a standard practice is established for testing of components and systems. A standard practice (SP) is a procedure or process that is established for testing. The document for the standard practice is called the standard practice (SP) document. A second practice is to establish an STM. The distinction between the standard practice (SP) and an STM is the STM procedure insures reproducibility and repeatability. In standards development, both standard practices (SP) and STM are established for the quality, reliability, and release of products to customers.

1.4.2 Repeatability

In STM development, repeatability is an important criterion in order to have a process elevate from a standard practice to an STM. It is important to know that if a test is performed, the experimental results are repeatable (Figure 1.5).

1.4.3 Reproducibility

In STM development, reproducibility is a second important criterion in order to have a process elevate from a standard practice (SP) to an STM. Reproducibility is key to verify that the experimental results can be reproduced (Figure 1.6).

1.4.4 Round Robin Testing

In order to determine if a standard practice can be elevated to an STM, reproducibility and repeatability are evaluated in a process known as Round Robin (RR) process. Statistical analysis is initiated to determine the success or failure of reproducibility and repeatability as part of the experimental methodology. RR is an interlaboratory test that can include measurement, analysis or performing an experiment. This process can include a number of

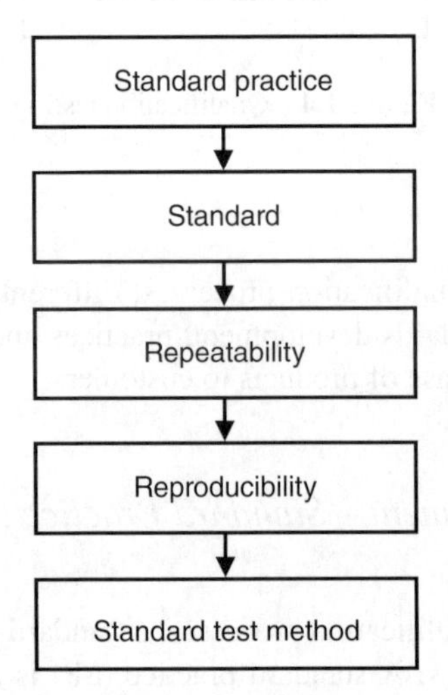

Figure 1.5 Repeatability and Reproducibility

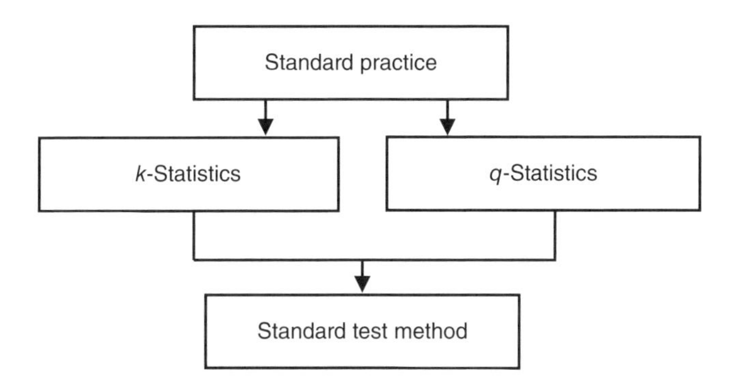

Figure 1.6 SP to STM Process

independent scientists and independent laboratories. In the case of ESD and EOS testing, different commercial test equipment is used in the process. To assess the measurement system, the statistics of analysis of variance (ANOVA) random effects model is used.

1.4.5 Round Robin Statistical Analysis – k-Statistics

In the RR process, the within-laboratory consistency statistics is known as the *k*-statistics. The *k*-statistics is the quotient of the laboratory standard deviation and the mean standard deviation of all the laboratories. These can be visualized using Mandel statistics and Mandel plots. Mandel's *k* is an indicator of the precision compared to the pooled standard deviation across all groups. Mandel's *k* plot is represented by a bar graph (Figure 1.7).

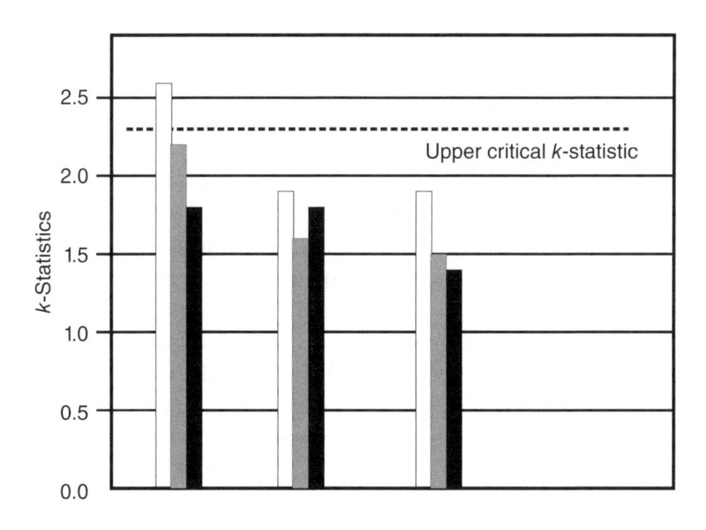

Figure 1.7 Mandel *k*-statistics plot

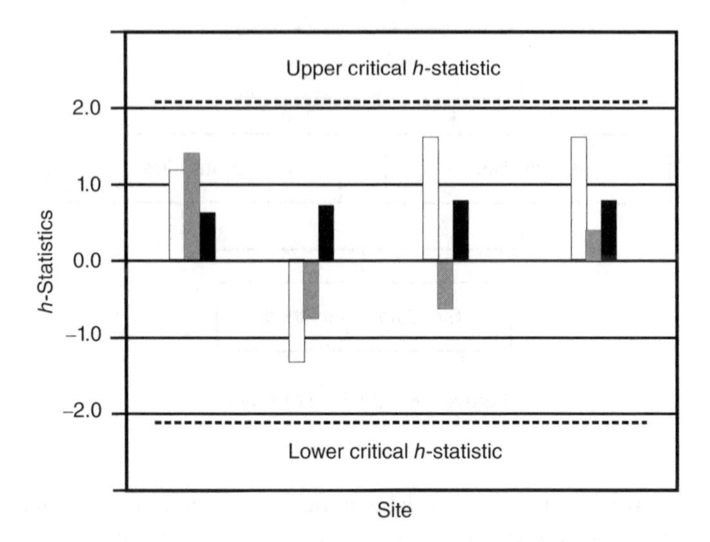

Figure 1.8 Mandel h-statistics plot

1.4.6 Round Robin Statistical Analysis – h-Statistics

In the RR process, the between-laboratory consistency statistics is known as the h-statistics. The h-statistics is the ratio of the difference between the laboratory mean and the mean of all the laboratories, and the standard deviation of the means from all the laboratories. These can be visualized using Mandel statistics and Mandel plots. Mandel's statistics are traditionally plotted for interlaboratory study data, grouped by laboratory to give a graphical view of laboratory bias and precision. Mandel's h-plots are bar graphs around a zero axis (Figure 1.8).

1.5 Component Level Standards

Today, in the semiconductor industry, components are tested to the HBM, MM, and CDM [1–5]. These tests are traditionally done on packaged components. In these tests, the components are also unpowered. For the qualification of semiconductor components for over 20 years, the HBM, MM, and CDM tests were completed prior to shipping components to a customer or system developer. In addition, latchup qualification was required in the shipping of components since the 1980s time frame [13]. A new test of components includes the HMM test to evaluate the influence of the components on system level tests.

In the 1990s, TLP testing became popular and is now a common characterization practice in the semiconductor industry. TLP testing did not have an established methodology until the year 2003 [6]. TLP testing is performed on test structures, circuits, to components. This was followed by a second test method known as the VF-TLP testing method [8]. VF-TLP testing is also completed on test structures to components. These TLP tests can also be performed on systems.

1.6 System Level Standards

System level standards are to address the sensitivity and susceptibility of electronic systems in shipping, handling, and usage environment. The objective is to simulate events that can occur. A distinction between many of the component standard tests and the system standards is the failure criteria. In system testing, the failure can be nondestructive and destructive. System level interrupts and disturbs can be regarded as a system failure.

System testing can be evaluated with all the modules of the system assembled or unassembled. System level testing can be with or without the cable connections between the system modules.

System level standards can include direct current (DC), alternating current (AC), pulse events, as well as transient phenomena. In the text, the system level tests known as IEC 61000-4-2 and IEC 61000-4-5 are discussed. In addition, CDE is discussed from charged cables.

1.7 Factory and Material Standards

In the semiconductor industry, there exist ESD and EOS standards for the factory and assembly environments [10]. ESD concerns in manufacturing are a combination of the materials, tooling, and the human factors. The materials influence the triboelectric charge transfer. The tooling used can lead to charge transfer, and operators can participate in this transfer process.

In the manufacturing area, the electric field between the ceiling and the floor is influenced by the height of the ceiling, air flow, and placement of the ionizers. The placement of the ionizers relative to the work surface where the sensitive parts are placed influences the effectiveness of the ionizers. The work surface material and its physical size is also a factor.

Operators in the manufacturing line can influence triboelectric charging process. All external surfaces of the operator, type of materials, and proximity of the operator to the item can influence the charge transfer, and the human-induced electric field imposed. The footwear, garments, wrist straps, and personnel grounding of the garments can all influence the impact of the operator on tribocharging and the ESD discharge. In addition, seating, position, and distance of the operator from the sensitive parts can also influence the electric fields.

The choice of ESD materials in a manufacturing environment can have a large effect on the ESD protected area (EPA). The material choice can influence its initial conductivity, as well as the conductivity as a function of time. Material coatings and cleaning processes can influence the material conductivity. The wearout of a floor or garment can influence its global conductivity as well as its spatial variation. It is these factors why it is important to qualify a manufacturing environment, establish a measurement set of procedures, and temporal audits of the items used in the manufacturing sector.

The manufacturing environment consists of the following categories (Figure 1.9) [23]:

- Grounding and bonding systems
- Work surfaces
- Wrist straps
- Monitors
- Footwear

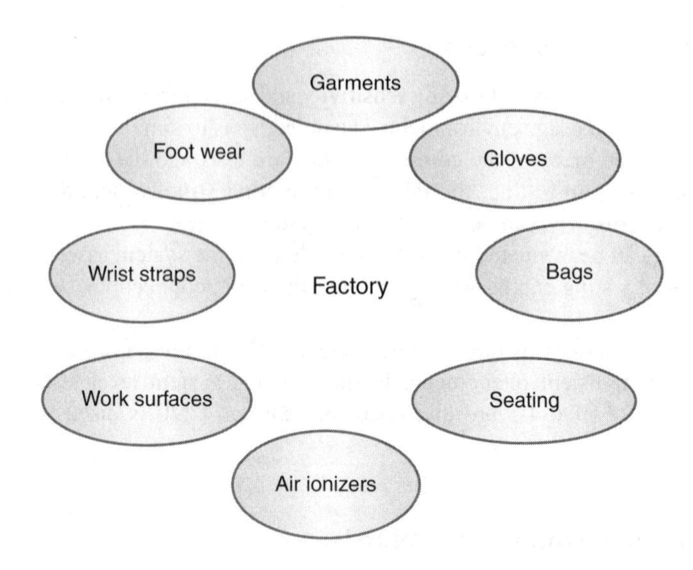

Figure 1.9 Factory and material standards

- Flooring
- Personnel grounding with garments
- Ionizers
- Seating
- Mobile equipment
- Packaging.

Manufacturing test equipment is needed for evaluation of compliance to specifications. ESD test equipment includes the following [23]:

- DC ohmmeter
- Electrodes
- Handheld electrodes
- Foot electrode
- AC outlet analyzer
- AC circuit tester (impedance meter)
- Insulative support surface
- Charged plate monitor

For all these items, it will be required to verify the electrical measurements. In order to verify compliance, electrical measurements will be the means of determining an "ESD safe" environment and compliance with objectives.

1.8 Characterization Testing

In component and system evaluation, characterization of the electrical characteristics is also done during the product development, assembly, and shipping [22].

1.8.1 Semiconductor Component Level Characterization

In component evaluation, characterization of the electrical characteristics is also done during the product development, assembly, and shipping [22]. These can be performed on a component level in packaged and unpackaged form.

1.8.2 Semiconductor Device Level Characterization

Characterization of the electrical characteristics can be completed on a semiconductor device level. ESD testing can be performed on individual devices in a semiconductor technology to evaluate the electrical characteristics, electrical response, failure mechanisms, and ESD robustness [15–25].

1.8.3 Wafer Level ESD Characterization Testing

Characterization of the electrical characteristics of semiconductor devices is typically performed on a wafer level [22, 23]. For wafer level ESD characterization, a probe station and adequate probes are required for ESD testing. Commercial test systems now accommodate wafer level testing. Wafer level ESD testing can be performed on individual devices in a semiconductor technology to evaluate the electrical characteristics, electrical response, failure mechanisms, and ESD robustness. Semiconductor devices can be integrated into bond pad sets to probe the components (Figure 1.10).

In semiconductor development, ESD networks can also be built into bond pad sets for development of ESD networks. These can be constructed with or without electrical circuits attached.

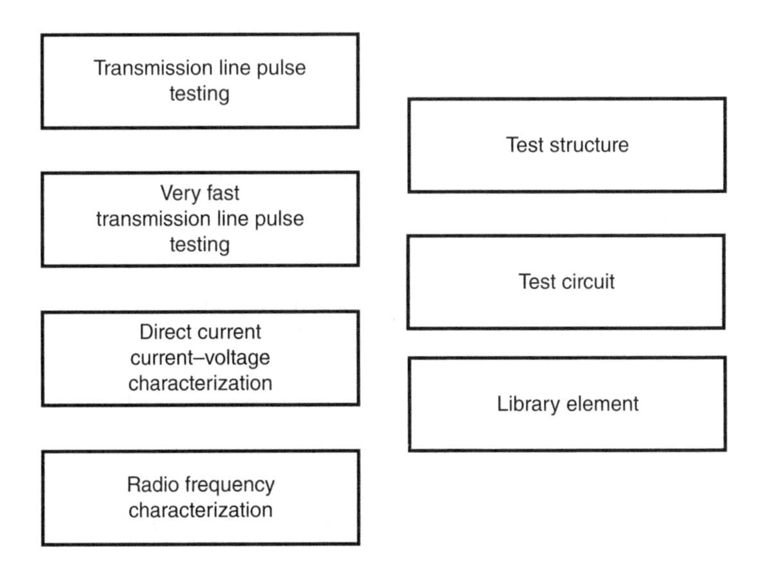

Figure 1.10 Semiconductor chip level characterization

1.8.4 Device Characterization Tests on Circuits

Characterization of the electrical characteristics of circuits can be evaluated on a wafer level. For wafer level ESD characterization, ESD networks integrated with input/output (I/O) circuits to evaluate the ESD network performance in protection of I/O circuits [16–18, 22–25]. Experiments can be performed with and without the I/O network or with and without the ESD network. Wafer level ESD testing can be performed on these test circuits in a semiconductor technology to evaluate the electrical characteristics, electrical response, failure mechanisms, and ESD robustness.

ESD network libraries and I/O libraries can be constructed on a wafer level for characterization or qualification for a given technologies [22, 23]. As part of the technology qualification, ESD networks can be evaluated, as well as integrated with the I/O library. For example, in an ASIC technology, the ESD networks are integrated with the I/O circuits. ESD and latchup testing of all the ASIC technology I/O libraries can be evaluated as part of an ASIC qualification.

1.8.5 Device Characterization Tests on Components

Device level characterization can be performed on components. In the testing on components, the pin type and pin combination play a role in the evaluation process.

1.8.5.1 Pin Combinations

In the testing of components, various pin combinations are established to test signal pins and supply pins. In the test of pins, various combinations exist:

- Signal pin to supply rail
- Supply rail to supply rail
- Signal pin to signal pin.

In the semiconductor chip, to avoid failure mechanisms from these cases, ESD networks are placed on signal pins, as well as bidirectional ESD networks on power rails and between power rails.

In multidomain semiconductor chips, combinations exist between domains. Domain-to-domain testing exists as follows:

- Signal pin of a first domain to supply rail of a second domain
- Supply rail of a first domain to supply rail of a second domain
- Signal pin of a first domain to signal pin of a second domain.

In all these cases, in order to avoid failure mechanisms, additional ESD networks are placed between the power domains.

1.8.5.2 Interrelation Between Standards, Pin Combination, and Chip Architecture

In the development of semiconductor chips, failure mechanisms are discovered for different modes of ESD testing. To avoid or eliminate these failure mechanisms, new ESD structures were placed between power supplies and between domains. Hence, there is a synergy between the pin combinations test, the standards, and the chip architecture.

1.8.6 System level Characterization on Components

System level testing is completed on printed circuit boards (PCB), shield assemblies, to assembled systems. Component failure can occur during system level testing for external ports or exposed pins coupled to the system. As a result, system developers have begun to request system level tests on semiconductor components. One standard practice that has been developed is the HMM. This is discussed later in the text.

1.8.7 Testing to Standard Specification Levels

ESD testing can be completed during the characterization and qualification process to different sensitivity levels. ESD testing levels can be used to quantify that components achieve different "classes" of protection levels. As a result, a method of testing is to test in large increments to the different ESD classification levels. For example, to qualify that a component achieves a 2000 V ESD protection level, ESD testing can be completed to see if the component survives a 2 kV HBM event without any pin failures in the test specification. In this case, testing to the standard specification levels demonstrates survivability to a fixed level without pin failures.

In some testing methods, testing continues to the next classification level to see if it is above the next ESD test level. In this manner, the test is continued to the next level and continues *ad infinitum* until the first indication of at least one pin failure occurs. Testing is stopped at the first level with at least one pin failure.

1.8.8 Testing to Failure

In some testing methods, testing is continued to ESD failure of all pins (e.g., all signal pins). In this method, small incremental steps (much less than the classification levels) are made and the testing continues until all the pins have demonstrated failure, based on some failure criteria. In this manner, actual failure levels of all pins are achieved [22, 23].

There are many advantages to this method. A first advantage of this methodology is that the experimental results of all pins (e.g., I/O circuits) can be quantified. In this manner, the quantification of the ESD robustness of each pin is evaluated to distinguish "weak pins" from "strong pins." In the case of multiple design passes of a component, the net improvement can be quantified when a circuit (or ESD network) is modified. Second, in the case of semiconductor manufacturing variations, the change in manufacturing variables can be evaluated.

A second advantage is that the failure mechanism of each pin is observable. The failure mechanism can be quantified and recorded for manufacturing control or future design improvements. A third advantage is that statistical analysis can be performed on the product to determine the mean, standard deviation, and nature of the distribution of pins (e.g., Gaussian, non-Gaussian).

Testing in this manner can lead to an acceleration of product learning for the technology and future technologies.

1.9 ESD Library Characterization and Qualification

In many release of new technologies, a "library" of ESD circuits are released to the circuit design teams to use for product applications. In the qualification of a technology, ESD libraries can be characterized to determine appropriateness of different I/O circuits. In this process, the ESD library characterization can be tested according to the following characterization methods and standards:

- TLP test [7]
- VF-TLP test [8]
- HBM test [1]
- MM test [2].

TLP and VF-TLP testing can provide high current *I–V* characteristics of the ESD elements. HBM and MM testing can provide survivability of the ESD circuits to different HBM and MM ESD levels.

1.10 ESD Component Standards and Chip Architectures

Chip architectures can significantly influence ESD results. Over many years of learning and evaluation of ESD failure mechanisms, ESD component standards were modified to capture specific failures related to the chip architecture. With these discoveries, ESD circuits were modified to allow passing of the new ESD standard tests within the ESD standard. Hence, there was a synergy between the ESD failure mechanisms, ESD circuits, and standards development (Figure 1.11).

1.10.1 Relationship Between ESD Standard Pin Combinations and Failure Mechanisms

A number of failure mechanisms occurred in semiconductor chips due to the chip architecture.

A common ESD failure was associated with a signal pin to a power rail within a second domain or that was electrically isolated from that signal pin. ESD standards required testing between a signal pin and a power rail, even when the power rail was in another domain or region of the semiconductor chip. The definition of what is the same power rail was also leading to a definition of how much resistance between "independent power rails." ESD networks were added between power rails to allow current flow (e.g., bidirectional ground-to-ground ESD networks). ESD power clamp networks were added between V_{DD} and V_{SS} within a domain to provide electrical connectivity.

A second common ESD failure was associated with signal pin to signal pin. ESD standards required "pin-to-pin" testing to capture the failure mechanism associated with the lack of a current path between two pins. ESD networks were added between power rails to allow current flow (e.g., bidirectional ground-to-ground ESD networks).

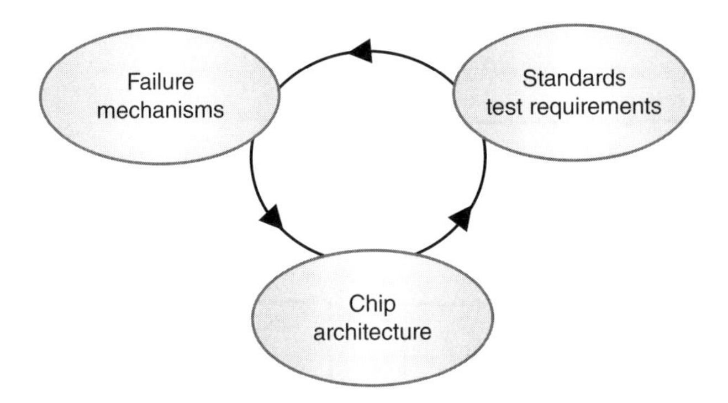

Figure 1.11 Failure mechanisms, standards, and chip architecture

1.10.2 Relationship Between ESD Standard Pin Combinations and Chip Architecture

Over many years of learning and evaluation of ESD failure mechanisms associated with these ESD standard pin combinations, the semiconductor chip architecture evolved to provide current paths between power rails of different power domains. Circuit design architecture separated power rails to provide noise isolation, but this led to ESD failures. Hence, the evolution of the semiconductor chip industry was to "recouple" the power grids and introduce ESD circuits to eliminate failure mechanisms associated with the chip architecture.

1.11 System Level Characterization

In system level analysis, system level characterization tests exist for the evaluation of system products. System level testing continues to be a growing field to address EOS and ESD failures in systems [21, 24]. Figure 1.12 provides examples of system level characterization tests.

1.12 Summary and Closing Comments

In this chapter, an introduction was given to the world of ESD testing. This includes a survey of different ESD tests, test procedures, devices, ESD networks, elements, to current paths. This will prepare the reader for the future chapters, which gets into specific tests in more detail.

In Chapter 2, one of the most important tests in the history of ESD testing is discussed. This chapter focuses on the HBM test, and in addition indicates many of the present and future anticipated changes that will occur on this test.

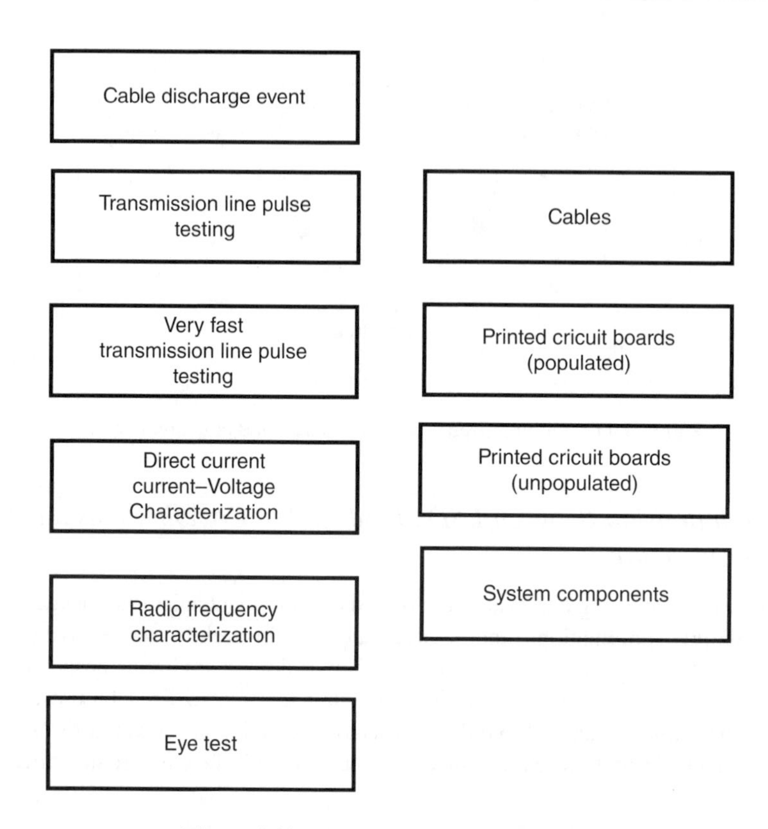

Figure 1.12 System level characterization

Problems

1. How do people charge that leads to a spark at a metal door knob? Explain the triboelectric charging process. Are people capacitors? Where is the charge stored?
2. Explain the discharge process when a charged person touches a metal door knob or object.
3. Is the process of discharging the same all the time? What influences the discharge process? How many issues influence the discharge process? List at least four of them.
4. Why is repeatability and reproducibility important for establishing a standard test method? What if a method is not repeatable or reproducible?
5. What are the advantages of wafer level testing? List four variables.
6. What cannot be achieved on a wafer level? What are the issues?
7. What is the reason for component level testing? What does it provide for the foundry that produces them?
8. Does component level testing help system level developers? What it cannot provide?
9. How do you determine the ESD robustness of a technology? How do you quantify the ESD robustness of one technology generation to a second generation? How do you determine the effect of technology scaling?

10. If you were an ESD engineer in a foundry, how would you structure the entire business process for delivering semiconductor components? How do you protect the interest of your foundry? How do you protect the interest of the customer?
11. How does one take into account manufacturing variation and distribution versus ESD latent damage? Can you tell the difference?

References

1. ANSI/ESD ESD-STM 5.1 – 2007. *ESD Association Standard Test Method for the Protection of Electrostatic Discharge Sensitive Items – Electrostatic Discharge Sensitivity Testing – Human Body Model (HBM) Testing – Component Level*. Standard Test Method (STM) document, 2007.
2. ESD Association. *ESD Association Standard Test Method for the Protection of Electrostatic Discharge Sensitive Items – Electrostatic Discharge Sensitivity Testing – Machine Model (MM) Testing – Component Level*, 1994.
3. ANSI/ESD ESD-STM 5.2 – 1999. *ESD Association Standard Test Method for the Protection of Electrostatic Discharge Sensitive Items – Electrostatic Discharge Sensitivity Testing – Machine Model (MM) Testing – Component Level*. Standard Test Method (STM) document, 1999.
4. ANSI/ESD ESD-STM 5.3.1 – 1999. *ESD Association Standard Test Method for the Protection of Electrostatic Discharge Sensitive Items – Electrostatic Discharge Sensitivity Testing – Charged Device Model (CDM) Testing – Component Level*. Standard Test Method (STM) document, 1999.
5. JEDEC. JESD22-C101-A. *A Field-Induced Charged Device Model Test Method for Electrostatic Discharge-Withstand Thresholds of Microelectronic Components*, 2000.
6. ANSI/ESD Association. ESD-SP 5.5.1-2004. *ESD Association Standard Practice for the Protection of Electrostatic Discharge Sensitive Items – Electrostatic Discharge Sensitivity Testing – Transmission Line Pulse (TLP) Testing Component Level*. Standard Practice (SP) document, 2004.
7. ANSI/ESD Association. ESD-STM 5.5.1 – 2008. *ESD Association Standard Test Method for the Protection of Electrostatic Discharge Sensitive Items – Electrostatic Discharge Sensitivity Testing – Transmission Line Pulse (TLP) Testing Component Level*. Standard Test Method (STM) document, 2008.
8. ESD Association. ESD-SP 5.5.2. *ESD Association Standard Practice for the Protection of Electrostatic Discharge Sensitive Items – Electrostatic Discharge Sensitivity Testing Very Fast Transmission Line Pulse (VF-TLP) Testing Component Level*. Standard Practice (SP) document, 2007.
9. ANSI/ESD Association. ESD-SP 5.5.2 – 2007. *ESD Association Standard Practice for the Protection of Electrostatic Discharge Sensitive Items – Electrostatic Discharge Sensitivity Testing – Very Fast Transmission Line Pulse (VF-TLP) Testing Component Level*. Standard Practice (SP) document, 2007.
10. International Electro-technical Commission (IEC). *IEC 61000-4-2 Electromagnetic Compatibility (EMC): Testing and Measurement Techniques – Electrostatic Discharge Immunity Test*, 2001.
11. ESD Association. ESD-SP 5.6 – 2008. *ESD Association Standard Practice for the Protection of Electrostatic Discharge Sensitive Items – Electrostatic Discharge Sensitivity Testing – Human Metal Model (HMM) Testing Component Level*. Standard Practice (SP) document, 2008.
12. International Electro-technical Commission (IEC). *IEC 61000-4-5 Electromagnetic Compatibility (EMC): Part 4–5: Testing and Measurement Techniques – Surge Immunity Test*, 2000.
13. JEDEC Standard. *IC Latch-Up Test*. JESD78A, 1997.
14. ANSI/ESD SP 5.1.2 – 2006 *ESD Association Standard Practice for the Protection of Electrostatic Discharge Sensitive Items – Human Body Model (HBM) and Machine Model (MM) Alternative Test Method: Split Signal Pin-Component Level*, 2006.
15. S. Voldman. *ESD: Physics and Devices*. Chichester, England: John Wiley and Sons, Ltd., 2004.
16. S. Voldman. *ESD: Circuits and Devices*. Chichester, England: John Wiley and Sons, Ltd., 2005.
17. S. Voldman. *ESD: Circuits and Devices 2nd Edition*. Chichester, England: John Wiley and Sons, Ltd., 2015.
18. S. Voldman. *ESD: RF Circuits and Technology*. Chichester, England: John Wiley and Sons, Ltd., 2006.
19. S. Voldman. *Latchup*. Chichester, England: John Wiley and Sons, Ltd., 2007.
20. S. Voldman. *ESD: Circuits and Devices*. Beijing, China: Publishing House of Electronic Industry (PHEI), 2008.

21. S. Voldman. *ESD: Failure Mechanisms and Models*. Chichester, England: John Wiley and Sons, Ltd., 2009.
22. S. Voldman. *ESD: Design and Synthesis*. Chichester, England: John Wiley and Sons, Ltd., 2011.
23. S. Voldman. *ESD Basics: From Semiconductor Manufacturing to Product Use*. Chichester, England: John Wiley and Sons, Ltd., 2012.
24. S. Voldman. *Electrical Overstress (EOS): Devices, Circuits, and Systems*. Chichester, England: John Wiley and Sons, Ltd., 2013.
25. S. Voldman. *ESD: Analog Design and Circuits*. Chichester, England: John Wiley and Sons, Ltd., 2014.

2

Human Body Model

The human body model (HBM) is the most widely established standard for the qualification and release of semiconductor components in the semiconductor industry. The HBM test is integrated into the qualification and release process of the quality and reliability teams for components in corporations and foundries. This chapter is dedicated to the HBM model [1–60]. The HBM standard is ANSI/ESD ESD-STM 5.1 – 2007 – ESD Association Standard Test Method for the Protection of Electrostatic Discharge Sensitive Items – Electrostatic Discharge Sensitivity Testing – Human Body Model (HBM) Testing – Component Level. Standard Test Method (STM) document, 2007 [1].

The HBM is regarded as an electrostatic discharge (ESD) event, not an electrical overstress (EOS) event because it is charge transfer-related source, with a short pulse event. The model was intended to represent the interaction of the electrical discharge from a human being, who is charged, with a component, or an object. The model assumes that the human being is the initial condition. The charged source then touches a component or an object using a finger. The physical contact between the charged human being and the component or object allows for current transfer between the human being and the object.

2.1 History

The HBM became of interest in early days in the mining industry in the 1950s. In the Bureau of Mines, investigation reports discussed the issue of electrostatic in the mining industry. A first publication was by P.G. Guest, V.W. Sikora, and B.L. Lewis as the *Bureau of Mines, Report of Investigation 4833*, U.S. Department of Interior, January 1952 [4]. A second article of interest was published by D. Bulgin, referred to as D. Bulgin. Static Electrification. *British Journal of Applied Physics*, Supplemental 2, 1953 [5]. Additional articles were published during this time frame [6, 7].

An early investigator of issues with the HBM standard was T.M. Madzy and L.A. Price II of IBM in 1979 discussed a test system titled "Module Electrostatic Discharge Simulator" [10]. This article discussed that the ESD simulator was used within IBM since 1974. In 1980, H. Calvin, H. Hyatt, H. Mellberg, and D. Pellinen proposed values for the resistance and capacitance for the human ESD event for the finger tip and field-enhanced discharges

ESD Testing: From Components to Systems, First Edition. Steven H. Voldman.
© 2017 John Wiley & Sons, Ltd. Published 2017 by John Wiley & Sons, Ltd.

in "Measurement of Fast Transients and Application to Human ESD," published in the 1980 Proceedings of the EOS/ESD Symposium [12]. The proposed resistance for the finger tip was averaged 1920 Ω and capacitance of 110 pF, whereas the field-enhanced discharge was a resistance of 550 Ω and 120 pF. In 1981, H. Hyatt, H. Calvin, and H. Mellberg investigated the human ESD event, published in the 1981 Proceedings of the EOS/ESD Symposium, titled "A Closer Look at the Human ESD Event [13]."

One of the first standards developed for the HBM was military standard MIL-STD 883B Method 3015.1 referred to as "Electrostatic Discharge Sensitivity Classification," published in January 1982.

In 1983, R.N. Shaw and R.D. Enoch of British Telecom Research Laboratories published one of the first publications on a programmable test equipment at the EOS/ESD Symposium, titled "A Programmable Equipment for Electrostatic Discharge Testing to Human Body Models [15]." The test system shown in the publication comprises of a high-voltage power source, a series charging resistor for charging the 100 pF capacitor, a 1500 Ω series resistor, and a single mercury wetted relay switch.

2.2 Scope

The scope of HBM ESD test is for testing, evaluation, and classification of components and micro- to nanoelectronic circuitry. The test is to quantify the sensitivity or susceptibility of these components to damage or degradation to the defined HBM test. The test is performed with the component in an unpowered state [1].

2.3 Purpose

The purpose of the HBM ESD test is for establishment of a test methodology to evaluate the repeatability and reproducibility of components to a defined pulse event in order to classify or compare ESD sensitivity levels of components [1].

2.4 Pulse Waveform

A characteristic time of the HBM event is associated with the electrical components used to emulate the human being. In the HBM standard, the circuit component to simulate the charged human being is a 100 pF capacitor in series with a 1500 Ω resistor [1]. This network has a characteristic rise time and decay time. The rise time is in the range of 17–22 ns. The characteristic decay time is associated with the time of the network, which is 150 ns. The pulse event is a single polarity event. To address HBM ESD qualification, the test is performed in both positive and negative polarity events. Figure 2.1 is an example of a positive polarity HBM pulse event [1].

HBM tests are performed in different pin combinations. Tests are performed between signal pin and referenced power rails. HBM test simulation also includes power-rail-to-power-rail testing, and signal-pin-to-signal-pin combinations. This is to evaluate all possible current

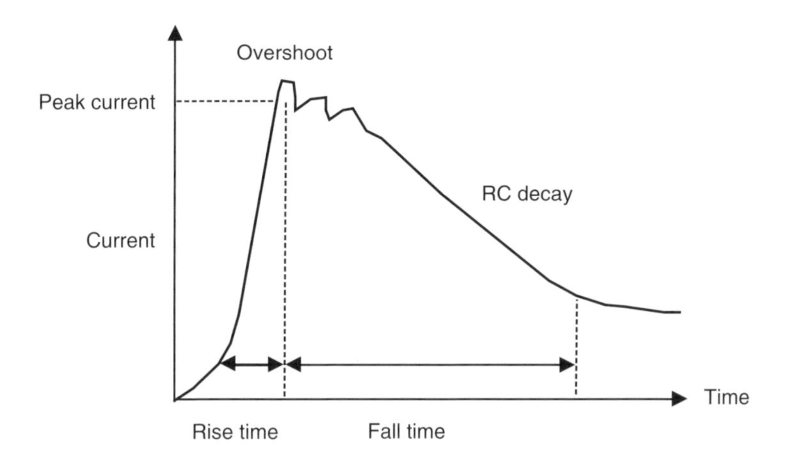

Figure 2.1 Human body model (HBM) pulse waveform

paths for the ESD event. The HBM test standard of pulse waveform and pin combination are integrated into commercial HBM test systems [1].

2.5 Equivalent Circuit

The HBM source can be represented as an equivalent circuit. Figure 2.2 is a circuit schematic representation of the HBM source. The equivalent circuit can be represented as a capacitor element and a resistor element. In the HBM equivalent circuit model, the circuit component to simulate the charged human being is a 100 pF capacitor in series with a 1500 Ω resistor [1]. The capacitance represents the human body and the resistor represents the resistance of an arm.

The HBM source can include the parasitic elements as well as the load in the equivalent circuit. Figure 2.3 is a circuit schematic representation of the HBM source including the parasitic elements and load [31]. In the equivalent circuit, the 1500 Ω resistor also has a parallel shunt

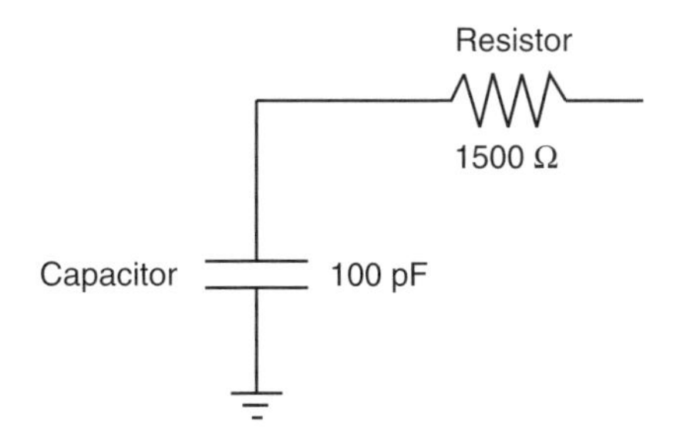

Figure 2.2 Human body model (HBM) equivalent circuit model

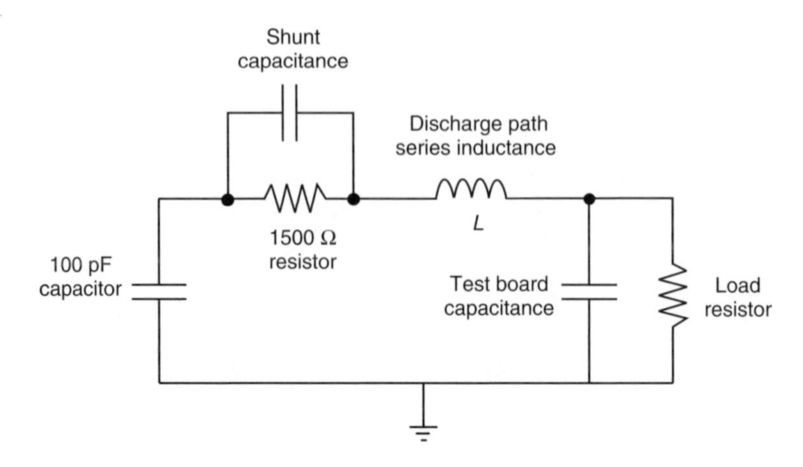

Figure 2.3 Human body model (HBM) equivalent circuit model with parasitics

capacitance. The discharge path also includes a series inductance. This element is followed by a test board capacitance and load resistance. The discharge path series inductance introduces damped oscillations in the waveform near the peak of the current.

2.6 Test Equipment

HBM commercial test equipment exists for the qualification and release of semiconductor components. Figure 2.4(a) shows an example of an HBM source for an early commercial HBM test system [69]. The HBM source comprises of a capacitor element and resistor element. These are encased in an insulating material to avoid parasitic discharge to other test system components.

Figure 2.4(b) shows an example of a second HBM source for another early commercial HBM test system [69]. In these sources, the capacitor and resistor elements are modules that can be placed in the commercial tester.

HBM commercial test equipment will be required to release an HBM pulse to the device under test (DUT). The commercial test equipment must have a high-voltage source used to charge the HBM source circuit. A switch is contained within the test system that is closed when the high-voltage source charges the equivalent circuit. A second switch must be open during the charging of the HBM source circuit. The second switch is closed when the HBM circuit current is discharged into the load or DUT.

Figure 2.5(a) shows an example of a present-day HBM ESD commercial test system. The system in the figure is the Hanwa S5000R HBM test system [72]. This test system is a 128- and 256-pin count package level system. The test system meets the JEITA, ESDA, and JEDEC international standards. This test system can test other ESD models (e.g., machine model (MM) and small charge model).

Figure 2.5(b) shows an example of a second HBM ESD commercial test system. This is the Hanwa HED-N5000 HBM tester for high pin count modules [73]. This test system is a fully automated ESD test system with a pin count of 1056 pin sockets. This system can also test up to eight chips in parallel on a given test board.

(a)

(b)

Figure 2.4 Human body model (HBM) source

Human body commercial systems were first developed for qualification and release of semiconductor components. In the 1980s, there was no wafer level commercial HBM test system for semiconductor development, ESD library test and verification. In the semiconductor environment, it was important for development to have a faster process to evaluate ESD robustness of devices and circuits. Wafer level ESD characterization was possible by extending the wires from the socket to a wafer level work station and wafer probes.

Today, wafer level HBM test systems exist for development. Figure 2.6 shows an example of a wafer level HBM ESD commercial test system.

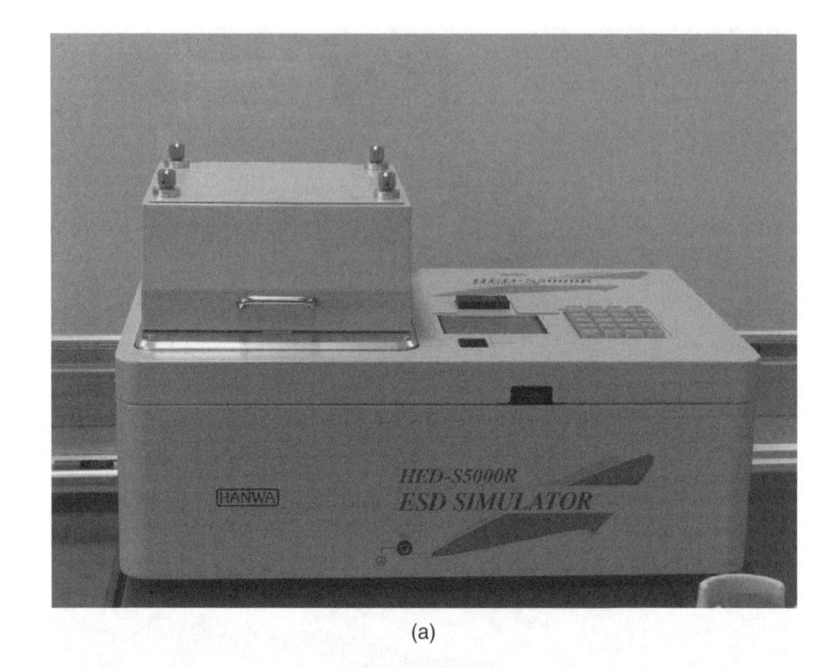

(a)

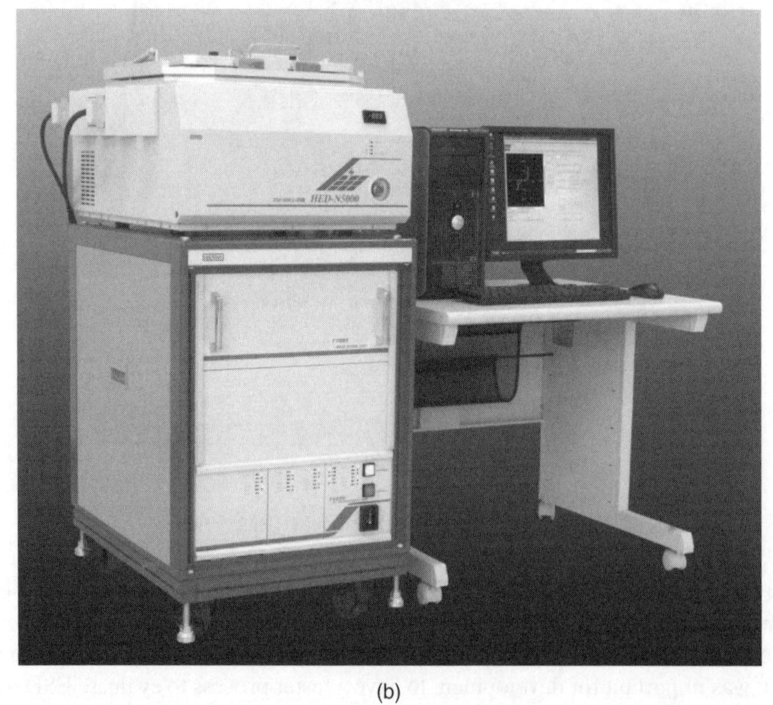

(b)

Figure 2.5 (a) Human body model (HBM) commercial test system – Hanwa S5000R. (b) Human body model (HBM) commercial test system – Hanwa HED-N5000

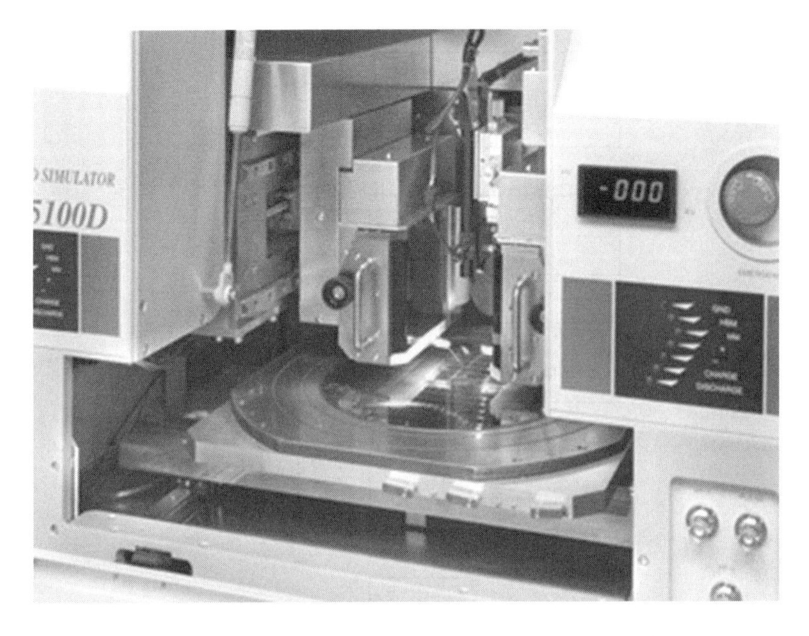

Figure 2.6 Human body model (HBM) wafer level commercial test system

2.7 Test Sequence and Procedure

In the testing of components, the commercial test systems have a defined test sequence and procedure. Figure 2.7 provides a test sequence flow for the commercial HBM test systems [1, 69].

In the testing of components to large increments for semiconductor device characterization, a grounded reference is established. An initial reading is performed prior to any applied voltage on the HBM source. A first step increment is defined on the capacitor element. All pins are pulsed with an ESD event, individually relative to the grounded reference. Each pin is evaluated based on a failure criterion (e.g., voltage shift, leakage). If the pin "passes," it remains in the test sequence for the next increment. Pins that "fail" are removed from the test sequence, where all other passing pins remain in the test sequence. The voltage is increased on the capacitor source to the next step level, and testing resumes for the surviving pins. In this case, after the desired steps are taken, the tester is stopped, and the failed pins versus surviving pins are recorded.

In the testing of components to small increments for semiconductor device characterization, a second test sequence is shown in Figure 2.8. A first precheck of the pins is performed prior to charging the capacitor source. In small increment testing, a grounded reference is established. A first increment is defined on the capacitor element. All pins are pulsed with an ESD event, individually relative to the grounded reference. Each pin is evaluated based on a failure criterion (e.g., voltage shift, leakage). If the pin "passes," it remains in the test sequence. Pins that "fail" are removed from the test sequence, where all other passing pins remain in the test sequence. The voltage is increased on the capacitor source to the next increment level, and testing resumes for the surviving pins. This continues until all the pins have failed the failure

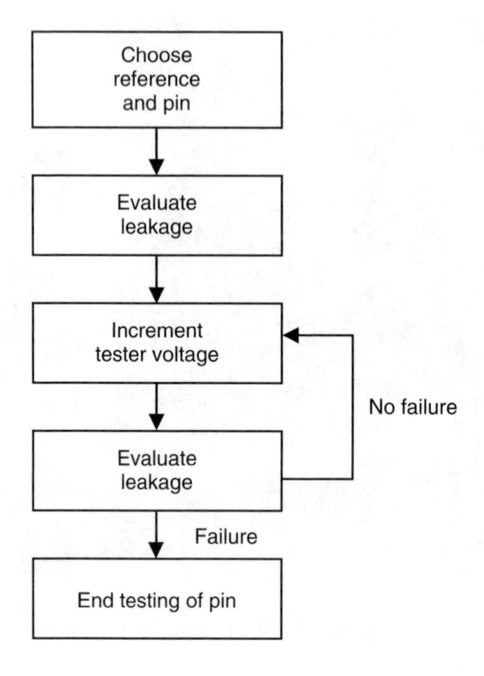

Figure 2.7 Human body model (HBM) test sequence

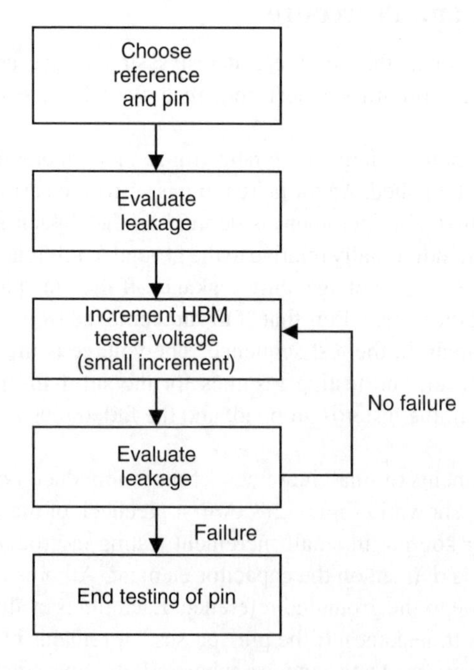

Figure 2.8 Human body model (HBM) incremental step stress test sequence

criterion. In this methodology, with small increments, the actual failure level is determined for all the pins that are tested, with an accuracy associated with the increment step size.

2.8 Failure Mechanisms

HBM failure mechanisms are typically associated with failures on the peripheral circuitry of a semiconductor chip that are connected to signal pins [61–64, 66–69]. HBM failures can occur on the power rails due to inadequate bus widths and ESD power clamps between the power rails. HBM failures can occur in both passive and active semiconductor devices. The failure signature is typically isolated to a single device, or a few elements in a given current path where the current exceeded the capability of the element. ESD circuits are designed to be responsive to HBM pulse widths; this is an issue for EOS events since they are not "tuned" for EOS events. Specifically, the RC-triggered metal oxide semiconductor field effect transistor (MOSFET) ESD power clamp is designed for HBM events [62–64, 66–69].

HBM ESD failures are also distinct from EOS events. HBM events will not typically cause failures in the package, printed circuit board (PCB), or single component devices mounted on a PCB.

Figure 2.9 is an example of an HBM failure of a diode structure. Integrated circuit diode structures fail at the contact interface, silicon surface, or junction region [61, 67].

Figure 2.10 is an example of an HBM failure of a MOSFET structure. Integrated circuit MOSFET structures failure occurs from MOSFET source-to-drain, or at the MOSFET gate. From HBM failures, typically, the failure is MOSFET source-to-drain failure [62–64, 66–69].

Figure 2.11 is an example of an HBM failure of a series cascode MOSFET structure [61–63]. Integrated circuit MOSFET structures failure in a series cascode structure can occur from the upper MOSFET drain to lower MOSFET source [61–63, 67].

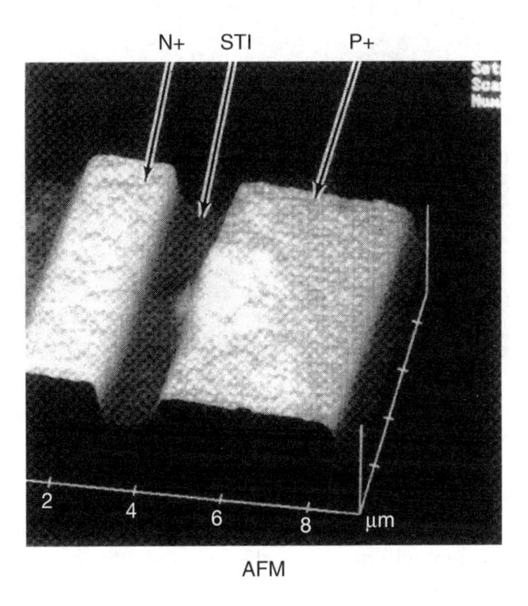

Figure 2.9 Human body model (HBM) failure mechanism – diode failure

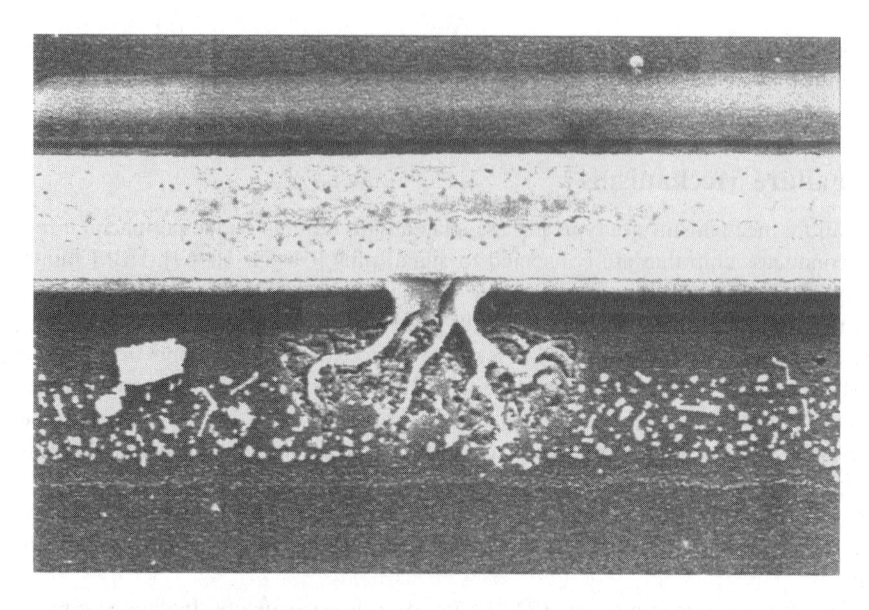

Figure 2.10 Human body model (HBM) failure mechanism – MOSFET failure

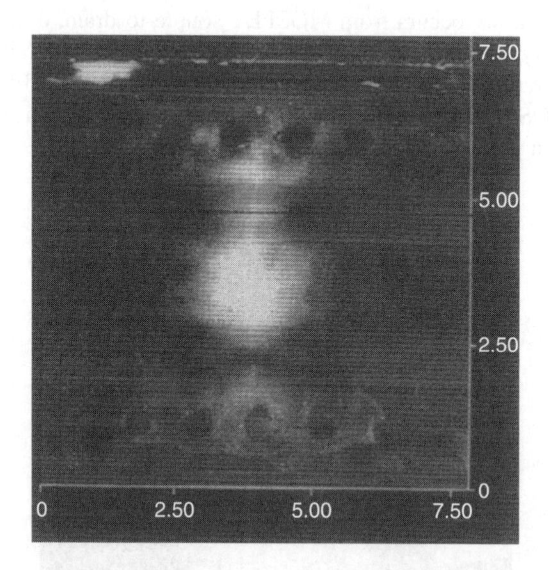

Figure 2.11 Human body model (HBM) failure mechanism – series cascode MOSFET failure

2.9 HBM ESD Current Paths

During ESD events, the current will flow from the signal pin that is pulsed to the grounded reference. Figure 2.12 is an example of the ESD current path from the signal pin to the grounded reference V_{DD}. In this case, the positive polarity ESD pulse leads to a positive current flow into

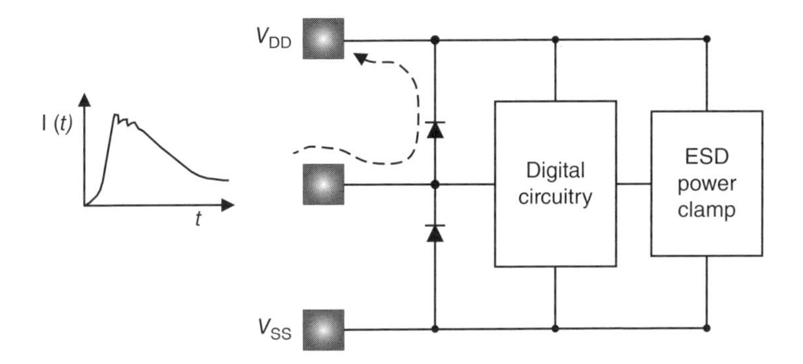

Figure 2.12 Human body model (HBM) ESD event current paths – signal pin to power (V_{DD})

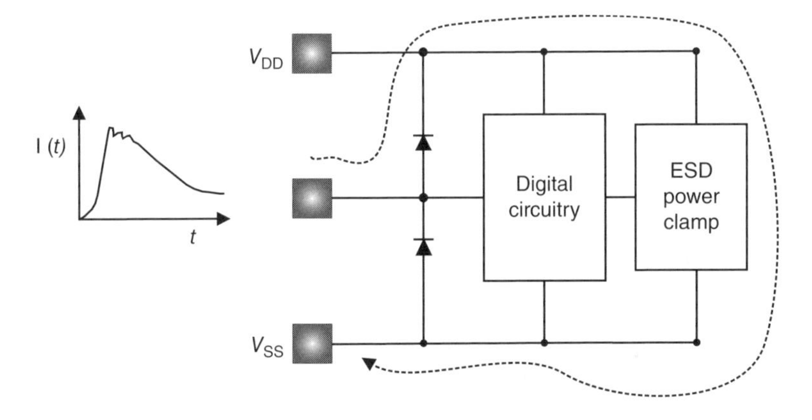

Figure 2.13 HBM ESD event current paths – signal pin to ground (V_{SS})

the bond pad. The ESD current flows through the ESD p–n diode structure connected to the grounded reference V_{DD}. Residual current also flows into the digital circuitry [62, 63, 68].

Figure 2.13 is an example of the ESD current path from the signal pin to the grounded reference V_{SS}. In this case, the positive polarity ESD pulse leads to a positive current flow into the bond pad. The ESD current flows through the ESD p–n diode structure connected to the power rail V_{DD}. The ESD current continues to flow through the ESD power clamp between the V_{DD} power rail and the V_{SS} power rail. Residual current also flows into the digital circuitry [68].

Figure 2.14 is an example of the ESD current path from the power rail V_{DD} bond pad to the signal pin to the grounded reference V_{SS}. In this case, the positive polarity ESD pulse leads to a positive current flow into the bond pad of the V_{DD} power supply. The ESD current continues to flow through the ESD power clamp between the V_{DD} power rail and the V_{SS} power rail [68].

Figure 2.15 is an example of the ESD current path from the signal pin to the grounded signal pin within a common power domain. In this case, the positive polarity ESD pulse leads to a positive current flow into the bond pad. The ESD current flows through the ESD p–n diode structure connected to the power rail V_{DD}. The ESD current continues to flow through the

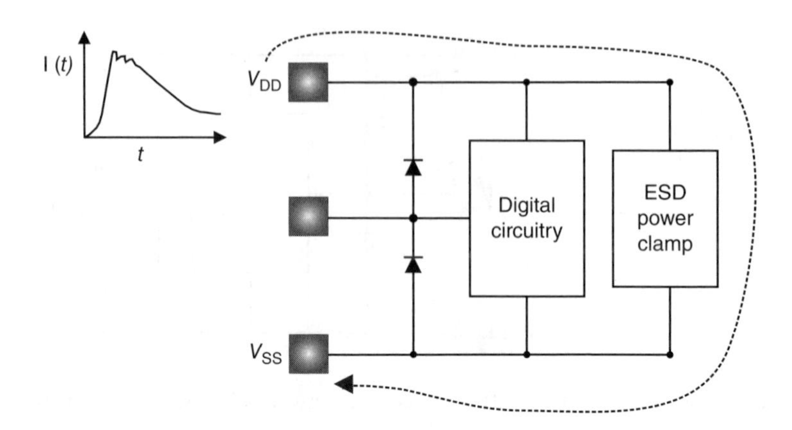

Figure 2.14 HBM ESD event current paths – power rail (V_{DD}) to power (V_{SS})

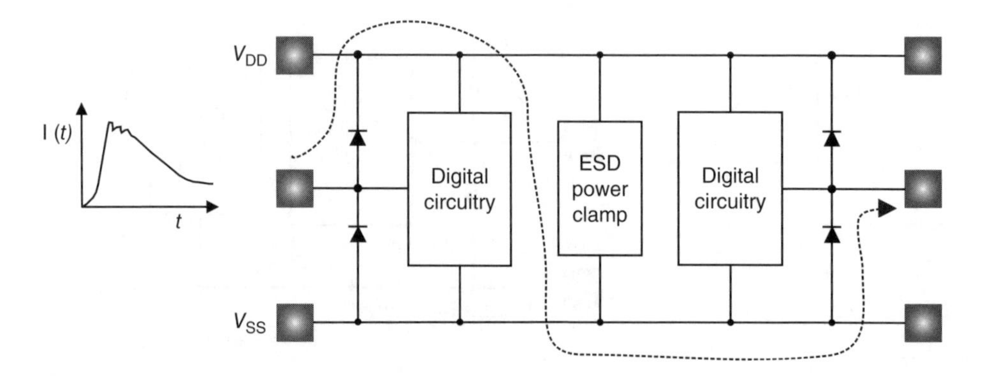

Figure 2.15 HBM ESD event current paths – signal pin to signal pin within a common power domain

ESD power clamp between the V_{DD} power rail and the V_{SS} power rail. Current flows through the ground rail, through the ESD p–n diode that is connected to the V_{SS} rail, and the second grounded signal pin [68].

Figure 2.16 is an example of the ESD current path from the signal pin to the grounded signal pin within an independent power domain. In this case, the positive polarity ESD pulse leads to a positive current flow into the bond pad. The ESD current flows through the ESD p–n diode structure connected to the power rail V_{DD}. The ESD current continues to flow through the ESD power clamp between the V_{DD} power rail and the V_{SS} power rail of its own domain. Current flows through the "rail-to-rail" ESD network placed between the two ground rails. The current flows to the grounded reference rail [68].

2.10 HBM ESD Protection Circuit Solutions

Figure 2.17 is an example of an ESD protection network known as a dual-diode network [61–64]. The dual-diode ESD network is a commonly used network for complementary metal

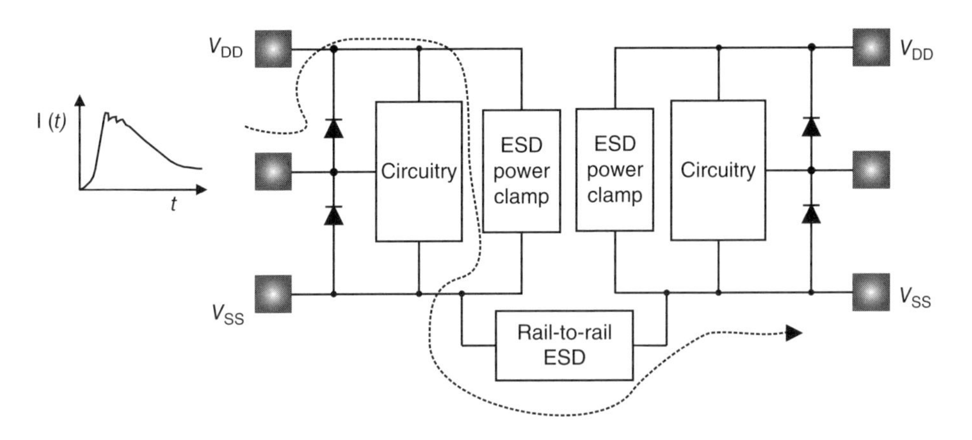

Figure 2.16 HBM ESD event current paths – signal pin to alternate ground rail domain to domain

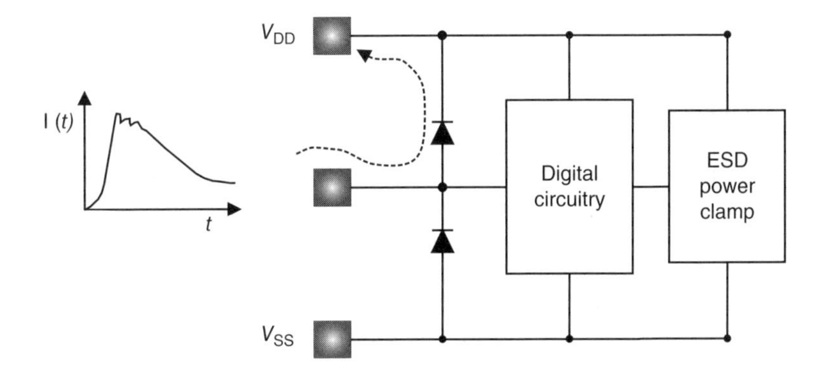

Figure 2.17 HBM signal pin ESD protection circuits – dual diode

oxide semiconductor (CMOS) technology. A first p–n diode element is formed in an n-well region where the p-anode is the p-diffusion implant of the p-channel MOSFET device and the n-cathode is the n-well region connected to the power supply V_{DD}. This is sometimes referred to as the "up diode." A second p–n diode element is formed in a p-well or p-substrate region where the n-cathode is the n-diffusion implant of the n-channel MOSFET device, or the n+/n–well implant and the p-anode is the p-well region or p-substrate region connected to the power supply V_{SS}. This is sometimes referred to as the "down diode." This circuit provides a "forward bias" ESD protection solution for positive and negative ESD pulse events to the two power rails V_{DD} and V_{SS}.

An advantage of the dual-diode ESD network is that it is easy to migrate from technology generation to technology generation. In shallow trench isolation (STI) technology, this structure is scalable. A second advantage is that it has a low turn-on voltage of 0.7 V. A third advantage is that it can be designed with low capacitance, making it suitable for CMOS, advanced CMOS, and RF technologies. A fourth advantage is that it does not contain MOSFET gate dielectric failure mechanisms [61–64].

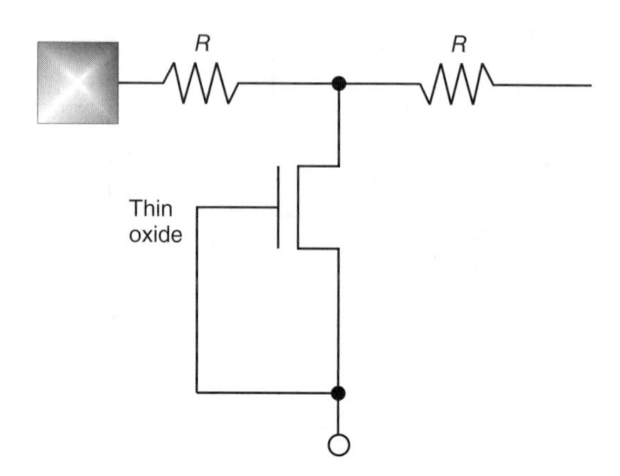

Figure 2.18 HBM signal pin ESD protection circuit – grounded gate NMOS (GGNMOS)

Figure 2.18 is an example of a signal pin ESD network consisting of a grounded gate n-channel MOSFET device [61–71]. The grounded gate NMOS (also referred to as GGNMOS) ESD network is a commonly used network for CMOS technology. Typically, it is an n-channel MOSFET whose MOSFET drain is connected to the signal pin and whose MOSFET source and gate are connected to the ground power rail. This circuit remains "off" in normal operation. When the signal pin exceeds the MOSFET snapback voltage, this circuit discharges to the V_{SS} power rail. When the signal pin is below the ground potential, the MOSFET drain forward biases to the p-well or p-substrate region. An advantage of the GGNMOS ESD network is that it is a natural scalable solution. As the technology scales, the MOSFET snapback voltage reduces, leading to an earlier turn-on of the MOSFET.

Figure 2.19 is an example of a signal pin ESD network consisting of a silicon controlled rectifier (SCR) network [61–64, 66–69]. The SCR ESD network is a commonly used network for CMOS technology. An SCR is a four-region pnpn device. The p-anode is the p-diffusion implant of the p-channel MOSFET device and the n-cathode is the n-well region. A third region is the p-substrate, and the n-cathode is the n-diffusion implant of the n-channel MOSFET device, or the n+/n− well implant There are a large variety of pnpn structure that can be constructed, from high-voltage SCR (HVSCR), medium-voltage SCR (MSCR), to low-voltage trigger SCR (LVTSCR) structures. These can be constructed as single- or bidirectional networks.

Figure 2.20 is an example of an ESD network that is placed between two power rails where the first power rail is a V_{DD} power rail and the second power rail is a V_{SS} power rail [62–64]. These types of networks are known as ESD power clamps. The most commonly used ESD networks are RC-triggered MOSFET ESD power clamps. This consists of a frequency trigger, a MOSFET drive stage, and an output MOSFET clamp device. The most commonly used network has a resistor–capacitor (RC) discriminator that is tuned to respond to an HBM ESD pulse event. The MOSFET drive stage consists of a series of inverter switches, with increasing width, to "turn-on" the output MOSFET clamp device. In recent years, these networks have been modified to respond to faster impulses, lower leakage, smaller area, and means to avoid false triggering [68].

Figure 2.21 is an example of an ESD network that is placed between two ground rails. ESD networks are placed between isolated ground rails to allow current flow between the

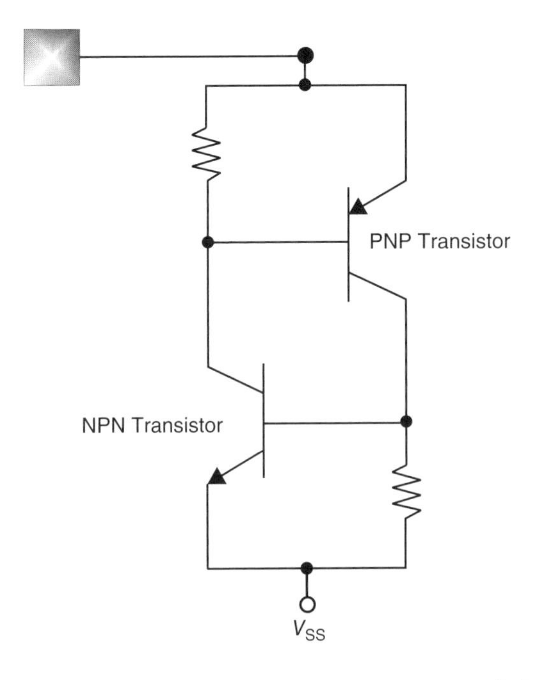

Figure 2.19 HBM ESD protection circuits – silicon controlled rectifier

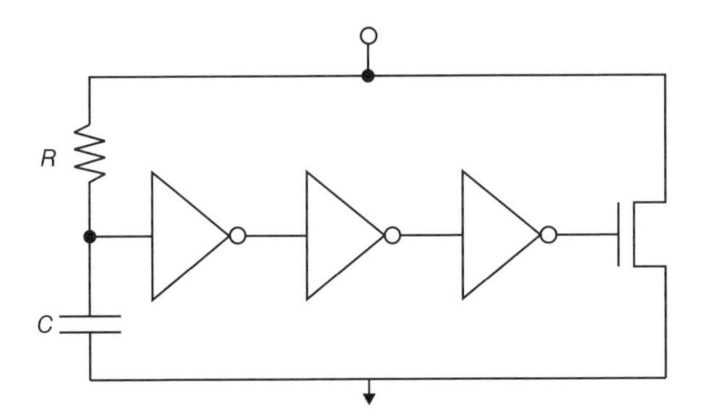

Figure 2.20 HBM ESD protection device – power rail to power rail protection

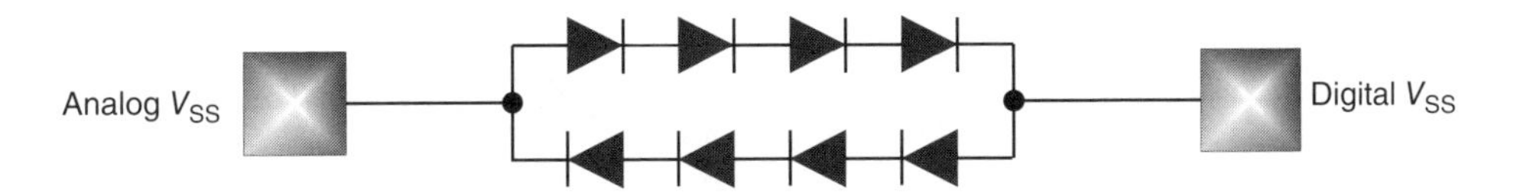

Figure 2.21 Human body model (HBM) ESD protection circuits – ground-to-ground network

grounds for signal pin-to-rail ESD testing where the signal pin is in a first domain, and the reference is contained in a second rail. In addition, it also assists pin-to-pin ESD HBM events and rail-to-rail HBM events.

2.11 Alternate Test Methods

Alternate test methods have been developed for HBM testing. Figure 2.22 shows a picture of alternate test methods. In the mid-1990s, the pin count of semiconductor chips continued to increase beyond 1000 pins. Yet, ESD test systems' largest pin count was 250 pins. ESD test systems that exceeded 1000 pins were also not available. In addition, with the increased pin count, the test time continued to increase.

One method of testing was known as "split fixture" testing; one method had multiple fixtures mounted onto a single test board, and some corporations, "rotated" the chip in the smaller socket. Also in this time frame, "sample testing" was proposed to address the cost and time to test a high pin count semiconductor chip. To accelerate "ESD learning," wafer level testing was introduced to provide characterization of ESD networks, and circuits on wafer, instead of a packaged environment. In addition, two-pin testing is now growing to avoid the complexity of relay matrix–based test systems. Other methods will also be discussed, such as small step stress testing to transmission line pulse (TLP)-like HBM testing.

2.11.1 HBM Split Fixture Testing

Alternate testing methods were at time developed because the test system pin count did not keep up with the number of pins on a component. The socket of the test system was not large enough for socketing the component into the ESD test system. This problem occurred when semiconductor chips were at 1000 pin designs for microprocessors and CPUs, yet the highest pin count ESD test system was 250 pins. As a result, to test and socket some HBM test systems,

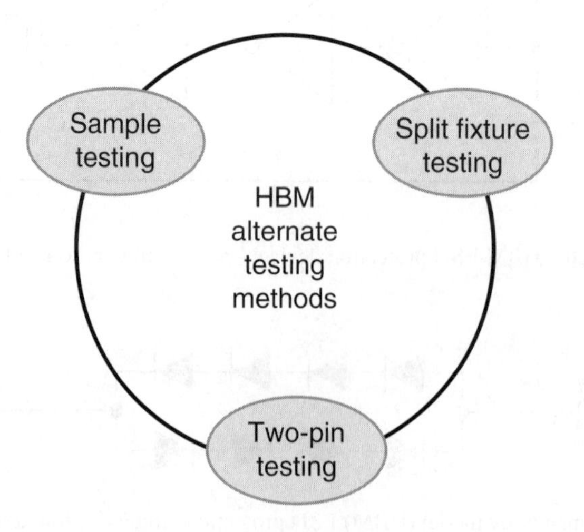

Figure 2.22 Alternate test methods

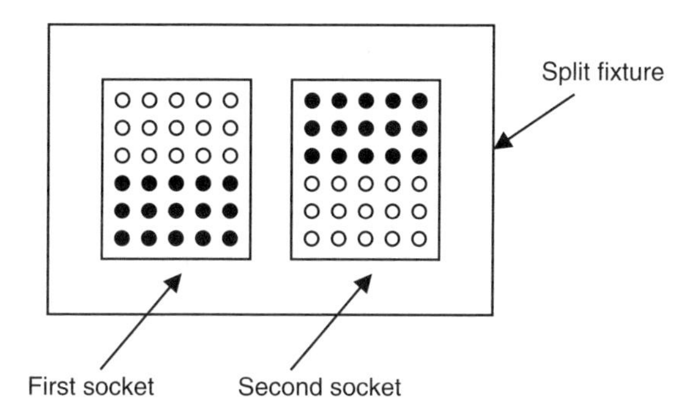

Figure 2.23 HBM split fixture testing

"split fixtures" were developed to test all pins on the component. Figure 2.23 is an example of a "split fixture."

2.11.2 HBM Sample Testing

As the number of pins increased on a semiconductor component, test time to test all pins on a component increased. As a result, it was proposed to introduce sampling methods to reduce the test time. With the increased pin count, the time to test an individual microprocessor was significant. With a large computer design system, which required eight semiconductor chips of 1000 pins each, this required a very large amount of test time. For example, in a 1000-pin semiconductor component, as many as 600 pins were the identical circuit. On testing this component, a Gaussian distribution of failures was observable. As a result, it is clear that by testing a smaller sample of "identical pins," the ESD robustness of the product can be qualified. The difficulty is the identification of "identical pins." Figure 2.24 shows an HBM sample testing method.

2.11.3 HBM Wafer Level ESD Testing

To improve turnaround time of semiconductor development, wafer level characterization was introduced. Figure 2.25 is an example of HBM wafer level test system. Wafer level testing can be used for ESD device characterization of active and passive elements in a technology library (e.g., resistors, capacitors, transistors) for ESD HBM technology benchmarking. Wafer level testing can be done on ESD networks that will be introduced into a technology ESD library. In addition, wafer level testing can be done on S-parameter extraction for RF technology library elements. HBM ESD testing of products can be done on a wafer to provide an early look prior to the first design pass.

2.11.4 HBM Test Extraction Across the Device Under Test (DUT)

One of the issues of the HBM test is that it does not provide the test engineer, ESD engineer, or circuit designer the true response of the pin under test. The HBM test only provides the

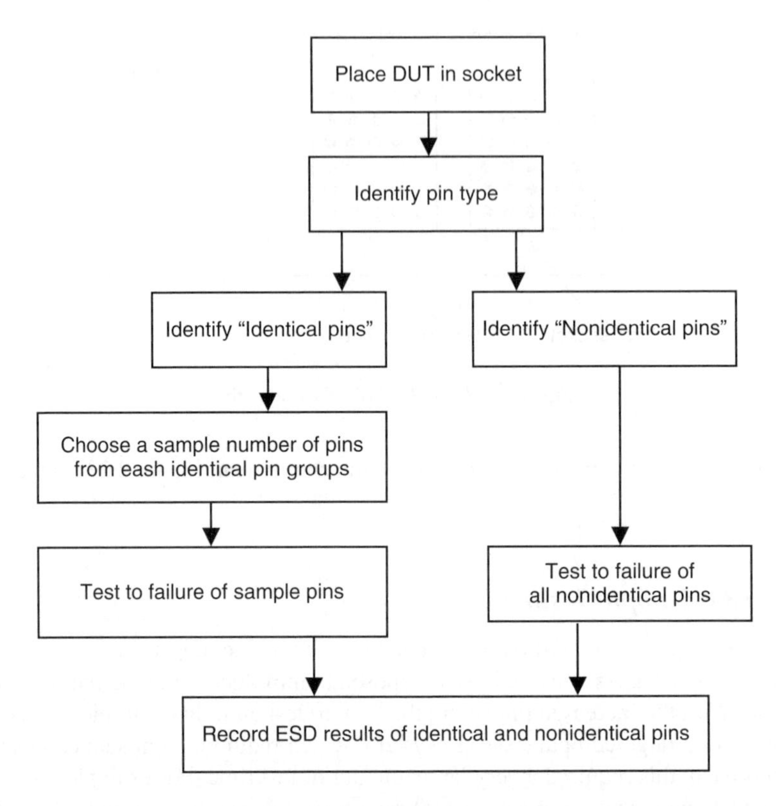

Figure 2.24 HBM sample testing

survivability of the pin to the voltage level on the capacitor of the test system; this limits the insight what is possible to understand about the actual current–voltage (*I–V*) response of the signal pin. A test system can be developed as a wafer level test system that extracts the voltage and current across the DUT during the pulse event. Using a voltage probe, and a current probe, the signals can be transferred to a two-channel oscilloscope during a pulse event. An HBM pulse can be generated by an RC-network source where the high-voltage supply charges the capacitor element, and a mercury wet switch can apply the pulse to the DUT. When the pulse is applied, sampling of the current and voltage must extract the conditions across the DUT for evaluation of the results. This is performed in TLP testing by establishing a measurement window where the voltage and current are sampled and averaged. This technique has the disadvantage that the measurement window is not flat and leads to some error in the HBM extraction process (extraction is performed near the peak of the HBM pulse).

2.12 HBM Two-Pin Stress

A present problem with HBM test systems is the relay systems used in the test system (Figure 2.26) [59]. Relay-based HBM test systems have reached their limitation for test channels in high-pin-count components. HBM tester designs with test fixture boards (TFB), and

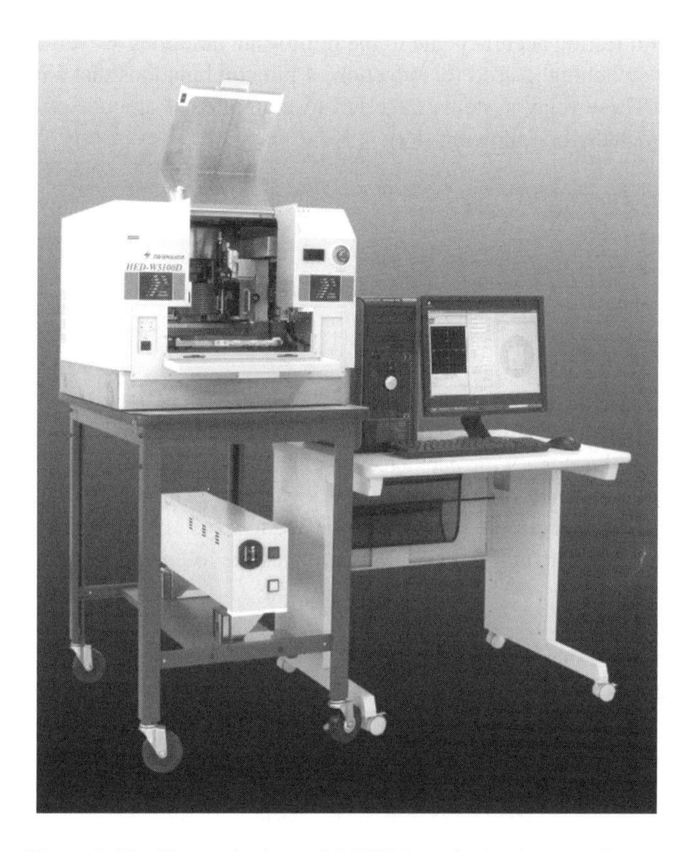

Figure 2.25 Human body model (HBM) wafer level test equipment

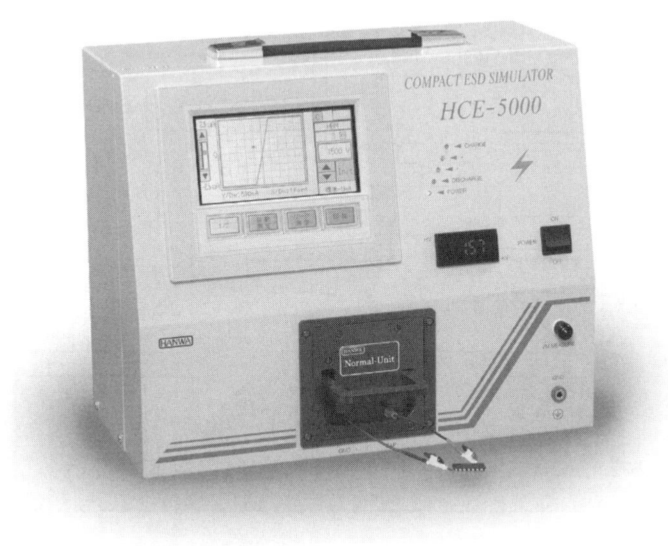

Figure 2.26 Human body model Two-Pin test system (Reproduced with permission of the Hanwa Corporation)

relay matrices limit testing accuracy due to the pin-to-pin variations as well as pulse-to-pulse variations. Relay switching is used for selection of pin combinations that are required by the HBM standards. These relay systems have led to a significant amount of tester problems. These problems include the following [59]:

- Recharging pulse events
- Relay switching noise
- Prepulse spikes
- Trailing pulse events
- Switching relay parasitics
- Tester costs.

HBM two-pin (2-pin) testing has been more recently introduced into testing methods. Two-pin testing can be done with packaged or unpackaged components. The advantage of two-pin testing is avoidance of the role of the switching matrix and socket on the HBM results. Figure 2.27 is an example of a two-pin HBM test system, the arcus HBM wafer and device test system [74]. The two-pin system eliminates the following [59]:

- Elimination of TFBs
- Elimination of test fixture board sockets
- Elimination of relay matrix.

Figure 2.27 shows the test system with the arcus HBM test system, HBM measurement pod, SMU, oscilloscope, and wafer probe station. The test system reduces parasitics, eliminates

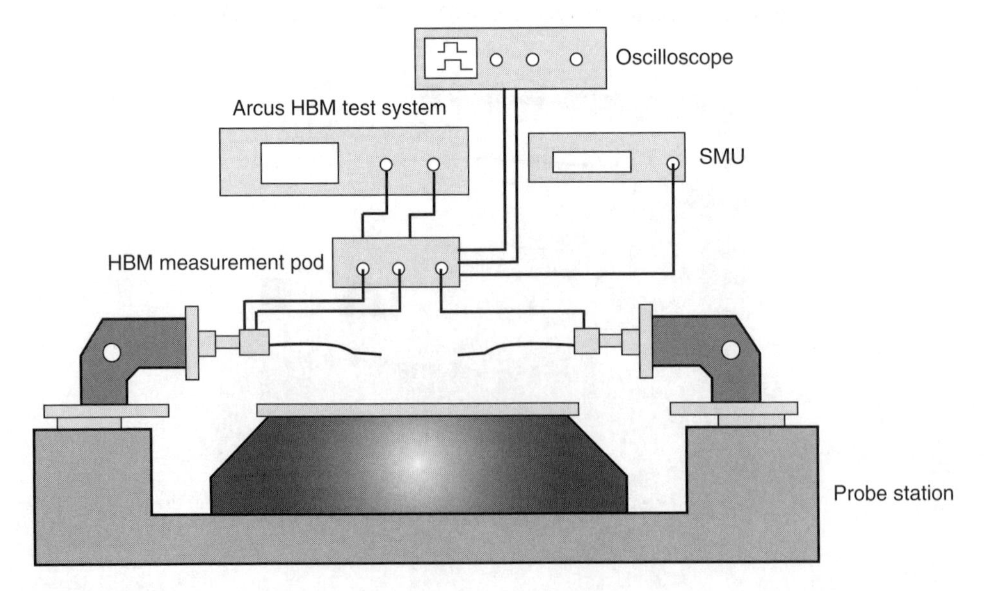

Figure 2.27 HBM two-pin stress test equipment – Grund Technical Solutions Arcus HBM wafer and device test system

anomalous stresses, avoids relay matrix problems, shortens test cycles, reduces cost, measures and verifies each pulse, and eliminates false failures [74].

2.12.1 HBM Two-Pin Stress – Advantages

There are a significant number of advantages using HBM two-pin testers. Two-pin testers provide the following [59]:

- Improved HBM accuracy
- Improved testing flexibility
- Low parasitic
- Elimination of TFBs
- Elimination of test fixture board sockets
- Elimination of TFB parasitic
- Elimination of test fixture board trace coupling
- Elimination of relay matrix
- Elimination of false failures from DUT-relay matrix tester interactions
- Cost reduction due to elimination of test fixture board, sockets, and relay matrix.

2.12.2 HBM Two-Pin Stress – Pin Combinations

One of the problems with two-pin testing is that it does not fulfill all the requirements of pin combinations in today's standards established in relay matrix–based test systems [59]. In the future, a solution will be needed to allow acceptance of HBM two-pin testing. These may include the following:

- Elimination of the requirement of I/O to I/O pin combination requirement
- Optional requirement of I/O to I/O pin combination.

2.13 HBM Small Step Stress

A method of testing is the application of a small step stress in the qualification procedure. In many corporations, the testing is performed as a single increment to the specification level. In the testing of components to small increments for semiconductor device characterization, a first precheck of the pins is performed prior to charging the capacitor source. In small increment testing, a grounded reference is established. A first increment is defined on the capacitor element. All pins are pulsed with an ESD event, individually relative to the grounded reference. Each pin is evaluated based on a failure criterion (e.g., voltage shift, leakage). If the pin "passes," it remains in the test sequence. Pins that "fail" are removed from the test sequence, where all other passing pins remain in the test sequence. The voltage is increased on the capacitor source to the next increment level, and testing resumes for the surviving pins. This continues until all the pins have failed the failure criterion. In this methodology, with small increments, the actual failure level is determined for all the pins that are tested, within an accuracy associated with the increment step size.

2.13.1 HBM Small Step Stress – Advantages

In this methodology, with small increments, there are a number of advantages as opposed to the method of a single increment.

A first advantage of small increment testing, is the avoidance of "ESD test windows." A disadvantage of a single value of ESD testing at the desired specification is, there are "ESD test windows" where the component can fail at a lower level. Some components may pass at a higher level but fail at lower levels. In this methodology, there are no "ESD windows" if the step stress has small increments.

A second advantage is that the actual failure levels of different pins and/or circuit types are determined. The actual failure levels of the different circuits can provide an understanding of the failure of specific circuits and which circuits are the "weak circuits." For example, in a microprocessor design, it was determined that the receiver network pin population was distinct from the off-chip driver (OCD) pin population.

A third advantage is that the recording of the failure levels of the component can provide quality control in a semiconductor manufacturing environment. Manufacturing semiconductor process shifts can be determined in a manufacturing line. An observed effect was epitaxial wafer control variations leading to ESD protection level variation. By knowing the protection level of specific pins, it was clear from the experimental data that there was a correlation of the ESD level versus well sheet resistance value. In a second example, silicide variation introduced ESD variation in ESD diode networks.

2.13.2 HBM Small Step Stress – Data Analysis Methods

In this methodology, with small increments, the statistical variation of the ESD pin population can be evaluated. By having the failure levels of all the pins, one can observe the Gaussian and non-Gaussian characteristics of the semiconductor chip ESD results. By having the experimental data for the mean ESD result, and the ESD level standard deviation, a complete understanding of the true robustness of the product chip can be determined.

2.13.3 HBM Small Step Stress – Design Optimization

In this methodology, with small increments, the statistical variation of the ESD pin population can be determined allowing for design optimization. By having the ESD failure levels of all the pins, semiconductor process experiments as well as layout improvements in design passes can be determined.

2.14 Summary and Closing Comments

In this chapter, one of the most important tests in the history of ESD testing is discussed. This chapter focuses on the HBM test, and in addition indicates many of the present and future anticipated changes that will occur on this test. The changes include sample testing, rotation testing, to two-pin test methods. In the future years, there will be continued change and improvements in this test method.

In Chapter 3, the MM test is discussed. The MM purpose, scope, and waveforms are shown. In addition, the discussion highlights the correlation between HBM and MM tests.

Problems

1. Given a resistor of 1500 Ω and a capacitor of 100 pF, calculate the RC time of the human body model (HBM) event.
2. In semiconductor chip development, RC-triggered ESD circuits are designed to respond to the human body model (HBM) pulse waveform. Assume an ESD HBM pulse is a double exponential waveform, and determine what the RC time should be to protect the semiconductor chip. Assume a single RC time, 2× RC time, and 3× RC time.
3. Given the ESD HBM is an RC source and a switch. Assume the ESD pulse is discharged to a semiconductor chip of capacitance C_{chip}. Derive the current and voltage assuming the HBM capacitor is initially charged, the chip is uncharged, and the switch is then closed instantaneously.
4. Given the ESD HBM is an RC source and a switch. Assume the ESD pulse is discharged to a semiconductor chip of capacitance C_{chip}. Derive the current and voltage assuming the HBM capacitor is initially charged, the chip is precharged to a given value, and the switch is then closed instantaneously.
5. Given the HBM source has a high-voltage charging source, and a 1 MΩ resistor element to charge the HBM RC capacitor of 100 pF. The charging source is to be isolated with a switch when applying the ESD pulse to the chip. Assume the gas relay switch has a leakage current (e.g., a resistance), which prevents the full isolation of the charging source. What is the outcome? How does it affect the ESD testing results? What would the waveform look like before and after the RC capacitor is discharged (at the device under test)?
6. As mentioned in Problem 5, an ESD pulse is applied to a chip. The RC pulse is discharged and the next step increment pulse is ready to be applied. Assume that the charging source is not fully isolated from the semiconductor chip of capacitance C_{chip}. Calculate the voltage and current assuming the high-voltage source is at voltage 2000 V, with a 1 MΩ resistor and switch resistance of 100 Ω.
7. Given that semiconductor chip has a large inductor load L, show the response of a chip to an RC-triggered circuit when testing V_{DD} to ground. Show the RLC response. What inductance magnitude is an issue?

References

1. ANSI/ESD ESD-STM 5.1 – 2007. *ESD Association Standard Test Method for the Protection of Electrostatic Discharge Sensitive Items – Electrostatic Discharge Sensitivity Testing – Human Body Model (HBM) Testing – Component Level.* Standard Test Method (STM) document, 2007.
2. ANSI/ESD SP 5.1.2 – 2006. *ESD Association Standard Practice for the Protection of Electrostatic Discharge Sensitive Items – Human Body Model (HBM) and Machine Model (MM) Alternative Test Method: Split Signal Pin-Component Level*, 2006.
3. W.N. Carr and J.P. Mize. MOS/LSI Design and Application. *Texas Instrument Electronics.* New York: McGraw-Hill, 1972.
4. P.G. Guest, V.W. Sikora, and B.L. Lewis. *Bureau of Mines, Report of Investigation 4833*, U.S. Department of Interior, 1952.
5. D. Bulgin. Static electrification. *British Journal of Applied Physics*, Supplemental 2, 1953.
6. C. Martin. *Duration of the Resistive Phase and Inductance of Spark Channels.* Atomic Weapons Research Establishment, SSWA/JCM/1065/25, 1996; 287–293.
7. R.C. O'Rourke. *Investigation of the Resistive Phase in High Power Gas Switching.* Science Applications, Inc., SAI-77-515-LJ. Jan 1997, Univ of California, Lawrence Livermore Lab, Master Thesis.
8. T.M. Madzy. FET circuit destruction caused by electrostatic discharge. *IEEE Transactions on Electron Devices*, Vol. 23, (9), Sept 1976.

9. W.M. King. Dynamic waveform characteristics of personal electrostatic discharge. *Proceedings of the Electrical Overstress/Electrostatic Discharge (EOS/ESD) Symposium*, 1979; 78–87.
10. T.M. Madzy, and L.A. Price II,. Module electrostatic discharge simulator. *Proceedings of the Electrical Overstress/Electrostatic Discharge (EOS/ESD) Symposium*, 1979; 36–40.
11. T.R. Schnetker. Human factors in electrostatic discharge protection. *Proceedings of the Electrical Overstress/Electrostatic Discharge (EOS/ESD) Symposium*, 1979; 122–125.
12. H. Calvin, H. Hyatt, H. Mellberg, and D. Pellinen. Measurement of fast transients and application to human ESD. *Proceedings of the Electrical Overstress/Electrostatic Discharge (EOS/ESD) Symposium*, 1980; 225–230.
13. H. Hyatt, H. Calvin, and H. Mellberg. A closer look at the human ESD event. *Proceedings of the Electrical Overstress/Electrostatic Discharge (EOS/ESD) Symposium*, 1981; 1–8.
14. H. Hyatt, H. Calvin, and H. Mellberg. Bringing ESD into the 20th century. *Proceedings of the IEEE International Electromagnetic Compatibility (EMC) Symposium,* 1982.
15. R.N. Shaw, and R.D. Enoch. A programmable equipment for electrostatic discharge testing to human body models. *Proceedings of the Electrical Overstress / Electrostatic Discharge (EOS/ESD) Symposium*, 1983; 48–55.
16. T.W. Lee. Construction and application of a tester for measuring EOS/ESD thresholds to 15 kV. *Proceedings of the Electrical Overstress / Electrostatic Discharge (EOS/ESD) Symposium*, 1983; 37–47.
17. P. Richman. ESD simulation – configuring a full performance facility. *Proceedings of the IEEE International Electromagnetic Compatibility (EMC) Symposium,* 1983.
18. H. Hyatt. Critical considerations for ESD testing. *Proceedings of the Electrical Overstress/Electrostatic Discharge (EOS/ESD) Symposium*, 1984; 104–111.
19. P. Richman. *Comparing Computer Models to Measured ESD Events*. Boston, MA: Electrical Overstress Exposition, 1985
20. R.G. Chemelli, and L. De Chairo. The characterization and control of leading edge transients from human body model ESD simulators. *Proceedings of the Electrical Overstress/Electrostatic Discharge (EOS/ESD) Symposium*, 1985; 155–162.
21. G.U. Schimrigk. *Influence of ESD pulse rise time on ESD test results. Texas Instruments Technical Journal*, Vol. 3, (5), Sept 1986; 72–74.
22. J.C. Martin. *Duration of Resistive Phase and Inductance of Spark Channels. Internal Technical Article. SSWA/JCM/1065/25*, Allermaston, England: *Atomic Weapons Research Establishment*, 1996.
23. D.L. Lin, M.S. Strauss, and T.L. Welsher. On the validity of ESD threshold data obtained using commercial human body model simulators. *Proceedings of the IEEE International Symposium on Reliability Physics*, 1987; 77–84.
24. T. Bilodeau. A novel high fidelity technique to view in-situ ESD stress voltage waveforms. *Proceedings of the Electrical Overstress/Electrostatic Discharge (EOS/ESD) Symposium,* 1988; 147–154.
25. R.J. Consiglio, and I.H. Morgan. A method of calibration for human body model testers to establish correlatable results. *Proceedings of the Electrical Overstress/Electrostatic Discharge (EOS/ESD) Symposium,* 1988; 155–161.
26. T. Bilodeau. Theoretical and empirical analyses of the effect of circuit parasitics on the calibration of HBM ESD simulators. *Proceedings of the Electrical Overstress/Electrostatic Discharge (EOS/ESD) Symposium,* 1989; 43–49.
27. J. Blakenagel and L. Lockwood. A high voltage pulse generator for ESD simulation. *Proceedings of the Electrical Overstress/Electrostatic Discharge (EOS/ESD) Symposium*, 1989; 50–54.
28. A. Nishimura, S. Murata, S. Kuroda, M. Yamaguchi, and H. Kitagawa. The arc problem and voltage scaling in ESD human body model. *Proceedings of the Electrical Overstress/Electrostatic Discharge (EOS/ESD) Symposium,* 1990; 111–113.
29. B. Bingold, TJ. Maloney, V. Wilson, and R. Levi. Package effects on human body and charged device ESD tests. *Proceedings of the Electrical Overstress/Electrostatic Discharge (EOS/ESD) Symposium,* 1991; 144–150.
30. L. Avery. Beyond MIL HBM testing: How to evaluate the real capability of protection structures. *Proceedings of the Electrical Overstress/Electrostatic Discharge (EOS/ESD) Symposium,* 1991; 120–126.
31. K. Verhaege, P.J. Rousell, G. Groeseneken, H. Maes, H. Gieser, C. Russ, P. Egger, X. Guggenmos, and F.G. Kuper. Influence of test board capacitance of HBM ESD testers and impact on HBM tester standard specification using a 4th order lumped element model. *Proceedings of the Electrical Overstress/Electrostatic Discharge (EOS/ESD) Symposium*, 1993; 385–393.
32. M. Kelly, G.E. Servais, T.V. Pfaffenbach. An investigation of human body model electrostatic discharge. *Proceedings of the International Symposium on Testing and Failure Analysis (ITSFA)*, 1993; 167–173.

33. C. Russ, H. Gieser, and K. Verhaege. ESD protection elements during HBM stress tests – further numerical and experimental results. *Proceedings of the Electrical Overstress/Electrostatic Discharge (EOS/ESD) Symposium*, 1994; 96–105.
34. K. Verhaege, C. Russ, D. Robinson-Hahn, M. Farris, J. Scanlon, D. Lin, J. Veltri, and G. Groeseneken. Recommendations to further improvements of HBM ESD component level test specifications. *Proceedings of the Electrical Overstress/Electrostatic Discharge (EOS/ESD) Symposium*, 1996; 40–53.
35. J. Barth, D. Dale, K. Hall, H. Hyatt, D. McCarthy, J. Nueble, and D. Smith. Measurements of ESD HBM events, simulator radiation, and other characteristics toward creating a more repeatable simulation, or; simulators should simulate. *Proceedings of the Electrical Overstress/Electrostatic Discharge (EOS/ESD) Symposium*, 1996; 211–222.
36. D. Lin, D. Pommerenke, J. Barth, L.G. Henry, H. Hyatt, M. Hopkins, G. Senko and D. Smith. Metrology and methodology of system level ESD testing, *Proceedings of the Electrical Overstress/Electrostatic Discharge (EOS/ESD) Symposium*, 1988; 29–39.
37. G. Notermans, P. de Jong, and F. Kuper. Pitfalls when correlating TLP, HBM and MM testing. *Proceedings of the Electrical Overstress/Electrostatic Discharge (EOS/ESD) Symposium*, 1998; 170–176.
38. SEMATECH. *Meeting Minutes of the SEMATECH ESD Work Group on Alternate Testing Methods*. SEMATECH ESD Work Group, 1997–2000.
39. J. Barth, K. Verhaege, L.G. Henry, and J. Richner. TLP Calibration, correlation, standards, and new techniques. *Proceedings of the Electrical Overstress/Electrostatic Discharge (EOS/ESD) Symposium*, 2000; 85–96.
40. J. Lee, M. Hoque, J.J. Liou, G. Croft, W. Young, and J. Bernier. A method for determining a transmission line pulse shape that produces equivalent results to human body model testing methods. *Proceedings of the Electrical Overstress/Electrostatic Discharge (EOS/ESD) Symposium*, 2000; 97–104.
41. M. Lee, C. Liu, C-C. Lin, J-T. Chou, H. Tang, Y. Chang, and K. Fu. Comparison and correlation of ESD HBM (Human Body Model) obtained between TLPG, wafer level, and package level tests. *Proceedings of the Electrical Overstress/Electrostatic Discharge (EOS/ESD) Symposium*, 2000; 105–110.
42. J. Barth, and J. Richner. Correlation considerations: Real HBM to TLP and HBM testers. *Proceedings of the Electrical Overstress/Electrostatic Discharge (EOS/ESD) Symposium*, 2001; 453–460.
43. J. Barth, K. Verhaege, L.G. Henry, and J. Richner. Calibration, correlation, standards, and new techniques. *IEEE Transactions on Electronics Packaging and Manufacturing* Vol. 24, (2), Apr 2001.
44. J. Barth, J. Richner, K. Verhaege, M. Kelly, and L.G. Henry. Correlation considerations II: Real HBM to HBM testers. *Proceedings of the Electrical Overstress/Electrostatic Discharge (EOS/ESD) Symposium*, 2002; 155–162.
45. J. Barth, J. Richner, L.G. Henry, and M. Kelly. Real HBM and MM – the dV/dt threat. *Proceedings of the Electrical Overstress/Electrostatic Discharge (EOS/ESD) Symposium*, 2003; 179–187.
46. T. Meuse, R. Barrett, D. Bennett, M. Hopkins, J. Leiserson, J. Schichl, L. Ting, R. Cline, C. Duvvury, H. Kunz, and R. Steinhoff. Formation and suppression of a newly discovered secondary EOS event in HBM test systems. *Proceedings of the Electrical Overstress/Electrostatic Discharge (EOS/ESD) Symposium*, 2004; 141–145.
47. R.A. Ashton, B.E. Weir, G. Weiss, and T. Meuse. Voltages before and after HBM stress and their effect on dynamically triggered power supply clamps. *Proceedings of the Electrical Overstress/Electrostatic Discharge (EOS/ESD) Symposium*, 2004; 153–159.
48. C. Duvvury, R. Steinhoff, G. Boselli, V. Reddy, H. Kunz, S. Marum, and R. Cline. Gate oxide failures due to anomalous stress from HBM ESD testers. *Proceedings of the Electrical Overstress/Electrostatic Discharge (EOS/ESD) Symposium*, 2004; 132–140.
49. H. Kunz, R. Steinhoff, C. Duvvury, G. Boselli, and L. Ting. The effect of high pin-count ESD tester parasitics on transiently-triggered ESD clamps. *Proceedings of the Electrical Overstress/Electrostatic Discharge (EOS/ESD) Symposium*, 2004; 146–152.
50. S. Voldman. Report to the ESDA Standards Committee. WG 5.1 (HBM) *2004*. ESD Standards Workgroup 5.1 Human Body Model, ESD Association, 2004.
51. J. Barth, and J. Richner. Voltages before and after current in HBM testers and real HBM. *Proceedings of the Electrical Overstress/Electrostatic Discharge (EOS/ESD) Symposium*, 2005; 141–151.
52. R. Gaertner, R. Aburano, T. Brodbeck, H. Gossner, J. Schaafhausen, W. Stadler, F. Zaengl. Partitioned HBM test – A new method to perform HBM test on complex tests. *Proceedings of the Electrical Overstress/Electrostatic Discharge (EOS/ESD) Symposium*, 2005; 178–183.
53. T. Brodbeck, and R. Gaertner. Experience in HBM ESD testing of high pin count devices. *Proceedings of the Electrical Overstress/Electrostatic Discharge (EOS/ESD) Symposium*, 2005; 184–189.
54. M. Chaine, T. Meuse, R. Ashton, L.G. Henry, M.I. Natarajan, J. Barth, L. Ting, H. Geiser, S. Voldman, M. Farris, E. Grund, S. Ward, M. Kelly, V. Gross, R. Narayan, L. Johnson, R. Gaertner, and N. Peachey. HBM tester

parasitic effects on high pin count devices with multiple power and ground pins. *Proceedings of the Electrical Overstress/Electrostatic Discharge (EOS/ESD) Symposium*, 2006; 353–362.
55. R. Ashton, and E. Worley. Pre-pulse voltage in the human body model. *Proceedings of the Electrical Overstress/Electrostatic Discharge (EOS/ESD) Symposium*, 2006; 325–333.
56. H. Kunz, C. Duvvury, J. Brodsky, P. Chakraborty, A. Jahanzeb, S. Marum, L. Ting, and J. Schichl. HBM stress of no-connect IC pins and subsequent arc-over events that lead to human-metal-discharge-like events into unstressed neighbor pins. *Proceedings of the Electrical Overstress/Electrostatic Discharge (EOS/ESD) Symposium*, 2006; 24–31.
57. M. Etherton, J. Miller, V. Axelrod, H. Marom, and T. Meuse. HBM ESD failures caused by a parasitic pre-discharge current spike. *Proceedings of the Electrical Overstress/Electrostatic Discharge (EOS/ESD) Symposium*, 2008; 30–39.
58. T. Brodbeck, R. Gaertner, W. Stadler, and C. Duvvury. Statistical pin pair combinations – A new proposal for device level HBM tests. *Proceedings of the Electrical Overstress/Electrostatic Discharge (EOS/ESD) Symposium*, 2008; 106–114.
59. E. Grund. Two-pin human body model testing. *Proceedings of the Electrical Overstress/Electrostatic Discharge (EOS/ESD) Symposium*, 2009; 125–134.
60. J. Lebon, G. Jenicot, P. Moens, D. Pogany, S. Bychikhin. IEC vs. HBM: How to optimize on-chip protections to handle both requirements. *Proceedings of the Electrical Overstress/Electrostatic Discharge (EOS/ESD) Symposium*, 2009; 358–363.
61. S. Voldman. *ESD: Physics and Devices*. Chichester, England: John Wiley and Sons, Ltd., 2004.
62. S. Voldman. *ESD: Circuits and Devices*. Chichester, England: John Wiley and Sons, Ltd., 2005.
63. S. Voldman. *ESD: Circuits and Devices 2nd Edition*. Chichester, England: John Wiley and Sons, Ltd., 2015.
64. S. Voldman. *ESD: RF Circuits and Technology*. Chichester, England: John Wiley and Sons, Ltd., 2006.
65. S. Voldman. *Latchup*. Chichester, England: John Wiley and Sons, Ltd., 2007.
66. S. Voldman. *ESD: Circuits and Devices*. Beijing, China: Publishing House of Electronic Industry (PHEI), 2008.
67. S. Voldman. *ESD: Failure Mechanisms and Models*. Chichester, England: John Wiley and Sons, Ltd., 2009.
68. S. Voldman. *ESD: Design and Synthesis*. Chichester, England: John Wiley and Sons, Ltd., 2011.
69. S. Voldman. *ESD Basics: From Semiconductor Manufacturing to Product Use*. Chichester, England: John Wiley and Sons, Ltd., 2012.
70. S. Voldman. *Electrical Overstress (EOS): Devices, Circuits, and Systems*. Chichester, England: John Wiley and Sons, Ltd., 2013.
71. S. Voldman. *ESD: Analog Design and Circuits*. Chichester, England: John Wiley and Sons, Ltd., 2014.
72. Hanwa Electronics. http://www.hanwa-ei.co.jp. Hanwa S5000R HBM test system, 2016.
73. Hanwa Electronics. http://www.hanwa-ei.co.jp. Hanwa HED-N5000 HBM test system, 2016.
74. Grund Technical Solutions. http://www.grundtech.com. Arcus HBM Wafer and Device Test System, 2016.

3

Machine Model

3.1 History

The machine model (MM) test is another electrostatic discharge (ESD) standard that was used in the industry for many years to qualify components [1]. The MM standard was supported by the ESD Association standards work group; the last version as a standard was ANSI/ESD ESD-STM5.2 – 1999 [2]. MM test simulation is integrated into commercial ESD test systems that support both the human body model (HBM) and MM test simulation [1–3].

The MM event was intended to represent the interaction of the electrical discharge from a conductive source, which is charged, with a component or object. The model assumes that the "machine" is charged as the initial condition. The charged source then touches a component or object. In this model, an arc discharge is assumed to occur between the source and the component or object, allowing for current transfer between the charged object and the component or object.

The MM event was incorporated into a majority of commercial test systems since it is compatible with the HBM event. In some systems, the pulse source was interchanged to test either HBM or MM events.

3.2 Scope

The scope of MM ESD test is for testing, evaluation, and classification of components and micro- to nanoelectronic circuitry. The test is to quantify the sensitivity or susceptibility of these components to damage or degradation to the defined MM test [1–26].

3.3 Purpose

The purpose of the MM ESD test is for establishment of a test methodology to evaluate the repeatability and reproducibility of components to a defined pulse event in order to classify or compare ESD sensitivity levels of components [1, 2].

ESD Testing: From Components to Systems, First Edition. Steven H. Voldman.
© 2017 John Wiley & Sons, Ltd. Published 2017 by John Wiley & Sons, Ltd.

3.4 Pulse Waveform

An MM characteristic time is associated with the electrical components used to emulate the discharge process. In the MM standard, the circuit component is a 200 pF capacitor; unlike the HBM standard, there is no resistor component [1–3]. An arc discharge fundamentally has a resistance on the order of 10–25 Ω. The characteristic RC decay time associated with the MM test simulation is the arc discharge resistor R and the charged capacitor C [1–3]. Figure 3.1 is an example of the MM equivalent circuit (e.g., without the arc resistance term).

3.4.1 Comparison of Machine Model (MM) and Human Body Model (HBM) Pulse Waveform

Unlike the HBM waveform, the MM event transitions through positive and negative polarities (Figure 3.2). The MM pulse oscillation switching from positive to negative currents leads to different current paths through the ESD network and circuitry. The MM event is higher peak current compared to the HBM event, leading to failure level ratio from 5 to 30× lower typically in a semiconductor component. Figure 3.3 shows an example of an MM source from a commercial MM test system.

MM events are faster than the HBM event due to the lower series resistance. As a result, MM pulse widths are different than long-pulse electrical overstress (EOS) events [25]. In an MM event, the pulse width is less than the thermal diffusion time, approaching adiabatic conditions. This is opposite of EOS events whose pulse widths are greater than the thermal diffusion time [25].

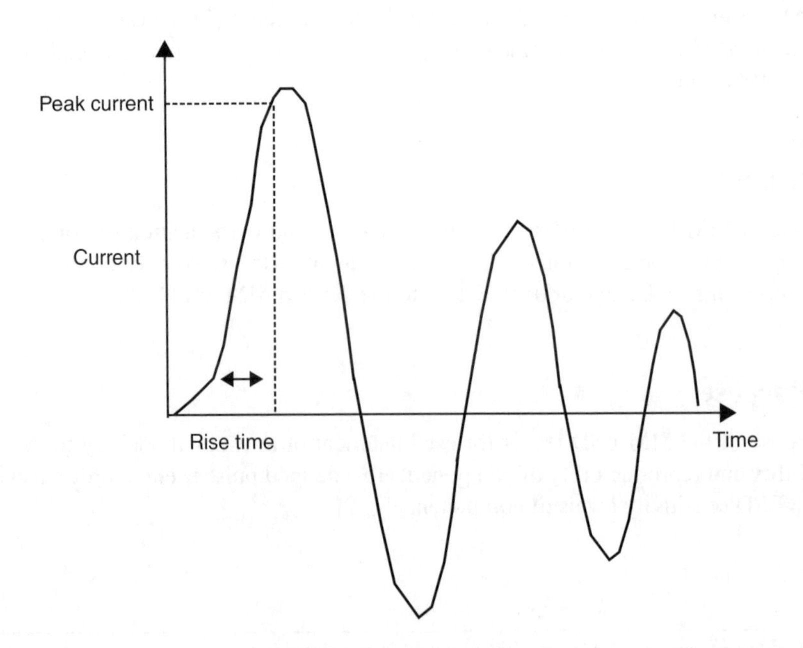

Figure 3.1 MM pulse waveform

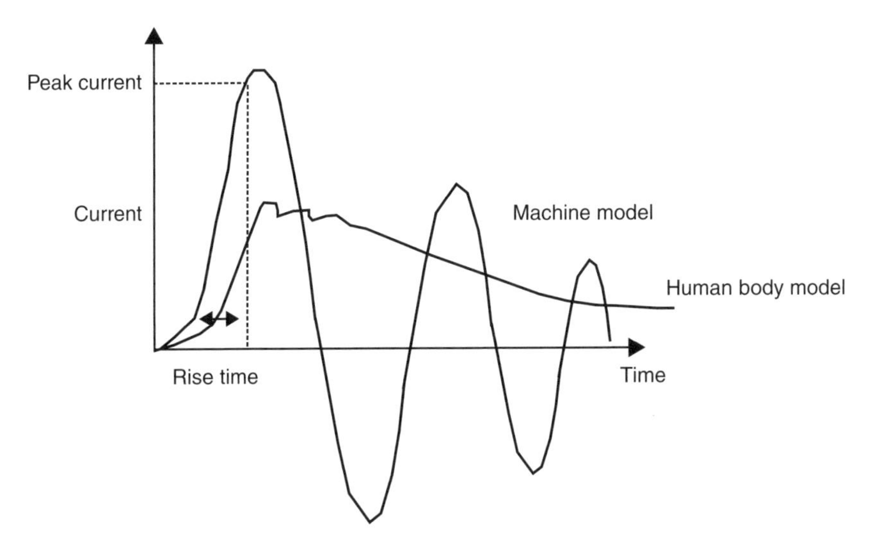

Figure 3.2 MM and HBM waveform comparison

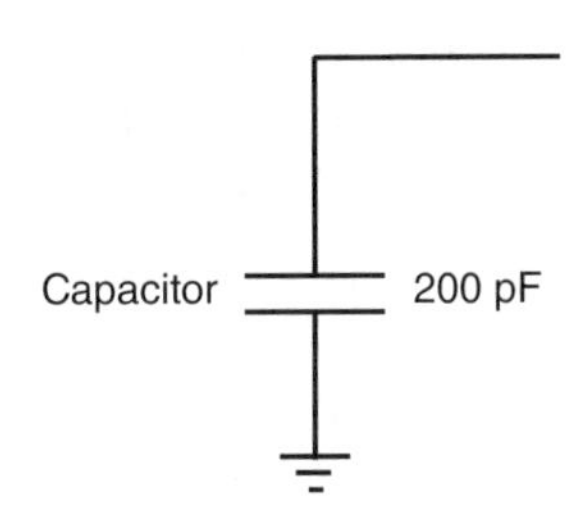

Figure 3.3 MM equivalent circuit model

3.5 Equivalent Circuit

The MM equivalent circuit can be represented as a single capacitor element. Figure 3.3 shows the MM equivalent circuit of a 200 pF capacitor [1, 2].

The MM is influenced by the discharge current, parasitic, and load conditions. Figure 3.4 addresses the MM equivalent circuit taking into account the parasitic and load components. In the figure, the discharge path series resistance, the discharge path series inductance, the test board capacitance, and load resistance are shown.

3.6 Test Equipment

MM commercial test equipment exists for the qualification and release of semiconductor components. Figure 3.5(a) shows an example of an MM source for an early commercial MM test system. In these sources, the capacitor element is a module that can be placed in the commercial tester [24].

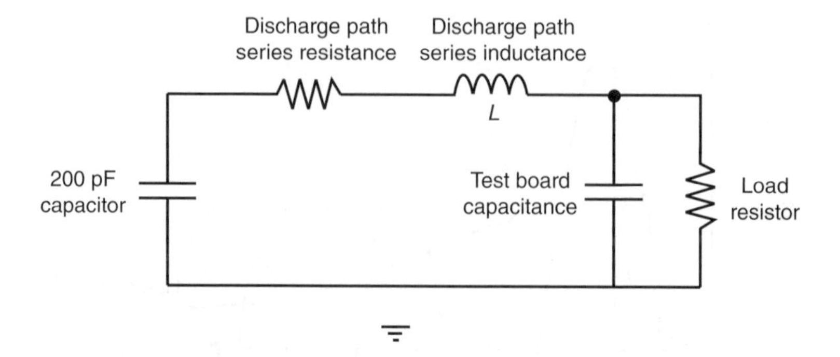

Figure 3.4 MM equivalent circuit model with parasitics

Figure 3.5 MM source

MM commercial test equipment will be required to release an MM pulse to the device under test (DUT). The commercial test equipment must have a high-voltage source used to charge the MM source circuit. A switch is contained within the test system that is closed when the high-voltage source charges the equivalent circuit. A second switch must be open during the charging of the MM source circuit. The second switch is closed when the MM circuit current is discharged into the load or DUT.

Figure 3.6(a) shows an example of a present-day MM ESD commercial test system. This test system is the Hanwa HCE-5000 ESD test system [27]. The HCE-5000 is designed for portability, improved use of space, and high cost performance. This test system can perform HBM and MM measurements.

The posttesting leakage measurement can be done without an independent computer or a curve tracer. The settings on the test system can be adjusted without an additional computer. This test system satisfies standards for JEDEC, ESDA, AEC, and JEITA standards organizations. Leakage current and curve tracing can be evaluated before and after testing. The methodology for failure can include percentage change (e.g., 50% shift) or absolute value

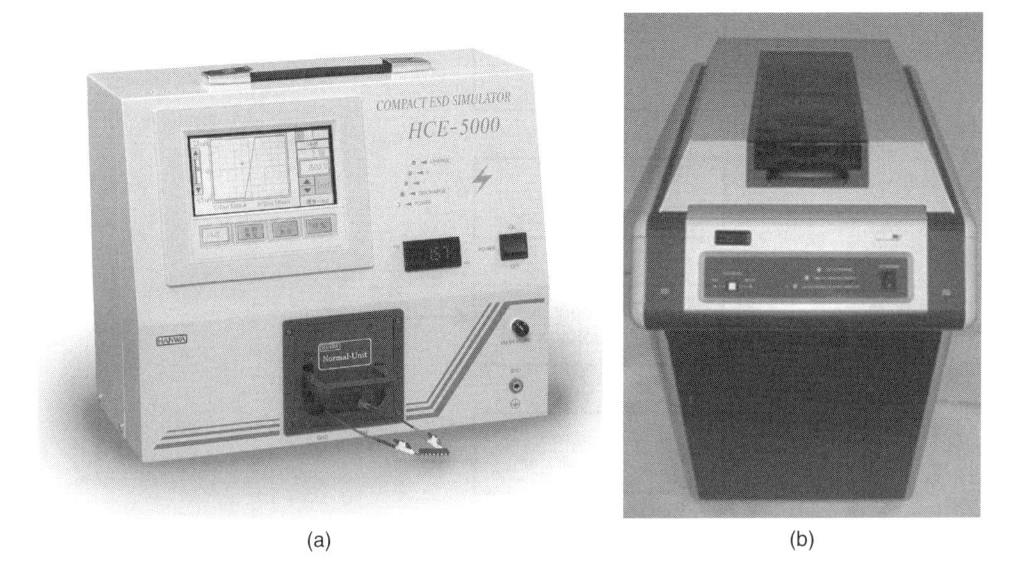

Figure 3.6 (a) MM commercial test system. (b) MM commercial test system – MK1 Thermo-KeyTek (Reproduced with permission of Thermo Fisher Scientific)

(e.g., 1 μA). The HCE-5000 charging voltage can be set from 10 V to 4 kV, in 5 V step increments. The test system can provide both a positive and negative polarity pulse. The leakage voltage accuracy is 1% – +50 mV [27].

Figure 3.6(b) is a second commercial test system used for MM testing. This system is the Thermo/KeyTek MK1 [28]. The Thermo Scientific™ MK1 system is an ESD simulator designed for high-yield testing of devices up to 256 pins. The system is a relay-based system that establishes pin combinations by a relay switching matrix. It is 5–10 times faster than robotic-driven ESD testers. The test system addresses the ESDA STM 5.2 specification, as well as JEDEC EIA/JESD22-A115 and AEC Q100-003 specifications. The test system has an MM voltage range from 30 V to 1 kV. The system provides a prestress, prefail, and postfail data with full curve tracing capability [28].

3.7 Test Sequence and Procedure

In the MM test, there is a test sequence and a procedure. Figure 3.7 is a flowchart of the test sequence and procedure for MM testing. In the testing of components to large increments for semiconductor device characterization, a grounded reference is established. An initial reading is performed prior to any applied voltage on the MM source. A first step increment is defined on the capacitor element. All pins are pulsed with an ESD event, individually relative to the grounded reference. Each pin is evaluated based on failure criteria (e.g., voltage shift, leakage). If a pin "passes," it remains in the test sequence for the next increment. Pins that "fail" are removed from the test sequence, where all other passing pins remain in the test sequence. The voltage is increased on the capacitor source to the next step level, and testing resumes for the surviving pins. In this case, after the desired steps are taken, the tester is stopped, and the failed pins versus surviving pins are recorded.

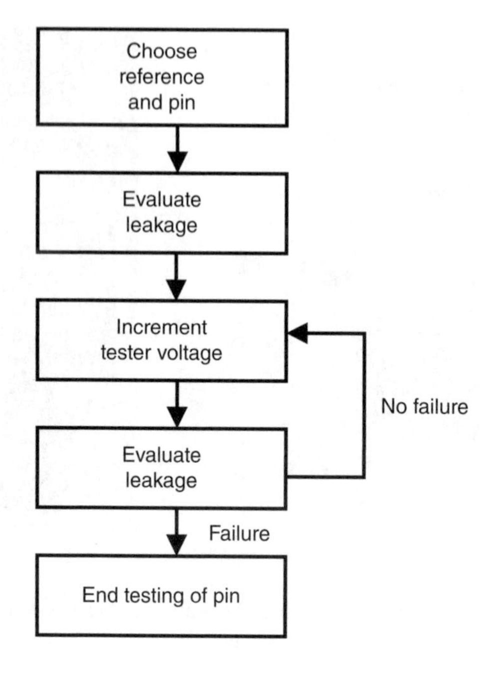

Figure 3.7 MM test sequence

In semiconductor corporations that test ESD in-house for characterization or qualification, sometimes a single increment at the test specification standard is completed. In this case, it is "pass–fail" criteria. If all the pins show no degradation after applying the MM pulse to the DUT, then the test is completed.

Figure 3.8 is a flowchart of the test sequence and procedure for MM testing where the testing is done in small increments. In the testing of components to small increments for semiconductor device characterization, a first precheck of the pins is performed prior to charging the capacitor source. In small increment testing, a grounded reference is established. A first increment is defined on the capacitor element. All pins are pulsed with an ESD event, individually relative to the grounded reference. Each pin is evaluated based on failure criteria (e.g., voltage shift, leakage). If the pin "passes," it remains in the test sequence. Pins that "fail" are removed from the test sequence, where all other passing pins remain in the test sequence. The voltage is increased on the capacitor source to the next increment level, and testing resumes for the surviving pins. This continues until all the pins have failed the failure criterion. In this methodology, with small increments, the actual failure level is determined for all the pins that are tested, with an accuracy associated with the increment step size.

In semiconductor corporations that in-house ESD test, test in small increments. In older MM test equipment, the smallest increment where accuracy was preserved was 30 V increments. In the newer test systems, the increments were reduced to 10 V increments. ESD testing would be initiated at 0 V as the first evaluation of the leakage characteristic prior to applying a pulse to the DUT. The first voltage increment was then applied to all pins relative to a given reference (e.g., V_{DD} or V_{SS} power pin). A leakage measurement was then taken after application of the pulse event on each pin. All pins that survived the pulse event will continue to be tested to the

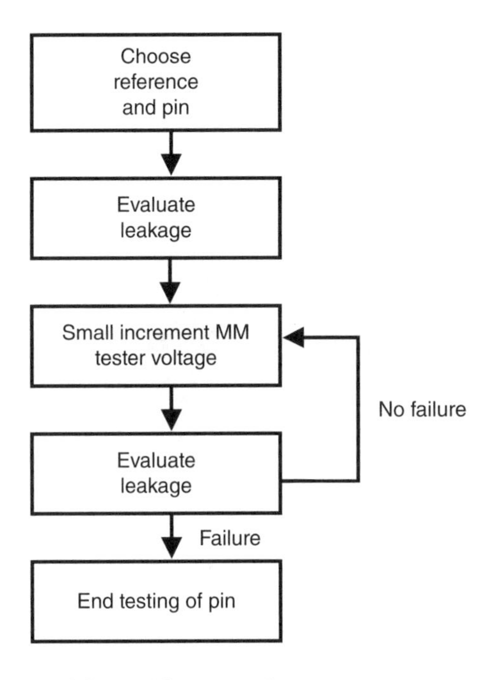

Figure 3.8 MM incremental step stress test sequence

next test increment. In the case where all the pins are tested to failure, the testing will continue until all the pins have failures. All the results are recorded.

3.8 Failure Mechanisms

MM failure mechanisms are typically associated with failures on the peripheral circuitry of a semiconductor chip that are connected to signal pins [17–22]. MM failures can occur on the power rails due to inadequate bus widths and ESD power clamps between the power rails. MM failures can occur in both passive and active semiconductor devices. The failure signature is typically isolated to a single device or a few elements in a given current path where the current exceeded the capability of the element.

ESD power clamp circuits are designed to be responsive to MM pulse widths; specifically, the RC-triggered MOSFET ESD power clamp is designed for MM events.

MM ESD failures are also distinct from EOS events. MM events will not typically cause failures in the package, printed circuit board (PCB), or single component devices mounted on a PCB (Figure 3.9).

3.9 MM ESD Current Paths

MM event current is oscillatory in nature [1, 2]. The MM waveform is a damped oscillation, which alternates in polarity. As a result, the current path through the circuitry is distinct from HBM, which is a single polarity event. Figure 3.10 illustrates an example of a signal pin, with

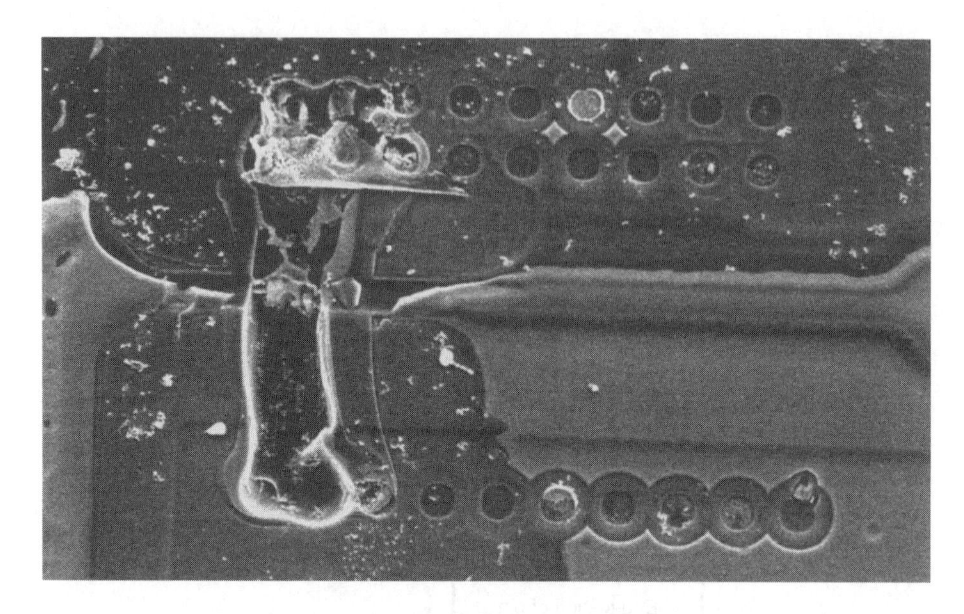

Figure 3.9 MM ESD failure mechanisms

an MM event where the VDD power rail is the grounded reference. During the positive part of the oscillation, the current will flow through the ESD element between the signal pad and the V_{DD} power rail signal pad. As illustrated, an ESD diode element will forward bias to allow current flow to the V_{DD} power pad. For the negative oscillation, the p–n diode between the signal pin and the ground rail V_{SS} will forward bias. ESD current will flow from the V_{DD} power pad, through the V_{DD} power grid, through the ESD power clamp, and back through the diode element connected to the signal pad. After this negative oscillation, it is followed by the second positive oscillation. Hence, during this event, the current is switching current paths as the polarity changes until the MM pulse is completed.

Figure 3.11 illustrates an example of a signal pin, with an MM event where the V_{SS} power rail is the grounded reference. During the positive part of the oscillation, the current will flow through the ESD element between the signal pad and the V_{DD} power rail signal pad. As illustrated, the ESD diode element will forward bias to allow current flow to the V_{DD} power rail. The current will flow through the V_{DD} power rail to the ESD power clamp placed between the V_{DD} and V_{SS} power rails. The current will flow through the ESD power clamp to the V_{SS} power rail and V_{SS} bond pad.

For the negative oscillation, the p–n diode between the signal pin and the ground rail VSS will forward bias. After this negative oscillation, it is followed by the second positive oscillation. Hence, during this event, the current is switching current paths as the polarity changes until the MM pulse is completed.

Figure 3.12 illustrates an example of an MM event where the V_{SS} power rail is the grounded reference and the pulse is applied to the V_{DD} power rail bond pad. During the positive part of the oscillation, the current will flow through the ESD power clamp and the power rails. The current will flow through the V_{DD} power rail to the ESD power clamp placed between the V_{DD}

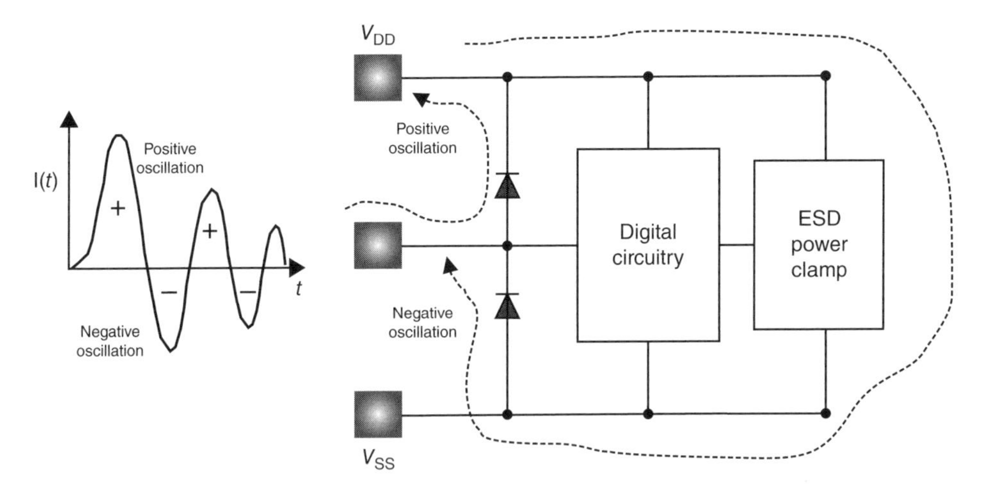

Figure 3.10 MM ESD event current paths – signal pin to power (V_{DD})

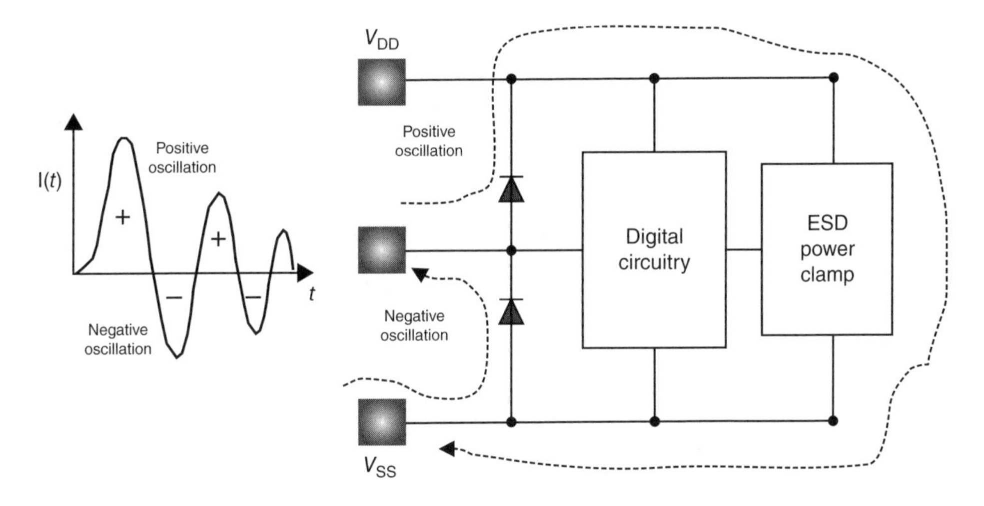

Figure 3.11 MM ESD event current paths – signal pin to ground power (V_{SS})

and V_{SS} power rails. The current will flow through the ESD power clamp to the V_{SS} power rail and V_{SS} bond pad.

For the negative oscillation, the current will flow in the reverse direction, through the ESD power clamp, as well as through the ESD diode network. After this negative oscillation, it is followed by the second positive oscillation repeating the current flow through the ESD power clamp. Hence, during this MM event, the current is switching current paths as the polarity changes until the MM pulse is completed.

Figure 3.13 illustrates an example of an MM event between two ground power rails. Because the ground-to-ground ESD network is bidirectional, current can flow between the two rails for both the positive and negative oscillations.

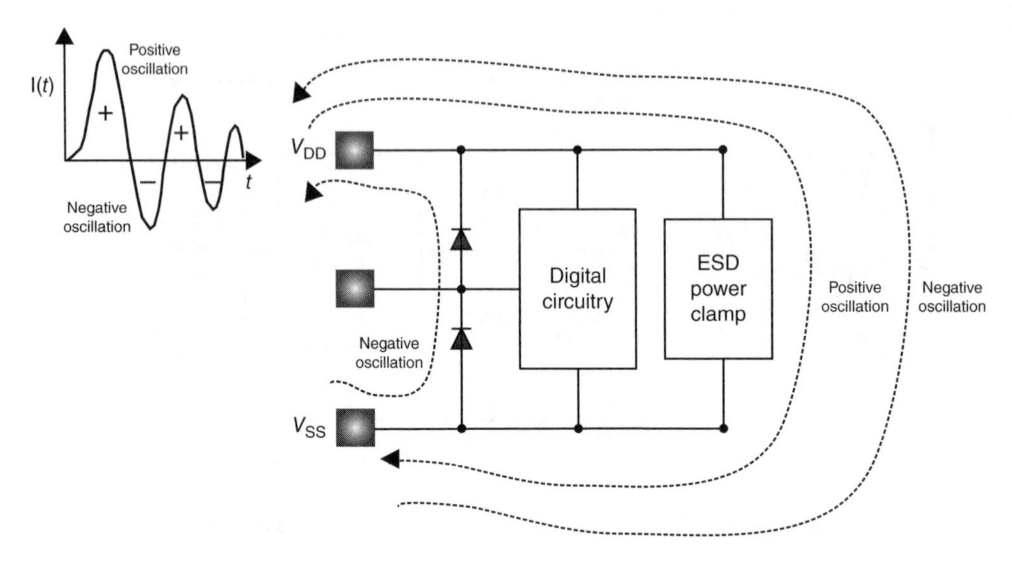

Figure 3.12 MM ESD event – power rail (V_{DD}) to ground power rail (V_{SS})

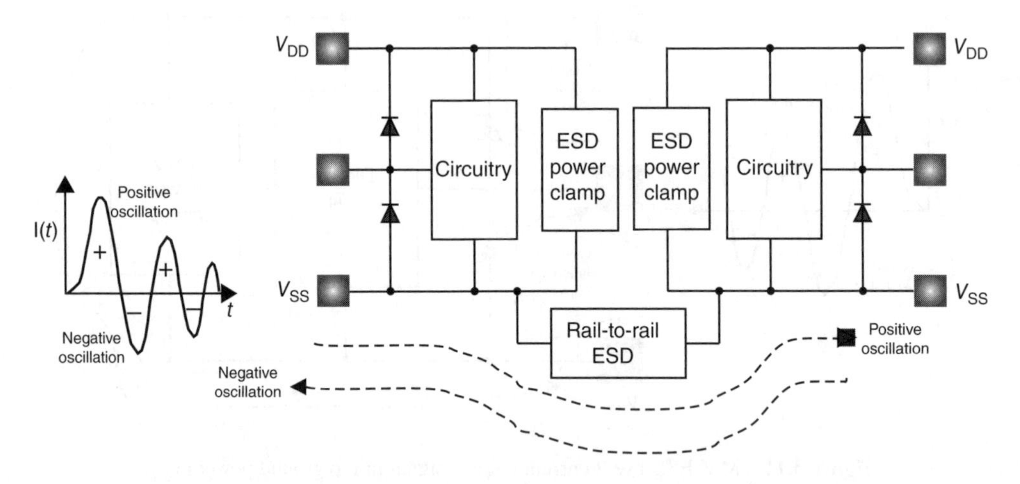

Figure 3.13 MM ESD event – power ground (V_{SS}) to ground power rail (V_{SS})

3.10 MM ESD Protection Circuit Solutions

To address MM events, ESD protection circuits provide an alternate current path to avoid the sensitive circuit elements [21–23]. Figure 3.14 shows an example of a dual-diode ESD network. For the MM event, the dual-diode circuit addresses both the positive and negative oscillations. For a positive MM test, the first peak is a positive sinusoidal oscillation, which turns on the first p–n diode connected to the power supply V_{DD}. For the first negative oscillation, the second p–n diode is turned on between the n-cathode and the p-substrate. The dual-diode ESD network is a natural solution for this type of event [18, 19].

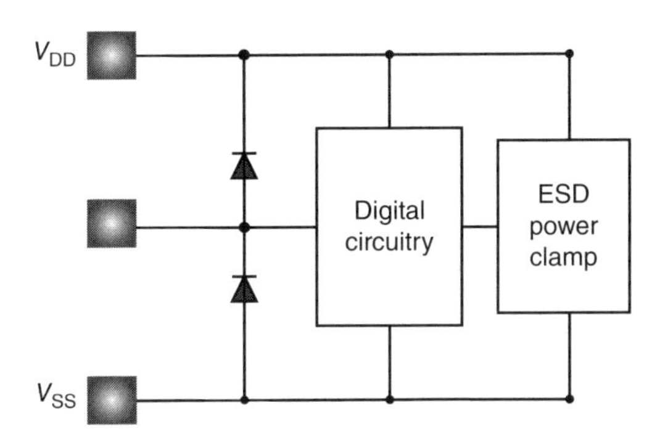

Figure 3.14 MM ESD protection circuits – dual diode

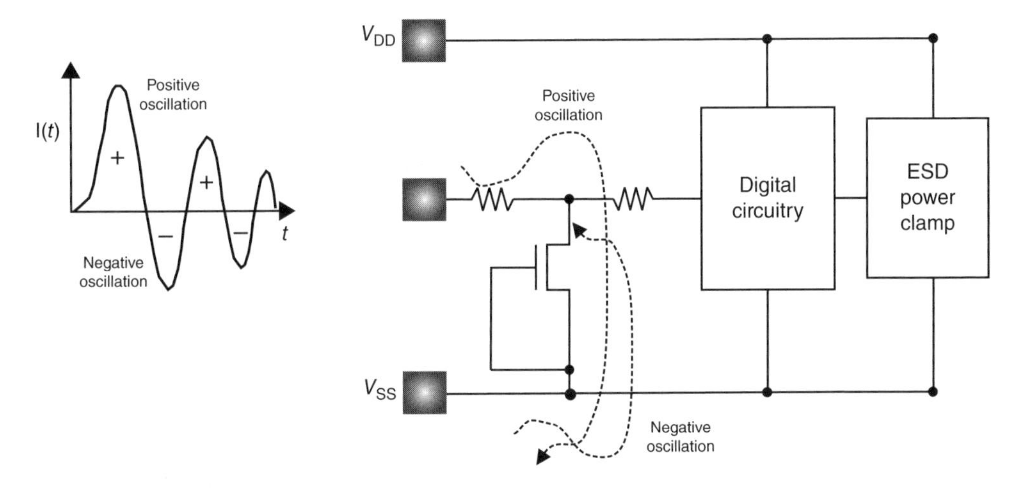

Figure 3.15 MM ESD protection circuits – grounded gate NMOS (GGNMOS)

Figure 3.15 shows an example of a grounded gate n-channel MOSFET (GGNMOS) ESD network. For the MM event, the GGNMOS circuit addresses both the positive and negative oscillations through MOSFET snapback and forward bias of the MOSFET drain. For a positive MM test, the first peak is a positive sinusoidal oscillation, which turns on the MOSFET through MOSFET snapback. For the first negative oscillation, the second p–n diode is turned on between the n-doped MOSFET drain and the p-substrate.

Figure 3.16 shows an example of a bidirectional silicon controlled rectifier (SCR) ESD network [19]. For the MM event, the bidirectional SCR circuit addresses both the positive and negative oscillations through pnpn turn-on. For a positive MM test, the first peak is a positive sinusoidal oscillation, which turns on the SCR from the signal pin to the ground rail. For the first negative oscillation, a second SCR is turned on between the ground rail and signal pad.

Figure 3.17 shows an example of an ESD network utilized between the V_{DD} and V_{SS} power rails. This commonly used circuit is known as an RC-triggered MOSFET ESD power clamp. The RC network is tuned to respond to an MM pulse to allow turn-on of the MOSFET between

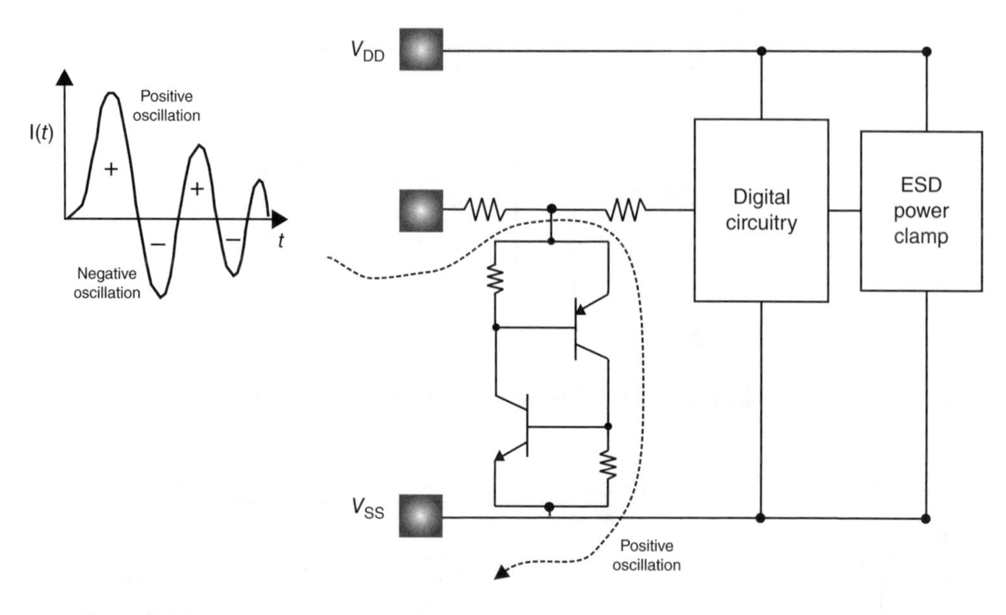

Figure 3.16 MM ESD protection circuits – bidirectional silicon controlled rectifier (SCR)

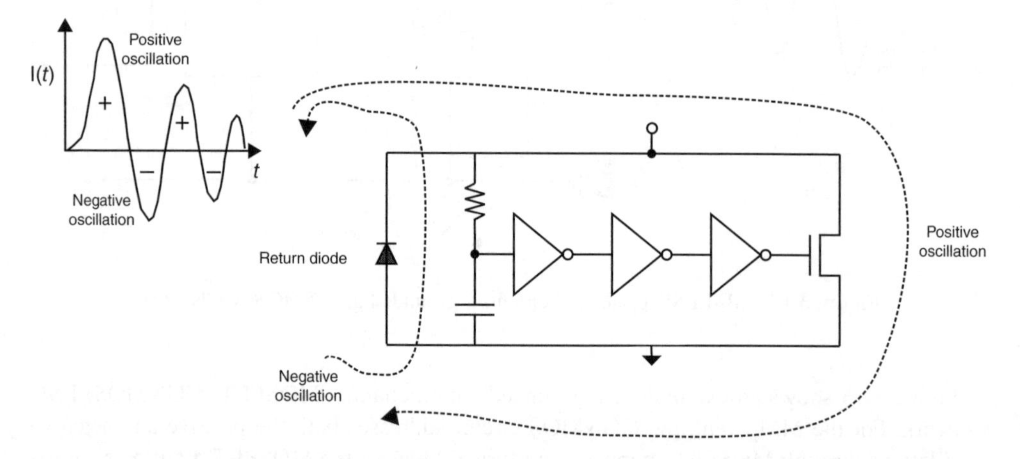

Figure 3.17 MM ESD protection circuits – V_{DD} to V_{SS} power rail RC-triggered power clamp network and return diode

V_{DD} and V_{SS}. For a positive MM test, the first peak is a positive sinusoidal oscillation, which turns on the RC-trigger network from the V_{DD} to V_{SS} rail. For a negative oscillation, the current flows through the "return diode" placed between the two power rails [23].

In many corporations, the ESD power clamp consists of two circuits for the different polarity pulse events. Hence, a "return diode" is used to clearly define an ESD solution for the current flowing from the V_{SS} power rail to the V_{DD} power rail. In the case of a MOSFET-based RC-triggered power clamp, there is a large n-diffusion drain on the MOSFET that forms a p–n diode between V_{DD} and V_{SS}.

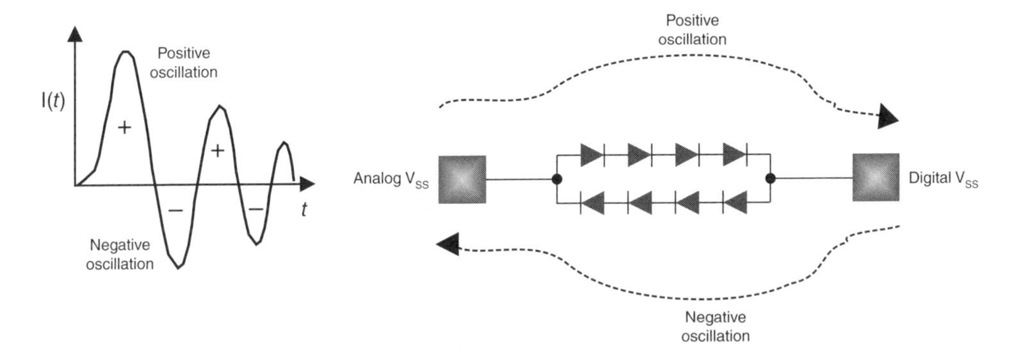

Figure 3.18 MM ESD protection circuits – V_{SS}-to-V_{SS} rail-to-rail network

Figure 3.18 shows an example of an ESD network utilized between two V_{SS} power rails [18–21]. This circuit is designed to be bidirectional. The bidirectional nature of this circuit makes it a natural solution for the MM event. For a positive MM test, the first peak is a positive sinusoidal oscillation, which turns on the p–n diode string from the V_{SS} to V_{SS} rail. For a negative oscillation, the current flows through the "return diode string" placed between the two power rails.

3.11 Alternate Test Methods

Alternate test methods exist in the industry for testing of components and systems. In this section, the small charge model (SCM) ESD test is discussed.

3.11.1 Small Charge Model (SCM)

An ESD test model is known as the small charge model (SCM) is used in many corporations. For some customers that use the SCM, it is also known as the Cassette Model. This is also used in the game industry, and many refer to it as the Nintendo model. Figure 3.19 shows an SCM source equivalent circuit. The small charge source is a 10 pF capacitor element that placed inside a commercial tester.

Figure 3.20 shows an SCM source. The small charge source is a 10 pF capacitor element that placed inside a commercial tester. For test systems and semiconductor suppliers that do

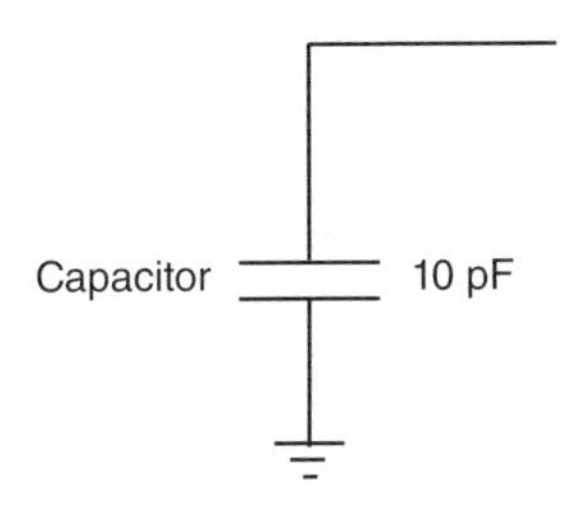

Figure 3.19 Small charge model (SCM) equivalent circuit (Reproduced with permission of Hanwa Corporation)

Figure 3.20 Small charge model (SCM) source (Reproduced with permission of Hanwa Corporation)

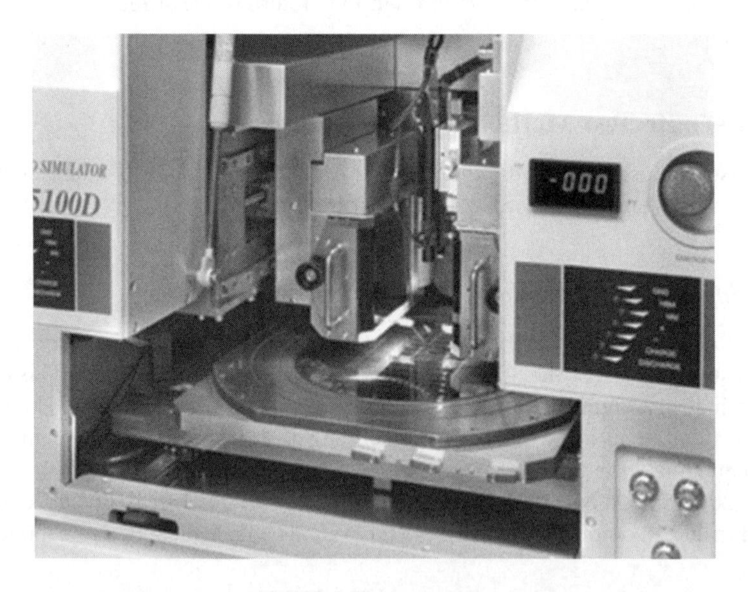

Figure 3.21 Small charge model (SCM) commercial tester

not have an SCM source, a "charge equivalency" is established for the MM test; a target is defined for the MM test that will establish equivalency.

Figure 3.21 shows an SCM commercial test system. Test system parasitic must be low due to the smaller capacitor element used for testing. The wafer level ESD tester the Hanwa HED-W51000D can perform MM and SCM tests on wafer [29]. This fully automatic MM and SCM test system can test wafers of 300 mm. This test system is in conformance with the JEITA, ESDA, and JEDEC test standards.

Since many suppliers do not have an SCM commercial tester, it is possible to use an MM ESD test system and evaluate the correlation between the SCM and MM ESD tests. This is possible, since there is a correlation between the SCM test and the MM test. For example, the MM can be used to verify the desired level needed for passing the SCM test requirement. As a result, corporations that desire passing the SCM test will request a certain magnitude that is achieved in the MM test for qualification.

3.12 Machine Model to Human Body Model Ratio

In the testing of components, there exists a relationship between the MM results and the HBM test results. A plot can be defined, which plots semiconductor chip results, with the HBM and MM results on the *x*-axis and *y*-axis. Experimental results show that the ratio of HBM-to-MM ESD results can be a value of 10×. Although this is the most often quoted HBM-to-MM ratio, this value can range between 5× and 20× [9].

Figure 3.22 shows an example of HBM versus MM plot for semiconductor components. The plot shows the HBM-to-MM ratio superimposed on the plot for the ratio of 5:1, 10:1, and 20:1. As can be observed from this sample of data, the HBM-to-MM ratio can vary from 5:1 to 20:1. For this semiconductor technology, and product type, the data falls mostly between 10:1 and 20:1. As a result, the MM results were lower than expected.

Another representation of the results is to plot the HBM-to-MM ratio versus HBM. Figure 3.23 shows an example of HBM-to-MM ratio versus HBM plot. From the results, it can be observed that the HBM-to-MM ratio can vary between superimposed on the plot from to over 20:1. In the case that a product must pass the ESD specification for HBM, and MM testing, an awareness of the ratio for a technology, design or corporation.

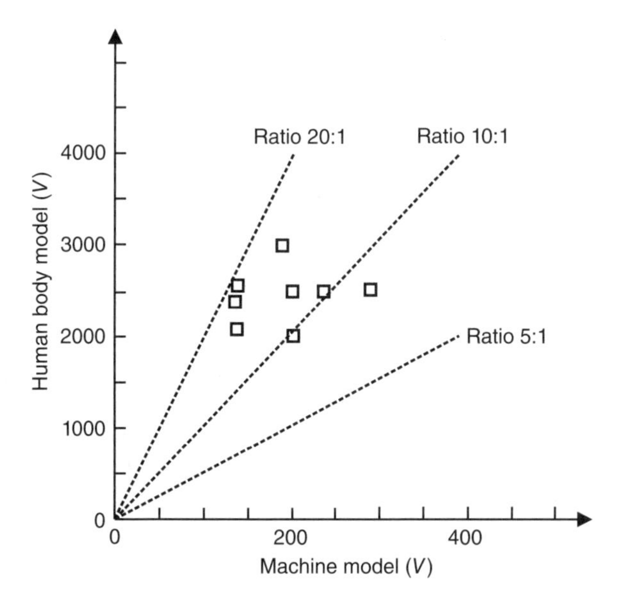

Figure 3.22 HBM to MM ratio

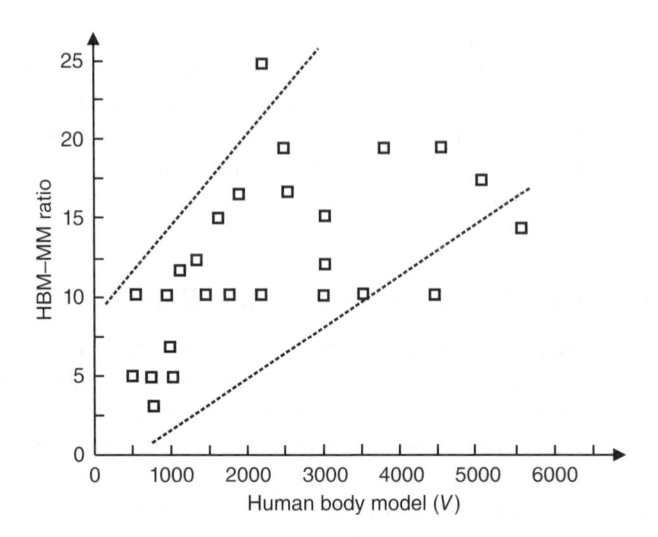

Figure 3.23 HBM to MM ratio versus HBM voltage

3.13 Machine Model Status as an ESD Standard

Today, there is an effort to downrate the MM test as a standard [30, 31]. Historically, this test was valuable to determine the response to an oscillatory nondamped waveform. Using this test in conjunction with the HBM, is important to determine the robustness of devices, components, and systems. As discussed in the preceding section, the ratio of HBM to MM ESD results can vary from 5:1 to 20:1. As a result, if the product has a 20:1 ratio, then the HBM results must be significantly high in order to achieve a good objective for MM results.

In relevance to EOS, the MM test has a higher peak current than that of the HBM standard test. In addition, for EOS analysis, the MM test is similar to the IEC 61000-4-5 surge test in its two polarities.

3.14 Summary and Closing Comments

In Chapter 3, one of the most important tests in the history of ESD testing, the MM standard is discussed. This chapter focused on model (HBM) purpose, scope, and waveform to test procedures and in addition indicated many of the present and future anticipated changes that will occur on this test. The changes include sample testing, rotation testing, to two-pin test methods. An important issue of the correlation of the HBM and MM test is discussed, as well as the ramification of this correlation issue.

In Chapter 4, the charged device model (CDM) test is discussed. The CDM is also an important test standard in today's semiconductor chip development. Chapter 4 highlights the test procedures as well as the current paths and failure mechanisms.

Problems

1. Assume the 200 pF capacitor of the MM source is charged to a given voltage, and assume a chip of capacitance C_{chip} is uncharged. The switch is closed. Calculate the net voltage on the chip, assuming the resistance of the switch is zero.
2. Assume the 10 pF capacitor of the SCM source is charged to a given voltage, and assume a chip of capacitance C_{chip} is uncharged. The switch is closed. Calculate the net voltage on the chip, assuming the resistance of the switch is zero.
3. Assume a semiconductor chip has a dual-diode ESD network on its signal pad, and an RC-triggered MOSFET power clamp between V_{DD} and V_{SS}. Assume that there is a diode between V_{DD} and V_{SS} in parallel with the RC-triggered circuit. Show the current paths through the first four oscillations of the MM pulse.
4. Assume a semiconductor chip has a dual-diode ESD network on its signal pad, and an RC-triggered MOSFET power clamp between V_{DD} and V_{SS}. Assume that there is no diode between V_{DD} and V_{SS} in parallel with the RC-triggered circuit. Show the current paths through the first four oscillations of the MM pulse. What is different from Problem 3?
5. Given the ratio of HBM to MM results of a product is 20:1, what HBM level should a product achieve, given that the MM specification is 200 V?
6. Given the ratio of HBM to MM results of a product is 10:1, what HBM level should a product achieve, given that the MM specification is 200 V?
7. A technology team's goal for qualification is 2000 V HBM and 200 V MM ESD specification. In order to pass both specifications, what is the maximum ratio of HBM to MM that is acceptable?
8. Two microprocessor chips are pasted together with polymide separated by 10 μm of polyimide for a system-on-chip (SOC) application. Assume the chip is 10 mm × 20 mm in size. A capacitor is formed between the two chips. What is the amount of capacitance between the two chips? Assume a 20 mm × 20 mm chip size. What is the capacitance? How does this compare to an MM source capacitance?

References

1. ESD Association. *ESD Association Standard Test Method for the Protection of Electrostatic Discharge Sensitive Items – Electrostatic Discharge Sensitivity Testing – Machine Model (MM) Testing – Component Level*, 1994.
2. ANSI/ESD ESD-STM 5.2 – 1999. *ESD Association Standard Test Method for the Protection of Electrostatic Discharge Sensitive Items – Electrostatic Discharge Sensitivity Testing – Machine Model (MM) Testing – Component Level.* Standard Test Method (STM) document, 1999.
3. ANSI/ESD ESD-STM 5.1 – 2007. *ESD Association Standard Test Method for the Protection of Electrostatic Discharge Sensitive Items – Electrostatic Discharge Sensitivity Testing – Human Body Model (HBM) Testing – Component Level.* Standard Test Method (STM) document, 2007.
4. ANSI/ESD SP 5.1.2 – 2006. *ESD Association Standard Practice for the Protection of Electrostatic Discharge Sensitive Items – Human Body Model (HBM) and Machine Model (MM) Alternative Test Method: Split Signal Pin – Component Level*, 2006.
5. J.C. Martin. *Duration of Resistive Phase and Inductance of Spark Channels. Internal Technical Article. SSWA/JCM/1065/25*, Allermaston, England: Atomic Weapons Research Establishment.

6. M. Kelly, T. Diep, S. Twerefour, G. Servais, D. Lin, and G. Shah. A comparison of electrostatic discharge models and failure signatures for CMOS integrated circuit devices. *Proceedings of the Electrical Overstress/Electrostatic Discharge (EOS/ESD) Symposium*, 1995; 175–185.
7. H. Ishizuka, K. Okuyama, K. Kubota, M. Komuro, and Y. Hara. A study of ESD protection devices for input pins: Discharge characteristics of diode, lateral-bipolar and thyristor under MM, and HBM tests. *Proceedings of the Electrical Overstress/Electrostatic discharge (EOS/ESD) Symposium*, 1997; 255–262.
8. Y. Fukuda. The relation between ESD current or voltage waveforms and damage phenomena on semiconductor device. RCJ Reliability Symposium, Japan, 8E-06, 1998.
9. G. Notermans, P. de Jong, and F. Kuper. Pitfalls when correlating TLP, HBM and MM testing. *Proceedings of the Electrical Overstress/Electrostatic Discharge (EOS/ESD) Symposium*, 1998; 170–176.
10. K. Yokoi, and T. Watanabe. A study of wafer level ESD testing. *Proceedings of the Electrical Overstress/Electrostatic Discharge (EOS/ESD) Symposium*, 2000; 1–6.
11. J. Walraven, J.M. Soden, E.I. Cole, D.M. Tanner, and R.E. Anderson. Human Body Model, Machine Model, and Charged Device Model ESD testing of surface micromachined microelectromechanical systems (MEMS). *Proceedings of the Electrical Overstress/Electrostatic Discharge (EOS/ESD) Symposium*, 2001; 238–248.
12. J. Barth, J. Richner, L.G. Henry, and M. Kelly. Real HBM and MM – The dV/dt threat. *Proceedings of the Electrical Overstress/Electrostatic Discharge (EOS/ESD) Symposium*, 2003; 179–187.
13. JEP155. Recommended Target Levels for HBM/MM Qualification. www.jedec.org and http://www.esdtargets.blogspot.com., March 2012.
14. K.T. Kaschani, M. Hofmann, M. Schimon, M.W. Wilson and R. Vahrmann. On the significance of the Machine Model. ESD Forum 2007, Munich, Germany, 4–5 December 2007.
15. M. Tanaka, JEITA/JEDEC Meetings, Tokyo, 2011.
16. M. Tanaka, K. Okada, and M. Sakimoto, Clarification of Ultra-high-speed electrostatic discharge and unification of discharge model. *Proceedings of the Electrical Overstress/Electrostatic Discharge (EOS/ESD) Symposium*, 1994; 170–181.
17. S. Voldman. *ESD: Physics and Devices*. Chichester, England: John Wiley and Sons, Ltd., 2004.
18. S. Voldman. *ESD: Circuits and Devices*. Chichester, England: John Wiley and Sons, Ltd., 2005.
19. S. Voldman. *ESD: Circuits and Devices 2nd Edition*. Chichester, England: John Wiley and Sons, Ltd., 2015.
20. S. Voldman. *ESD: RF Circuits and Technology*. Chichester, England: John Wiley and Sons, Ltd., 2006.
21. S. Voldman. *ESD: Circuits and Devices*. Beijing, China: Publishing House of Electronic Industry (PHEI), 2008.
22. S. Voldman. *ESD: Failure Mechanisms and Models*. Chichester, England: John Wiley and Sons, Ltd., 2009.
23. S. Voldman. *ESD: Design and Synthesis*. Chichester, England: John Wiley and Sons, Ltd., 2011.
24. S. Voldman. *ESD Basics: From Semiconductor Manufacturing to Product Use*. Chichester, England: John Wiley and Sons, Ltd., 2012.
25. S. Voldman. *Electrical Overstress (EOS): Devices, Circuits, and Systems*. Chichester, England: John Wiley and Sons, Ltd., 2013.
26. S. Voldman. *ESD: Analog Design and Circuits*. Chichester, England: John Wiley and Sons, Ltd., 2014.
27. Hanwa Electronics. http://www.hanwa-ei.co.jp. Hanwa HED-N5000 HBM test system, 2016.
28. Hanwa Electronics. http://www.hanwa-ei.co.jp. Hanwa HCE 5000 test system, 2016.
29. Thermo Scientific. http://www.thermofisher.com. Thermo Scientific™ MK1 test system, 2016.
30. Hanwa Electronics. http://www.hanwa-ei.co.jp. Hanwa HED-W51000D test system, 2016.
31. C. Duvvury, R. Ashton, A. Righter, D. Eppes, H. Gossner, T. Welsher, M. Tanaka. Discontinuing use of the Machine Model for device ESD qualification. https://www.esda.org/assets/The-Machine-Model2012.pdf, 2016.

4

Charged Device Model (CDM)

4.1 History

The charged device model (CDM) is an electrostatic discharge (ESD) test method that is part of the qualification of semiconductor components [1–65]. The CDM standard is supported by ESD Association as ANSI/ESD ESD-STM5.3.1 – 1999 [1, 2]. Presently, there are four CDM test standards (ESDA S5.3.1, JEDEC JESD22-C101, AEC-Q100-011 Rev. C, and JEITA ED-4701-300). Each requires different test platform, testing, waveform, and calibration requirements. The CDM event is associated with the charging of the semiconductor component through different charging processes. Charging of the package can be achieved through direct contact charging or field-induced charging processes. The field-induced charging method is called the field-induced charged device model (FICDM).

There is presently an effort to align the CDM standards between the ESD Association and the JEDEC organization by establishing a joint ESDA/JEDEC standard. The ESDA/JEDEC joint standard (JS-002 2014) will replace existing CDM ESD standards JEDS22-C101 and ANSI/ESD S5.3.1. The new joint standard will preserve test systems in the field and improve the waveform measurement process.

In addition, there are multiple types of CDM-like tests. Figure 4.1 shows examples of different CDMs. These include the CDM, the socketed device model (SDM) [60, 61], and the charged board model (CBM) [48]. The SDM was developed into a standard but was eliminated as a standard after years of research and development [60, 61]. It was found that there was inductive coupling of signals in the test system, leading to faulty response. The CBM is presently under development toward a standard [48].

4.2 Scope

The scope of CDM ESD test is for the testing, evaluation, and classification of components and micro- to nanoelectronic circuitry. The test is to quantify the sensitivity or susceptibility of these components to damage or degradation to the defined CDM test [1, 2].

ESD Testing: From Components to Systems, First Edition. Steven H. Voldman.
© 2017 John Wiley & Sons, Ltd. Published 2017 by John Wiley & Sons, Ltd.

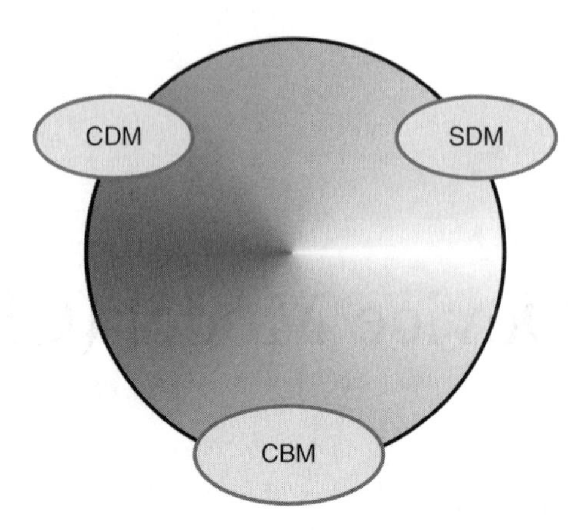

Figure 4.1 Charged device models – CDM, SDM, and CBM

4.3 Purpose

The purpose of the CDM ESD test is for establishment of a test methodology to evaluate the repeatability and reproducibility of components to a defined pulse event in order to classify or compare ESD sensitivity levels of components [1, 2].

4.4 Pulse Waveform

The CDM pulse waveform and its comparison to other ESD models are shown in the following sections. Presently, there are four CDM test standards (ESDA S5.3.1, JEDEC JESD22-C101, AEC-Q100-011 Rev. C, and JEITA ED-4701-300) with different waveform requirements.

4.4.1 Charged Device Model Pulse Waveform

The CDM pulse is regarded as the fastest event of all the ESD events. Figure 4.2 shows a CDM pulse waveform [1, 2]. Note that the CDM pulse waveform is influenced by the test platform and measurement metrology. The test platform is influenced by the field plate, field plate dielectric thickness and material type, and the probe assembly (e.g., test head and ground plane). The metrology is influenced by the oscilloscope and verification module specifications.

First, the event is oscillatory. The CDM current pulse rise time is on the order of 250 ps, and with peak currents in the range of 10 A. The energy spectrum of the CDM pulse event extends to 5 GHz frequency. The CDM pulse waveform has a fast current pulse. The timescale of the CDM event is significantly lower than the thermal diffusion time; hence, CDM events are in the "adiabatic regime" of a Wunsch–Bell power-to-failure curve. CDM events are thus at the opposite direction of long-pulse electrical overstress (EOS) events.

In the calibration and verification procedure, the JEDEC standard requires a 1 GHz oscilloscope, whereas the ESDA standard requires 3 GHz. Both standards today are bandwidth-limited signal since the CDM waveform is faster than 1 GHz. These oscilloscopes were chosen based on availability at the time. It is well known that the energy spectrum of the CDM pulse waveform can extend into the 5 GHz frequency.

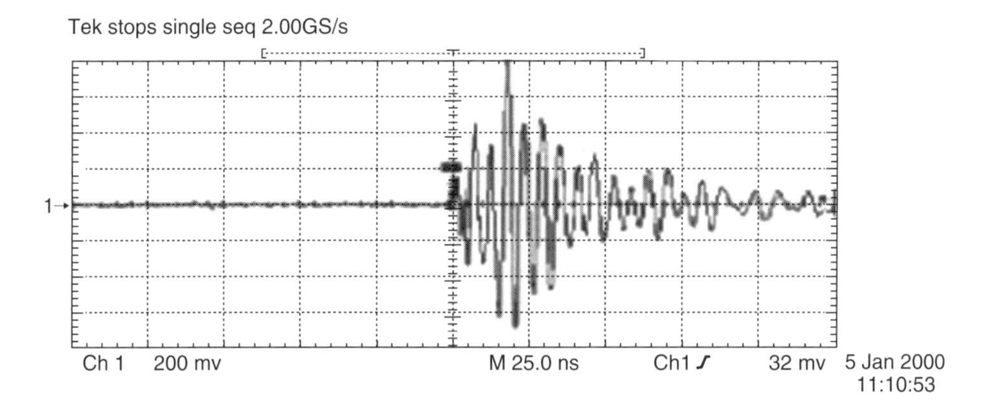

Figure 4.2 Charged device model pulse waveform

The new joint specification JS-002 will increase the bandwidth requirement to over 6 GHz. In addition, new "test conditions" (TC) of 125, 250, 500, 750, and 1000 V will be specified. Full width half maximum (FWHM) values for minimum and maximum magnitudes will also be specified.

4.4.2 Comparison of Charged Device Model (CDM) and Human Body Model (HBM) Pulse Waveform

The CDM pulse is significantly different from the human body model (HBM). Figure 4.3 shows a comparison of the CDM and HBM events. First, the CDM event is oscillatory, whereas primarily the HBM event is a damped double exponential decay response. Second, the rise time of the CDM event is significantly faster than that of the HBM event. The CDM event rise time

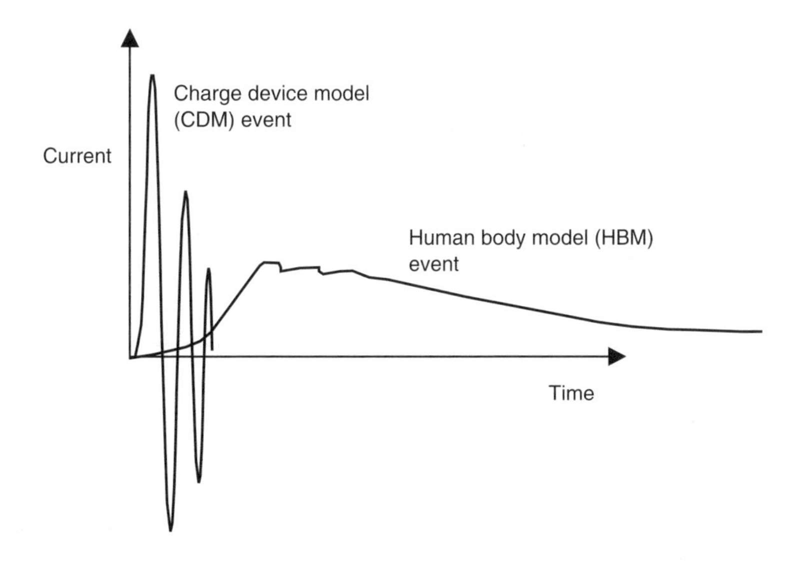

Figure 4.3 Charged device model (CDM) and HBM waveform comparison

is 250 ps, whereas the HBM event rise time is 17–22 ns. Third, the decay time of the CDM event is faster than that of the HBM event. The HBM event is associated with the RC time of the HBM network (e.g., 150 ns).

From a Wunsch–Bell power-to-failure curve perspective, the CDM event is in the adiabatic regime, whereas the HBM event is in the thermal diffusion time regime.

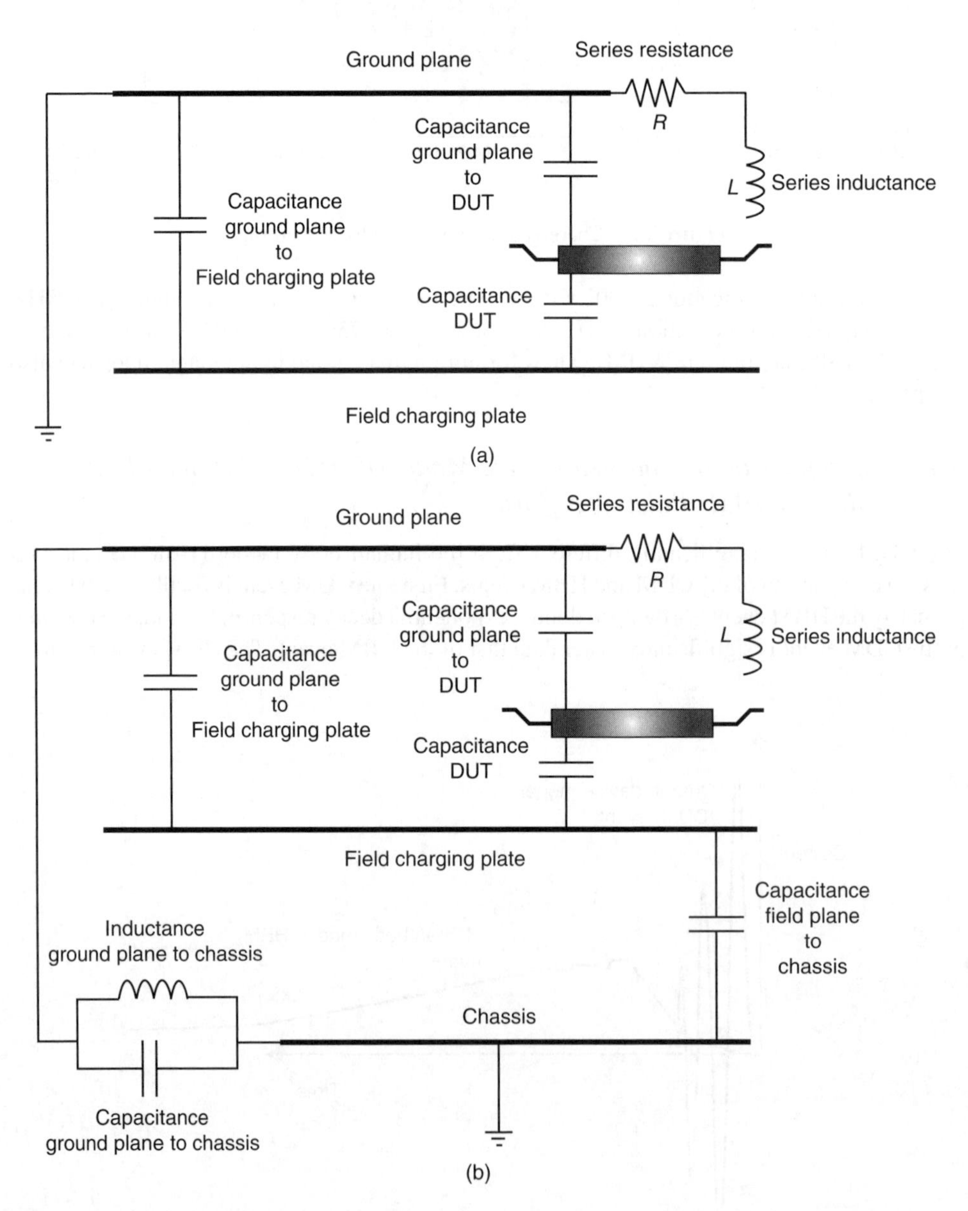

Figure 4.4 (a) Charged Device Model equivalent circuit. (b) Charged Device Model equivalent circuit with parasitics associated with the chassis

4.5 Equivalent Circuit

The equivalent circuit for the CDM is highlighted in Figure 4.4. In the equivalent circuit, the capacitances have the largest influences on the CDM. The test system consists of a field charging plate, the ground plane, and the device under test (DUT). A primary capacitance is the capacitance formed between the DUT and the field charging plate (C_{DUT}). A second primary capacitance is the capacitance of the DUT to the ground plane (C_g). There is also a capacitor formed between the field charging plate and the ground plane (C_{gpd}). In the discharge process, the discharge is initiated by the pogo pin between the DUT and the ground plane. The equivalent circuit is an inductor and a resistor, shown as a resistor, R_S, and inductance L_S.

In the equivalent circuit, additional parasitic capacitances can be shown to address the chassis in the test system [43] (Figure 4.5). Figure 4.4b illustrates three additional circuit elements in the equivalent circuit associated with the chassis. A first capacitor is formed between the field charging plate and the chassis (C_{fgc}). Between the ground plane and the chassis is an inductor (L_C) and capacitance (C_{gpc}).

4.6 Test Equipment

Commercial test equipment has been developed to perform the CDM test. Early CDM test equipment was not commercial but practiced by early developers at AT&T. The first CDM tester was developed at AT&T and was called the "Happy Zapper." Without the standardization of the CDM event, there were issues of the test system's repeatability and reproducibility. The early test system could not verify the release of a pulse event, and hence performed the same

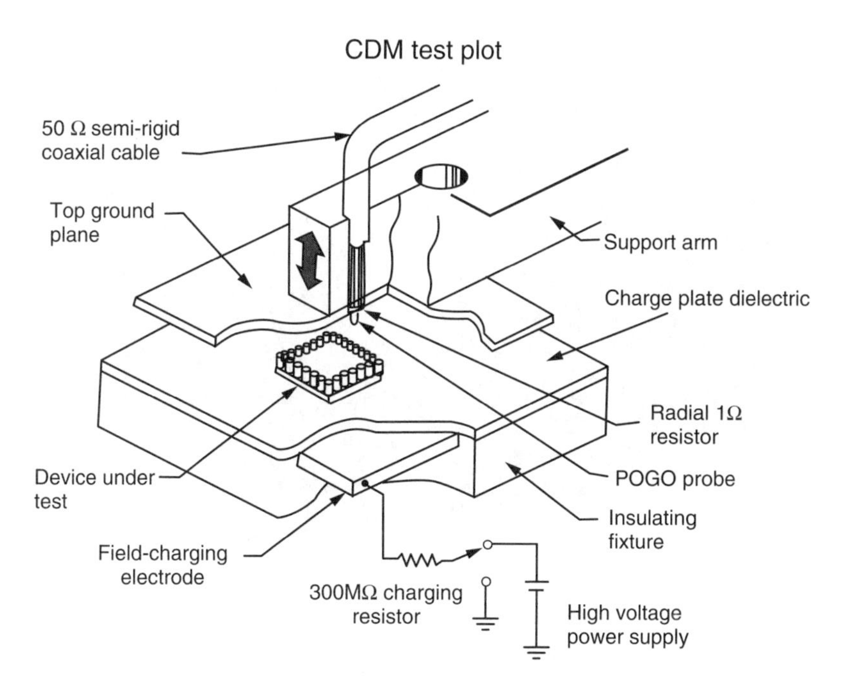

Figure 4.5 Charged device model tester

pulse five times. Second, the early test systems pulse event was not consistent but showed variation. Third, the CDM response was a function of the insulator thickness and insulator material properties; this led to variation of the CDM event results. These problems were solved with new commercial test systems allowing for repeatable and reproducible results.

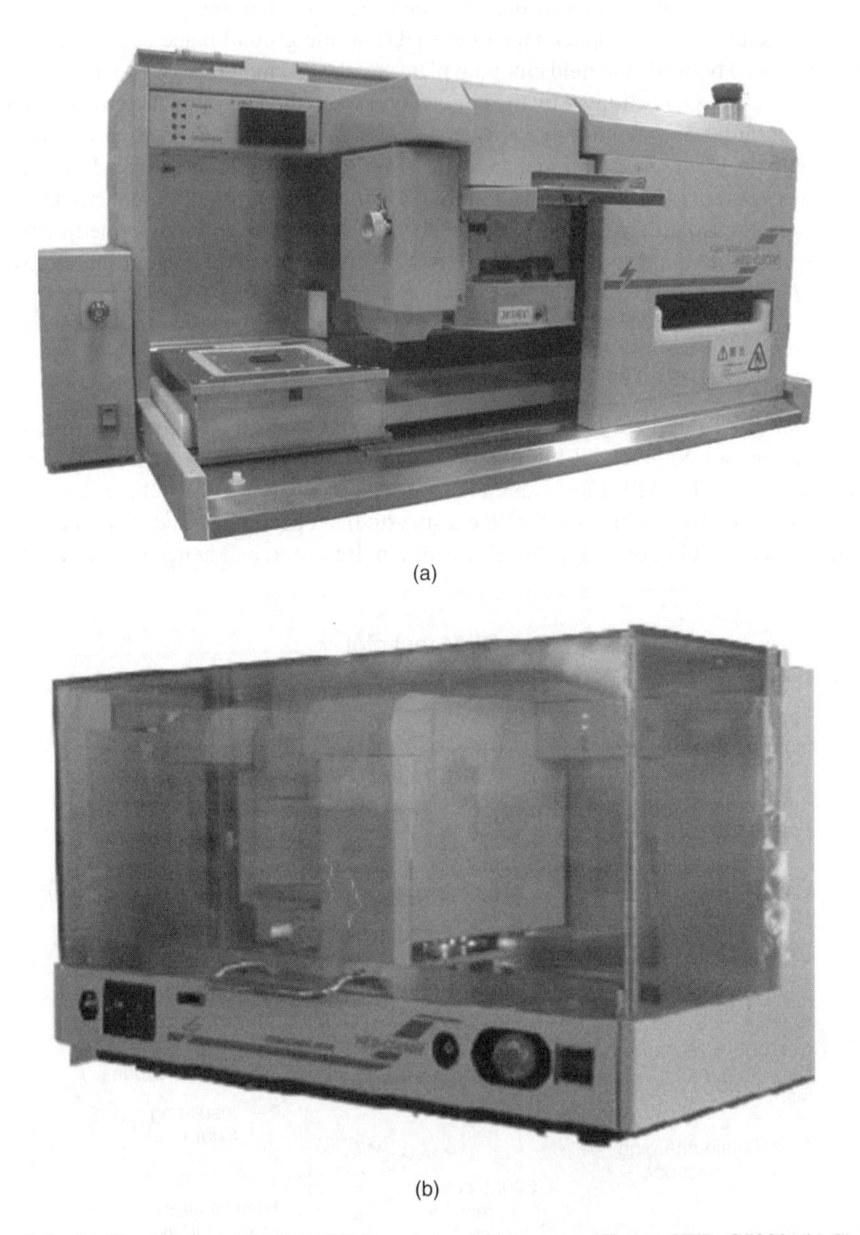

(a)

(b)

Figure 4.6 (a) Charged device model (CDM) commercial test system Hanwa HED-C5000. (b) Charged device model (CDM) commercial test system Hanwa HED-C5000R (Reproduced with permission of the Hanwa Corporation)

Figure 4.6(a) shows a commercial test system for CDM testing developed in Japan. The figure shows a test system from the Hanwa HED-5000 series [62]. Figure 4.6(b) shows a second image of the commercial test system for CDM testing developed in Japan by the Hanwa Corporation. The HED-C5000R CDM test system is in conformance with the JEDEC, ESDA, AEC, and JEITA standard organizations. The system has built-in humidity control and N_2 enclosed chamber. The test system provides a wide motion test area (e.g., 80 mm × 200 mm) allowing for large semiconductor chips to be placed in the test system. The test capability of the HED-C5000R CDM system can test 1024 pin components. The testing voltage can range from 0 to 4000 V in step voltage of 5 V. The test voltage accuracy is 1% $\pm$10 V. The number of pulses can be set for one pulse, or up to 100 times. The interval time between pulses can be changed from 0 to 10 s [62].

Figure 4.7 shows a different commercial test system for CDM testing developed by the Thermo KeyTek Corporation. This test system is the Orion™ CDM tester [63]. This test system is configurable to different ESD test standards (e.g., JEDEC, ESDA, AEC, JEITA, and CDM2). The system has a high-resolution display to allow for discharge pin alignment during the testing setup. This system has an enclosed test area for moisture control. By controlling the environment for testing, this will insure repeatability when performing an air discharge event (e.g., field-induced CDM (FICDM)). One of the issues with CDM testers is whether the CDM event actually occurred. In this system, an "event detector" is used to confirm the CDM ESD discharge by internal detection and waveform capture. In this system, software exists for evaluation of the waveform performance during device testing. The software package in the test system allows for programming the test voltage, polarity, delay time between pulses, as well as alignment of the chip in the system (e.g., pin coordinates) [63].

4.7 Test Sequence and Procedure

A test procedure and sequence is required for CDM testing to assure proper testing [1, 2]. Figure 4.8 shows a flowchart of the CDM testing procedure. The CDM event is associated with the charging of the semiconductor component through different charging processes. Charging of the package can be achieved through direct contact charging or field-induced charging processes. The field-induced charging method is called the FICDM. Today, the commercial test system has the sequence of steps for the charging process, as well as for the discharging process (Figure 4.8). In the CDM test simulation, discharging is achieved using a pogo pin touching one of the semiconductor component package pins [1, 2].

Each requires different test platform, testing, waveform, and calibration requirements. Presently, there are four CDM test standards (ESDA S5.3.1, JEDEC JESD22-C101, AEC-Q100-011 Rev. C, and JEITA ED-4701-300). The JEDEC standard calibration, a 55 pF large coin (metal only) is used for the test system calibration. For the ESDA standard, it uses a 30 pF target metal disk encased in a 0.88-mm-thick FR4 insulator. For the small coin calibration, JEDEC uses a 6.8 pF coin, whereas ESDA uses a 4.0 pF coin encased in 0.8-mm-thick FR4 insulator. In the calibration and verification procedure, the JEDEC standard requires a 1 GHz oscilloscope, whereas the ESDA standard requires 3 GHz. Both standards today are bandwidth-limited signal since the CDM waveform is faster than 1 GHz. The new specification will increase the bandwidth requirement to over 6 GHz.

Additional differences exist between the JEDEC and JS-002 standards. The JEDEC standard uses a 0.388 mm FR-4 dielectric, and in the discharge path, a ferrite and 1 Ω resistor element.

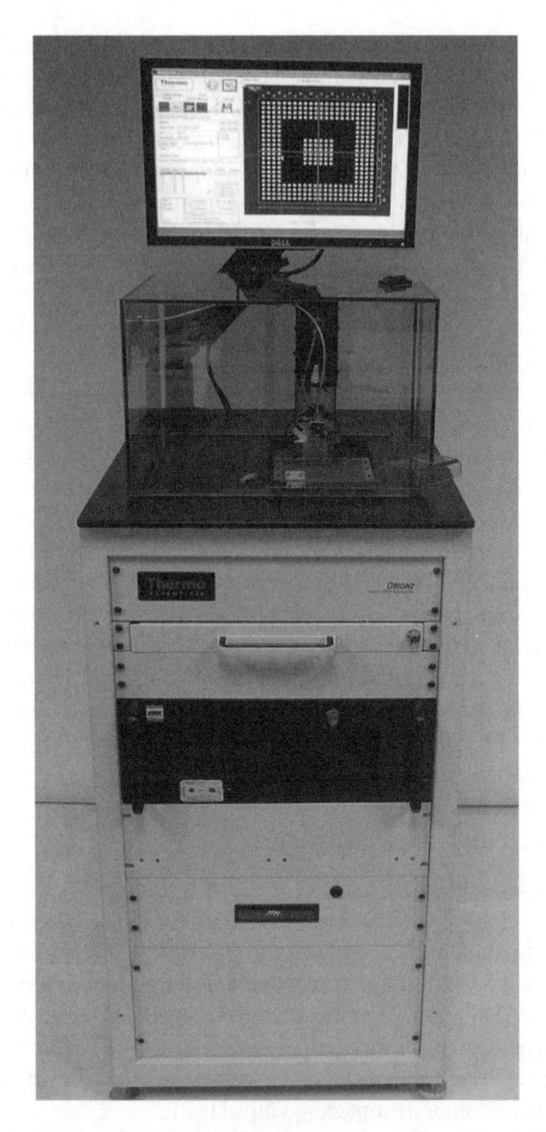

Figure 4.7 Charged device model (CDM) commercial test system – Thermo KeyTek Orion2 (Reproduced with permission of Thermo Fisher Scientific)

In the new JS-002 standard, there is no ferrite specified. The new joint specification JS-002 will increase the bandwidth requirement to over 6 GHz. In addition, new "test conditions" of TC 125, 250, 500, 750, and 1000 V will be specified. FWHM values for minimum and maximum magnitudes will also be specified.

For the new test standard, JS-002, the nominal field plate charging resistor is 1000 MΩ. In the standard, there will be an option for a lower value for large products, to avoid

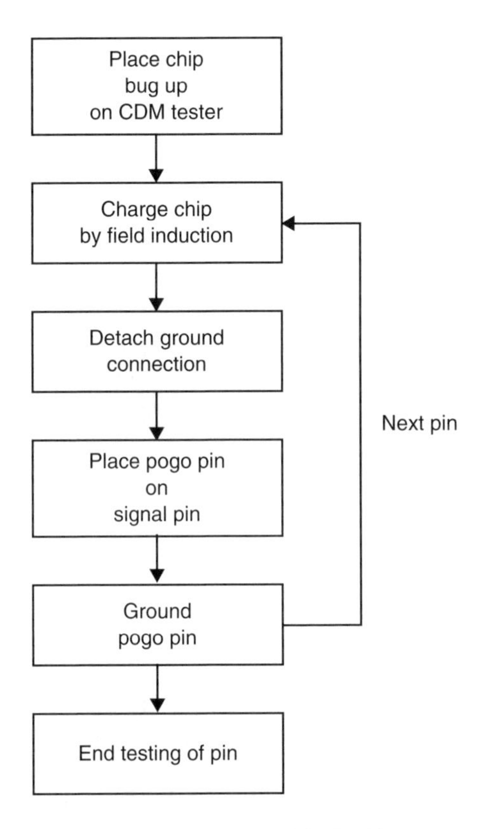

Figure 4.8 Charged device model (CDM) test sequence

charging resistor delay effects, as well as to improve the current peak measurement for large area/high-capacitance components. The procedure to evaluate the charging resistor effect on large modules is as follows:

- Average of 10 current peak measurements with the charging delay setting to 0 ms.
- Average of a second set of current peaks with charging delay of 500 ms.
- If the two averages are the same, there is no requirement for a change of charging resistor value.

4.8 Failure Mechanisms

CDM event damage can occur in the semiconductor chip through the substrate or the power supply. The charge stored on the substrate or the power supply rapidly discharges through the pin that the pogo pin is grounding. CDM event can be small "pin-hole" in a MOSFET gate structure, small transistors in receiver networks, and metal interconnects. Figure 4.9 shows an example of a MOSFET gate dielectric failure from a CDM event [55].

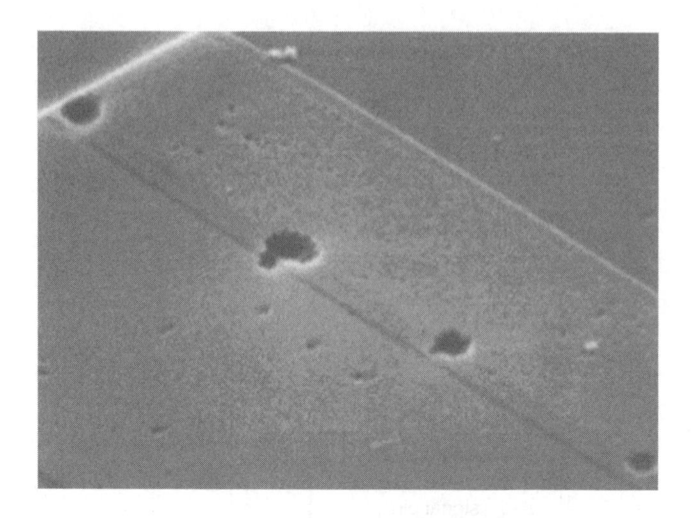

Figure 4.9 Charged device model (CDM)-induced failure mechanism – receiver gate failure

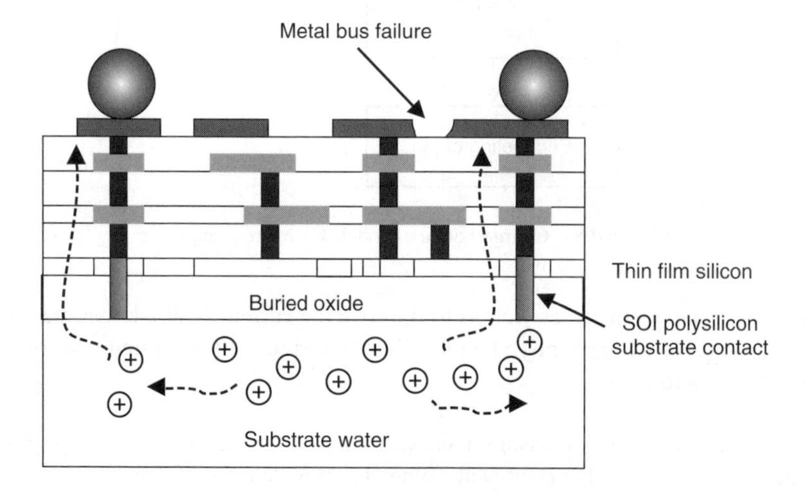

Figure 4.10 Charged device model (CDM)-induced failure mechanism – interconnects

The CDM metal interconnect failure can be significant due to the high current magnitude of the CDM event. Figure 4.10 shows an interconnect failure from a CDM event [55].

4.9 CDM ESD Current Paths

The current path for CDM in components is significantly different from other ESD events [55]. In the case of the CDM, the package and/or chip substrate is charged through a power or ground rail. The component itself is charged slowly to a desired voltage state. As a result, the current flows from the component itself to the grounded pin during ESD testing. This is

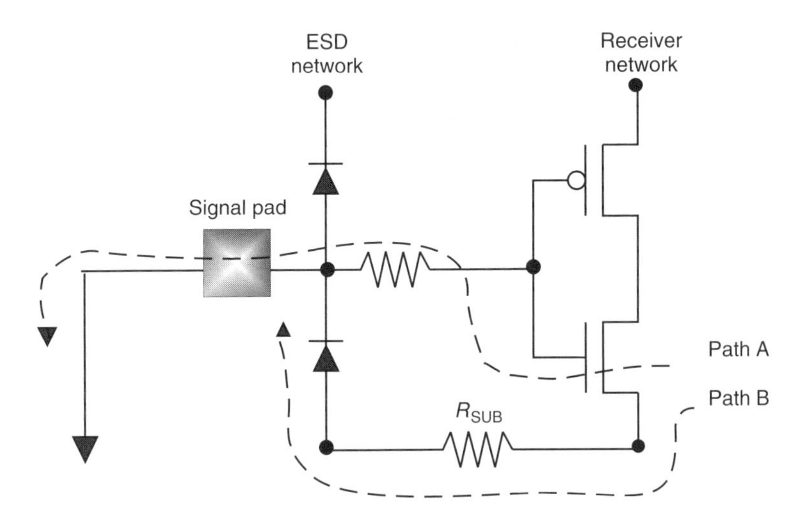

Figure 4.11 Charged device model (CDM) ESD event current path – charged V_{SS} positive polarity and grounded receiver pin

significantly different from other ESD tests that ground a reference, and then apply an ESD event to a signal or power pin. As a result, the current path that a CDM event follows is from inside the component to the pin that is grounded during test.

Figure 4.11 is an example showing the CDM current path from a charged ground rail (e.g., p-substrate) to the grounded receiver pin. For a positive charging of the substrate, the current flows from the substrate to any possible path that will reach the grounded signal pad node. The figure shows two paths: Path A is through the n-channel MOSFET receiver circuit and Path B is through the substrate and the ESD diode.

In the case of the ESD protection circuit is far from the signal pad, the substrate resistance can be significant. For Path B, the total resistance from the grounded location to the grounded signal path is the sum of the substrate resistance and the ESD diode series resistance. In the case that the receiver network is adjacent to the ESD network, or close to the signal to the signal pad, a majority of the current will flow through the ESD network. But, when the ESD network is far from the receiver network, the substrate resistance will be significant. When the impedance of the n-channel MOSFET receiver and the series resistor (e.g., Path A) is low compared to that of Path B, the current will flow through the receiver gate structure, causing rupturing of the MOSFET gate dielectric [55].

Figure 4.12 is an example showing the CDM current path from a charged power rail (e.g., V_{DD}) to the grounded receiver pin. For a positive charging of the power rail, the current flows from the power rail to any possible path that will reach the grounded signal pad node. In the figure, two paths are shown. Path A is from the power supply rail to the substrate and then through the n-channel MOSFET receiver circuit. Path B is from the power supply rail, through the substrate, and the ESD diode. When the power supply is charged up to voltages above the n-well-to-substrate breakdown voltage, current will flow to the p-substrate.

As in the last case, if the ESD protection circuit is far from the signal pad, the substrate resistance can be significant. For Path B, the total resistance from the grounded location to the grounded signal path is the sum of the impedance of the n-well-to-substrate diode, the substrate

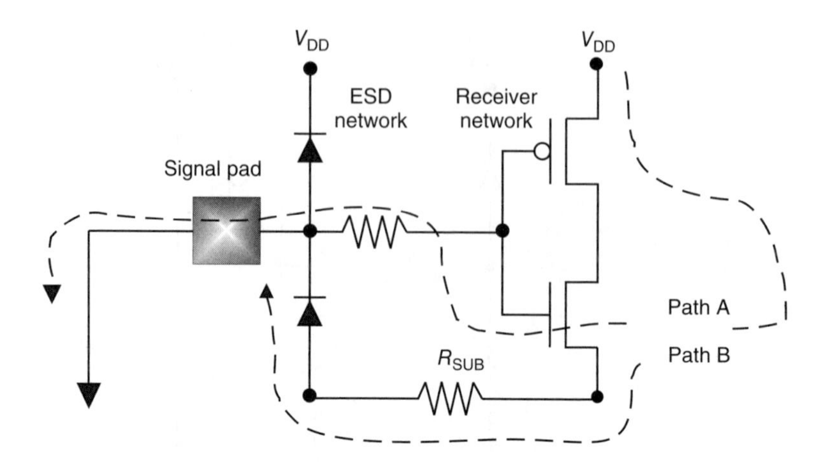

Figure 4.12 CDM ESD current paths – charged V_{DD} positive polarity and grounded receiver pin

resistance and the ESD diode series resistance. In the case that the receiver network is adjacent to the ESD network, or close to the signal to the signal pad, a majority of the current will flow through the ESD network. But, when the ESD network is far from the receiver network, the substrate resistance will be significant. When the impedance of the n-channel MOSFET receiver and the series resistor (e.g., Path A) is low compared to Path B, the current will flow through the receiver gate structure causing rupturing of the MOSFET gate dielectric. Also note in this case, an additional path is through the reverse-bias p–n diode of the ESD network; this has never been observed as a failure mechanism [50–52, 54–59].

4.10 CDM ESD Protection Circuit Solutions

For the CDM failures, circuit solutions exist to address low-level failure mechanisms. Additional ESD circuit elements can be added to existing circuitry to solve the CDM to acceptable results.

Figure 4.13 is an example of a commonly used CDM ESD network [50, 51]. The ESD network comprises of a first-stage dual-diode network placed adjacent or in proximity to the signal pad. A second set of diodes (e.g., second-stage network) are placed adjacent to the receiver circuit. A resistor is placed between the first and second stage. Figure 4.13 shows the possible paths of the CDM event for the case of a charged substrate. Three paths are shown for the CDM current path from a charged ground rail (e.g., p-substrate) to the grounded receiver pin. For a positive charging of the substrate, the current flows from the substrate to any possible path that will reach the grounded signal pad node. Path A is through the n-channel MOSFET receiver circuit and to the second-stage ESD diode. Path B is through the substrate to the second-stage diode network. Path C is through the substrate to the first-stage ESD network.

In the case that the first-stage ESD protection circuit is far from the signal pad, the substrate resistance can be significant. For Path C, the total resistance from the grounded location to the grounded signal path is the sum of the substrate resistance and the ESD diode series resistance.

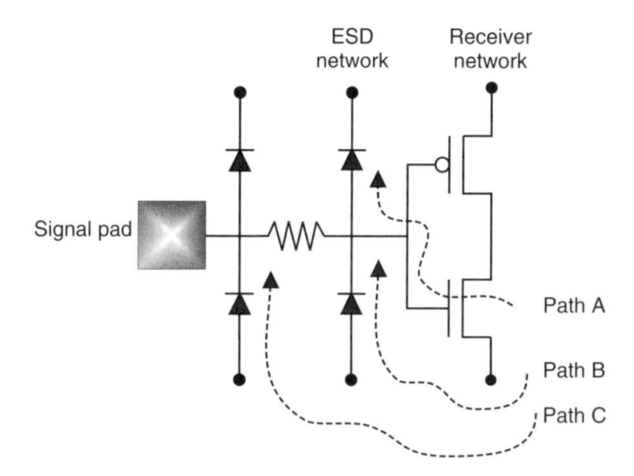

Figure 4.13 Charged device model (CDM) ESD protection circuits – dual-diode–resistor–dual-diode network

In the case that the receiver network is adjacent to the second-stage ESD network, the current will prefer to follow Path B instead of Path A. When the impedance of the n-channel MOSFET receiver (e.g., Path A) is higher than resistance through Path B, the receiver gate structure can avoid rupturing of the MOSFET gate dielectric. To insure that the current flows through Path B through the second-stage CDM ESD network, the circuit must be close to the receiver, and a low series resistance diode element.

Figure 4.14 is an example of a second CDM ESD network [50, 51]. In this network, a first-stage dual-diode network is placed adjacent or in proximity to the signal pad. The second-stage network consists of a series resistor, and a grounded gate n-channel MOSFET placed adjacent to the receiver circuit. In the case of HBM ESD testing, the second stage serves as a resistor divider network, increasing the HBM protection level significantly. For CDM ESD protection, the n-channel MOSFET drain serves as a p–n diode element. Placement of the n-channel MOSFET adjacent to the receiver network provides a low impedance path through the MOSFET parasitic p–n diode (Path B). In addition, if the local substrate increases positively, the MOSFET turn-on occurs (Path C). As in the prior discussion, Path B prevents current flow through Path A given that the impedance of the MOSFET gate receiver is higher compared to the series resistance of the substrate and the impedance of the p–n junction of the CDM GGNMOS structure. The current flow through Path D is a function of the proximity of the receiver network to the signal pad and the substrate resistance.

For silicon-on-insulator (SOI) technology, with the separation of the MOSFET from the substrate, the current paths will be different [49–51, 55]. Figure 4.15 shows an SOI circuit and CDM failure mechanism. The CDM current flowed from the p-substrate to the power supply rail $V_{DD.}$ The CDM current path went from the pass transistor MOSFET gate to grounded signal pad, causing gate-to-diffusion failure. The solution to improve the CDM robustness was placement of a series resistor in series with the pass transistor MOSFET gate, increasing the impedance between the pass transistor and signal pad. In this manner, current flowed through the first-stage ESD elements instead of through the receiver network [49–51, 55].

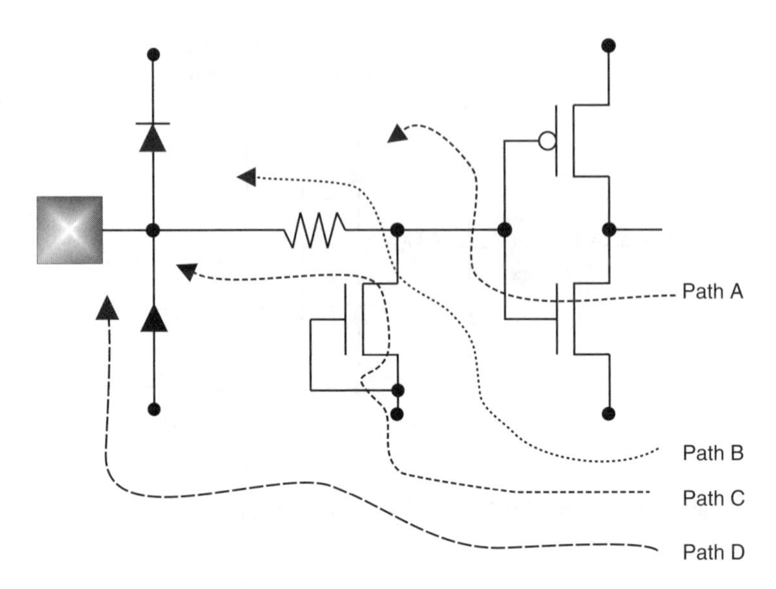

Figure 4.14 Charged device model (CDM) ESD protection circuits – dual-diode–resistor–grounded gate NMOS

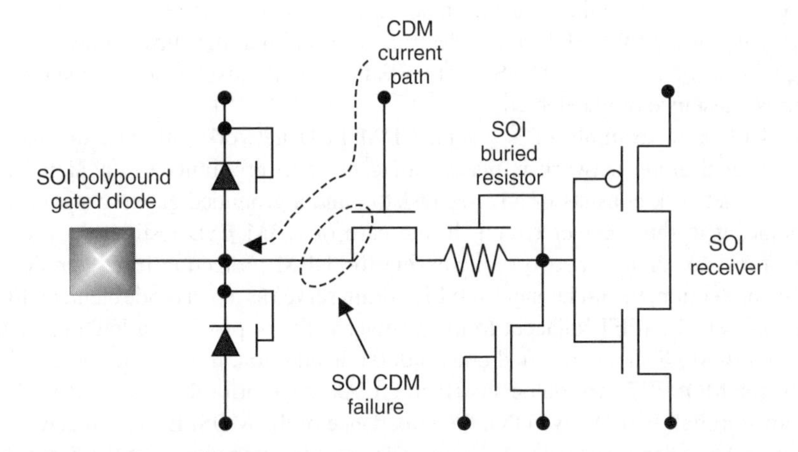

Figure 4.15 Charged device model (CDM) failure mechanisms in silicon-on-insulator (SOI) technology

4.11 Alternative Test Methods

An alternative test method for CDM evaluation has been proposed. In the following section, the SDM and the CBM are discussed.

4.11.1 Alternative Test Methods – Socketed Device Model (SDM)

An alternative test method for CDM evaluation was known as the SDM [60, 61]. This test was performed with the semiconductor component socketed into a test board. The semiconductor component was charged to the substrate (e.g., V_{SS}) or power supply (e.g., V_{DD}) by a power supply. The packaged semiconductor chip was then discharged through one of the signal pins. An SDM standard was established for this test.

The standard practice is ESD Association Standard Practice for Electrostatic Discharge Sensitivity Testing – SDM – Component Level, released in 2013 [60]. This standard practice defines a method to verify the correct operational state of the test simulator. It describes a procedure for measuring acceptable SDM discharge waveforms, determines critical waveform parameters and their acceptable values, and provides recommendations for device testing and ESD stressing levels.

In addition, there is a technical report, referred to as ESD Association Technical Report – Socket Device Model (SDM) Tester – An Applications and Technical Report [61]. This technical report provides a detailed description of existing SDM test systems. It defines a waveform confirmation test method, as well as explains why a derating stress voltage method must be used to account for differences in SDM test equipment. The technical report describes the side effects of background tester parasitic RLC transmission line (TL) components and proposes design requirements to be achieved by the next-generation test system. Since this report, the SDM test has been abandoned and is not used nowadays for qualification of semiconductor components.

4.12 Charged Board Model (CBM)

An alternative test method for CDM evaluation was known as the CBM (as well as the charged board event (CBE)). In the following sections, the charge board model is described [48] (Figure 4.16).

4.12.1 Comparison of Charged Board Model (CBM) and Charged Device Model (CDM) Pulse Waveform

The CBM and the CDM have different waveforms due to the different electrical characteristics involved in the discharge process. First, the field-induced charged board model (FICBM) has a larger capacitance compared to the FICDM due to the printed circuit board (PCB) [48]. Second, the PCB resistance and inductance characteristics are distinct from a component's resistance and inductance characteristics. In the case of the FICBM, the PCB traces have lower resistance than that of the interconnects within the DUT.

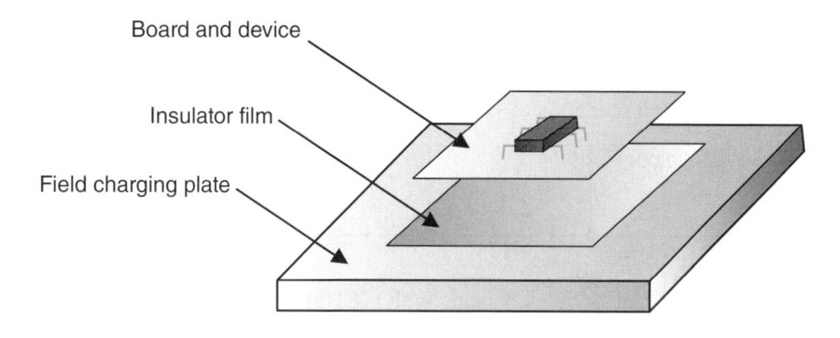

Figure 4.16 Charged board model

Figure 4.17 shows a comparison of the waveforms of the FICBM and the FICDM. First, the rise time of the FICBM waveform is faster than that of the FICDM waveform due to the lower inductance and lower resistance of the discharge path of the PCB. Second, the FICBM peak current is significantly higher than the FICDM peak current. The FICBM has a larger effective capacitance due to the addition of the PCB capacitance. The increase in the peak current will lead to lower FICBM ESD protection levels of protection in comparison to the FICDM ESD protection levels.

The increase in the FICBM capacitance is evident from the circuit schematic in Figure 4.18. Figure 4.18 shows the capacitance elements from the DUT as well as the additional capacitors from the PCB.

The circuit schematic contains the capacitance of the input diodes to both the power supply and ground rails. In addition, in parallel to these DUT capacitors are the board capacitors to the power supply and ground power planes. In the schematic, the DUT capacitance between

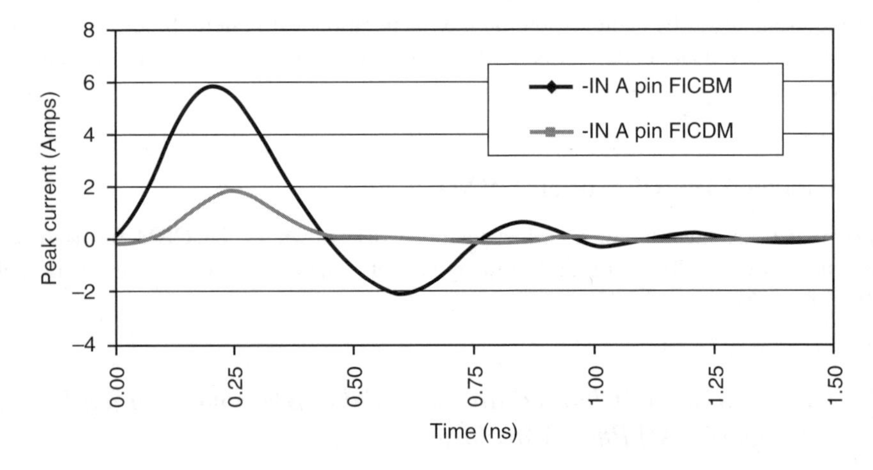

Figure 4.17 FICDM and field-induced charged board model (FICBM) waveform comparison

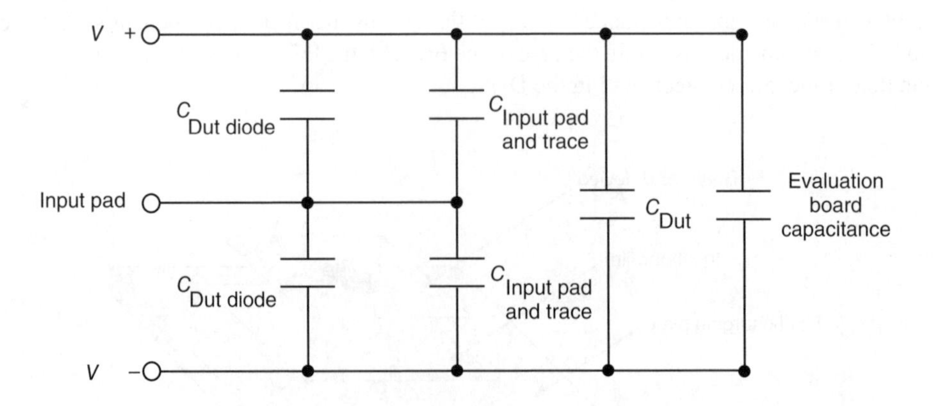

Figure 4.18 Equivalent circuit schematic with FICDM and FICBM capacitive elements

Figure 4.19 Field-induced charged board model (FICBM) testing of populated printed circuit board

the power supply and the ground rail is in parallel with the evaluation board capacitance (Figure 4.19).

Figure 4.20(a) shows an example of an FICBM failure of an input protection diode. The element in the center of the scanning electron microscope (SEM) image is the anode connected to the negative power supply. The damage in the perimeter is the cathode electrically connected to the cathode. This damage was evident at a −500 V FICBM level.

Figure 4.20(b) shows an example of an FICBM failure of an input protection diode at a higher test level. The damage in the perimeter is the cathode electrically connected to the cathode. This damage was evident at a −625 V FICBM level.

Figure 4.20(c) shows interconnect damage from an FICBM event. It is clear that the large damage pattern is a CDM, CBM [48], or EOS. In this case, it is known that this damage pattern was from an FICBM ESD test.

4.12.2 Charged Board Model (CBM) as an ESD Standard

Today, the CBM is not a supported ESD standard. There is a strong possibility that this will be a standard in the near future. This is most critical to small systems, such as cell phones.

4.13 Summary and Closing Comments

In this chapter, the CDM test was discussed. The CDM is also one of the most important test standards in today's semiconductor chip development. To combat CDM failures, CDM-specific circuitry was integrated into the power grid and input circuitry of semiconductor chips. In addition, new tests are being developed to address chips mounted on the PCB.

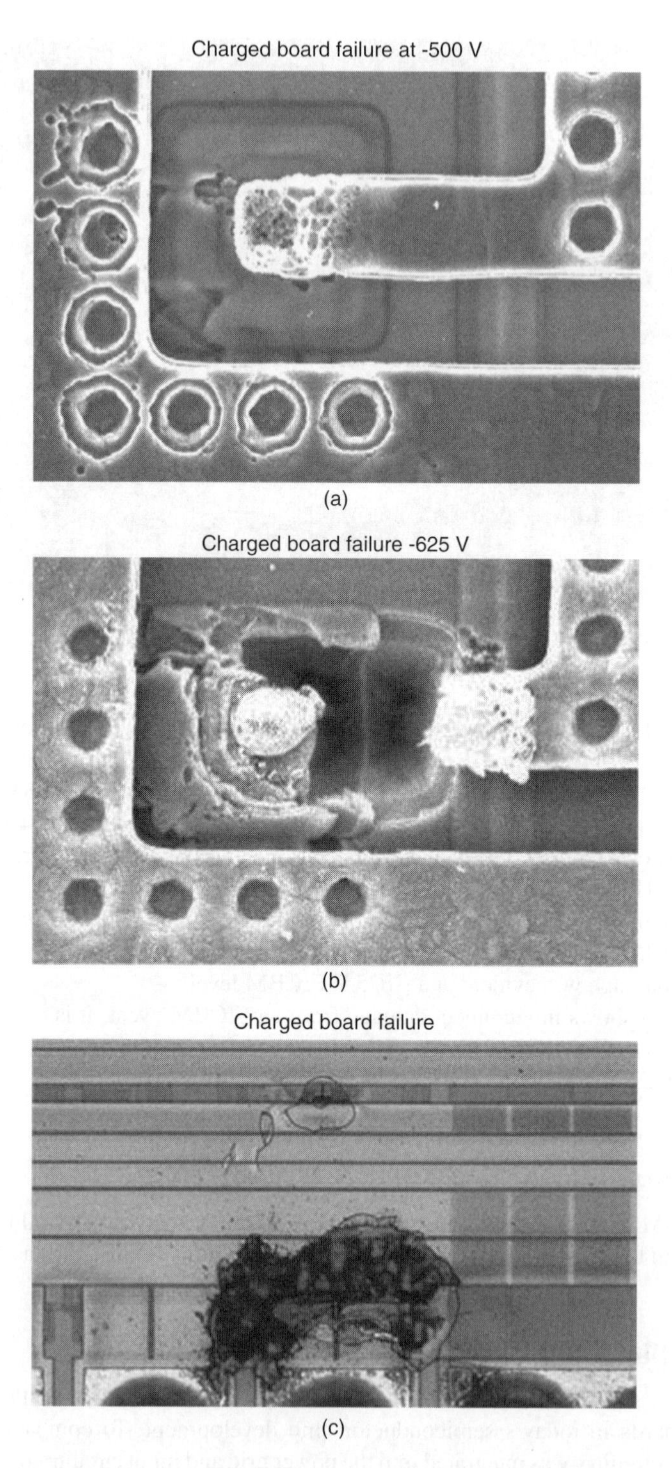

Figure 4.20 (a) CBM failure of an ESD diode. (b) CBM failure of an ESD diode. (c) CBM failure of interconnects

In Chapter 5, again another large step forward in the history of ESD testing is the development of transmission line pulse (TLP) testing standard. An important issue of the correlation of the TLP test to the HBM test is discussed.

Problems

1. In manufacturing, semiconductor chips are placed on an insulating conveyer belt to avoid scratching of the paint off the metal package of a microprocessor. As the chips are picked up by a manufacturing person, the microprocessor is damaged by CDM. Out of 300 microprocessors, 200 were destroyed due to CDM failures. Explain the complete charging process that leads to this failure. Explain a solution to the ESD failures.
2. In shipping, DRAM chips are placed into a shipping tube. When the DRAMs are removed from the shipping tube, they come in contact with a table top. Explain the charging and discharge process of the chip. Explain how the charge transfers internally in the chip that leads to the interior failure. Explain the circuitry and which pins would fail. Explain the complete charging process that leads to this failure.
3. Magnetic recording heads are placed in a plastic tray that contains carbon to improve the conductivity of the shipping tray. Unfortunately, the extrusion of the carbon was not uniformly distributed in the shipping tray. This problem leads to a high yield loss of magnetic recording heads. Show the charging process within the tray. Explain how to measure the conductivity to determine the problem.
4. Silicon germanium (SiGe) radio frequency (RF) chips are placed in a plastic tray that contains carbon to improve the conductivity of the shipping tray. RF chips were picked up with metal tweezers prior to measuring on a bench station. Fifty percent of the chips were destroyed by the test engineer. Show the charging process and discharging process. Explain the RF chip failures. Explain how to change the tweezers and manufacturing steps to test these RF components.
5. A semiconductor chip is 20 mm × 20 mm. How much charge is stored in the chip compared to a 1 mm × 1 mm chip? What are the CDM ramifications?
6. A semiconductor substrate of a CMOS chip is charged using a 1 MΩ resistor to 1000 V to the substrate V_{SS} signal pin. What is the voltage of the V_{DD} power supply?
7. A semiconductor chip is a CMOS chip with a single n-well in a p-substrate. The chip size is 20 mm × 20 mm. What is the size of the n-well to the substrate diode area?
8. A semiconductor chip is a silicon-on-insulator (SOI) technology. The technologist did not place a substrate contact between the backside of the wafer and the top surface. The substrate region is charged by the testing chuck. Explain the possible CDM failure mechanisms.
9. A semiconductor SRAM chip is a silicon-on-insulator (SOI) technology. The technologist placed a polysilicon substrate contact between the backside of the wafer and the top surface on the edges of the SRAM array. The buried oxide (BOX) region was destroyed by the CDM event. Explain the interrelationship between the charge distribution in the substrate, the buried oxide thickness, the size of the chip, the substrate resistance, and polysilicon substrate contact resistance.
10. CDM circuitry is typically placed local to a MOSFET receiver network. Explain the current paths and all parasitic components of the circuitry and substrate assuming a dual-diode–resistor–dual-diode ESD network. Where are the CDM circuit elements to

be placed? What is the interrelationship between the substrate resistance, the placement near the receiver, and size of the CDM circuitry.

11. Given a dual-diode–resistor–dual-diode ESD network, show all current paths from the CDM event to the signal pad when the resistor value is 200 Ω. How does the current get back to the signal pad? What should be the on-resistance of the CDM diode (e.g., second-stage network)?
12. Given a dual-diode–resistor–grounded gate MOSFET ESD network, show all current paths from the CDM event to the signal pad when the resistor value is 40 Ω. How does the current get back to the signal pad?
13. Given a dual-diode–resistor–grounded gate MOSFET ESD network, what advantages does this circuit have over the dual-diode–resistor–dual-diode network for CDM, for HBM, and for MM?
14. Show the CDM failure mechanisms of a CMOS receiver network that contains a dual-diode ESD network, followed by a MOSFET half-pass network, and a MOSFET receiver. This circuit is mapped into SOI with no body contacts on the MOSFETs. Show the CDM failure mechanisms now after mapping into SOI. Show the current paths back to the signal pin.

References

1. ANSI/ESD ESD-STM 5.3.1 – 1999. *ESD Association Standard Test Method for the Protection of Electrostatic Discharge Sensitive Items – Electrostatic Discharge Sensitivity Testing – Charged Device Model (CDM) Testing – Component Level.* Standard Test Method (STM) document, 1999.
2. JEDEC. JESD22-C101-A. *A Field-Induced Charged Device Model Test Method for Electrostatic Discharge-Withstand Thresholds of Microelectronic Components*, 2000.
3. T. Speakman. A model for the failure of bipolar silicon integrated circuits subjected to electrostatic discharge. *Proceedings of the International Reliability Physics Symposium*, 1974; 60–67.
4. P.R. Bossard, R.G. Chemelli, and B.A. Unger. ESD damage from triboelectrically charged IC-pins. *Proceedings of the Electrical Overstress/Electrostatic Discharge (EOS/ESD) Symposium*, 1980; 17–22.
5. R.D. Enoch, and R.N. Shaw. An experimental validation of the field-induced ESD model. *Proceedings of the Electrical Overstress/Electrostatic Discharge (EOS/ESD) Symposium*, 1986; 224–231.
6. R.N. Shaw. A programmable equipment for ESD testing to the charged device model. *Proceedings of the Electrical Overstress/Electrostatic Discharge (EOS/ESD) Symposium*, 1986; 232–237.
7. L. Avery. Charged device model testing: trying to duplicate the reality. *Proceedings of the Electrical Overstress/Electrostatic Discharge (EOS/ESD) Symposium*, 1987; 88–92.
8. R. Renninger, M. Jon, D. Lin, T. Diep, and T. Welsher. A microwave-bandwidth waveform monitor for charged device model simulators. *Proceedings of the Electrical Overstress/Electrostatic Discharge (EOS/ESD) Symposium*, 1988; 162–172.
9. R. Renninger, M. Jon, D. Lin, T. Diep, and T. Welsher. A field-induced charged device model simulator. *Proceedings of the Electrical Overstress/Electrostatic Discharge (EOS/ESD) Symposium*, 1989; 59–71.
10. R.G. Renninger. Mechanisms of charged device electrostatic discharges. *Proceedings of the Electrical Overstress/Electrostatic Discharge (EOS/ESD) Symposium*, 1991; 127–143.
11. B. Bingold, T.J. Maloney, V. Wilson, and R. Levi. Package effects on human body and charged device ESD tests. *Proceedings of the Electrical Overstress/Electrostatic Discharge (EOS/ESD) Symposium*, 1991; 144–150.
12. H. Hyatt. The resistive phase of fast risetime ESD pulses. *Proceedings of the Electrical Overstress/Electrostatic Discharge (EOS/ESD) Symposium*, 1992; 55–67.
13. D.L. Lin, and T.L. Welsher. From lightning to charged-device model electrostatic discharges. *Proceedings of the Electrical Overstress/Electrostatic Discharge (EOS/ESD) Symposium*, 1992; 68–75.
14. M. Tanaka, M. Sakimoto, I. Nishimae, and K. Ando. An advanced ESD test method for charged device model. *Proceedings of the Electrical Overstress/Electrostatic Discharge (EOS/ESD) Symposium*, 1992; 76–87.

15. D. Pommerenke. The influence of speed of approach, humidity, and arc length on the ESD breakdown. *ESD Forum Number 3*, ISBN 7-9802845-2-2, 1993; 103–111.
16. H. Gieser, P. Egger, J.C. Reiner, and M.R. Herrmann. A CDM-only reproducible field degradation and its reliability aspects. *Proceedings of the European Symposium on Reliability of Electron Devices, Failure Physics and Analysis (ESREF)*, 1993; 371–378.
17. H. Gieser, and P. Egger. Influence of tester parasitics on "charged device model" failure thresholds. *Proceedings of the Electrical Overstress/Electrostatic Discharge (EOS/ESD) Symposium*, 1994; 69–84.
18. M. Chaine, C. Liong, and H. San. A correlation study between different types of CDM testers and "real" in-line leakage failures. *Proceedings of the Electrical Overstress/Electrostatic Discharge (EOS/ESD) Symposium*, 1994; 63–68.
19. K. Verhaege, G. Groeseneken, H. Maes, P. Egger, and H. Gieser. Influence of tester, test method, and device type on CDM testing. *Proceedings of the Electrical Overstress/Electrostatic Discharge (EOS/ESD) Symposium*, 1994; 49–62.
20. D. Pommeranke. ESD: transient fields, arc simulation, and risetime limits. *Journal of Electrostatics*, Vol. 36, Nov 1995; 31–54.
21. P. Eggers, R. Kropf, and H. Gieser. ESD monitor circuit – a tool to investigate the susceptibility and failure mechanisms of the CDM. *Proceedings of the European Symposium on Reliability on Electron Devices, Failure Physics and Analysis (ESREF)*, 1995; 223–228.
22. H. Gieser, and M. Haunschild. Very-fast transmission line pulsing of integrated structures and the charged device model. *Proceedings of the Electrical Overstress/Electrostatic Discharge (EOS/ESD) Symposium*, 1996; 85–94.
23. M. Tanaka, and K. Okado. CDM ESD test considered phenomena of division and reduction of high voltage discharge in the environment. *Proceedings of the Electrical Overstress/Electrostatic Discharge (EOS/ESD) Symposium*, 1996; 54–61.
24. L.G. Henry, H. Hyatt, J. Barth, M. Stevens, and T. Diep. Charged device model (CDM) metrology: limitations and problems. *Proceedings of the Electrical Overstress/Electrostatic Discharge (EOS/ESD) Symposium*, 1996; 167–179.
25. A. Kager. Einfluss der Bausteinkapazitat auf die ESD-Festigkeit beim Charged Device Model (CDM). PhD Thesis, 1997.
26. R. Carey, and L. DeChairo. An experimental and theoretical consideration of physical design parameters in field-induced charged device model ESD simulators and their impact upon measured withstand voltages. *Proceedings of the Electrical Overstress/Electrostatic Discharge (EOS/ESD) Symposium*, 1998; 40–53.
27. A. Kagerer, and T. Brodbeck. Influence of device package on the results of the CDM tests – consequences for tester characterization and test procedure. *Proceedings of the Electrical Overstress/Electrostatic Discharge (EOS/ESD) Symposium*, 1998; 320–327.
28. L.G. Henry, M. Kelly, T. Diep, and J. Barth. Issues concerning CDM ESD verification modules – the need to move to alumina. *Proceedings of the Electrical Overstress/Electrostatic Discharge (EOS/ESD) Symposium*, 1999; 203–211.
29. S. Frei, M. Senghaas, R. Jobava, and W. Kalkner. The influence of speed of approach and humidity on the intensity of ESD. *Proceedings of the International Symposium on Electromagnetic Compatibility*, Zurich Switzerland, 1999; 105–110.
30. L.G. Henry, M. Kelly, T. Diep, and J. Barth. The importance of standardizing CDM ESD test head parameters to obtain data correlation. *Proceedings of the Electrical Overstress/Electrostatic Discharge (EOS/ESD) Symposium*, 2000; 72–84.
31. S. Bonich, W. Kalkner, and D. Pommeranke. Broadband measurements of ESD risetimes to distinguish between different discharge mechanisms. *Proceedings of the Electrical Overstress/Electrostatic Discharge (EOS/ESD) Symposium*, 2001; 373–384.
32. Automotive Electronic Council, AEC-Q100-011, *Revision A: Charged Device Model Electrostatic Discharge Test*, 2001.
33. W.D. Greason, Z. Kucerovsky, and C. Zaharia. Development of an experimental platform to study the effect of speed of approach on the electrostatic discharge (ESD) event. *Proceedings of the Electrical Overstress/Electrostatic Discharge (EOS/ESD) Symposium*, 2001; 408–414.
34. H. Geiser, H. Wolf, W. Soldner, H. Reichl, A. Andreini, M. Natarajan, and W. Stadler. A traceable method for the arc-free characterization and modeling of CDM testers and pulse metrology chains. *Proceedings of the Electrical Overstress/Electrostatic Discharge (EOS/ESD) Symposium*, 2003; 328–337.

35. H. Wolf, H. Geiser, W. Stadler, and W. Wilkening. Capacitively coupled transmission line pulsing (CC-TLP) – a traceable and reproducible stress method in the CDM-domain. *Proceedings of the Electrical Overstress / Electrostatic Discharge (EOS/ESD) Symposium*, 2003; 338–345.
36. S. Bonich, W. Kalkner, and D. Pommeranke. Modeling of short-gap ESD under consideration of different discharge mechanisms. *IEEE Transactions on Plasma Science*, Vol. 31, Aug 2003; 736–744.
37. C. Goeau, C. Richier, P. Salome, J.P. Chante, and H. Jaouen. Impact of the CDM tester ground plane capacitance on the DUT stress level. *Proceedings of the Electrical Overstress/Electrostatic Discharge (EOS/ESD) Symposium*, 2005; 170–177.
38. S. Sowariraj. *Full chip modeling of IC's under CDM stress*. PhD Dissertation. University of Twente, Netherlands, 2005.
39. C. Ito, and W. Loh. A new mechanism for core device failure during CDM ESD events. *Proceedings of the Electrical Overstress/Electrostatic Discharge (EOS/ESD) Symposium*, 2006; 8–13.
40. C. Ito, W. Loh, T.W. Chen, and R.W. Dutton. A frequency domain VF-TLP pulse characterization methodology and its application to CDM ESD modeling. *Proceedings of the Electrical Overstress/Electrostatic Discharge (EOS/ESD) Symposium*, 2007; 317–324.
41. L.G. Henry, R. Narayan, L. Johnson, M. Hernandez, E. Grund, K. Min, and Y. Huh. Different CDM ESD simulators provide different failure thresholds from the same device even though all the simulators meet the CDM standard specifications. *Proceedings of the Electrical Overstress/Electrostatic Discharge (EOS/ESD) Symposium*, 2007; 342–352.
42. T. Brodbeck, K. Esmark, and W. Stadler. CDM tests on interface test chips for the verification of ESD protection concepts. *Proceedings of the Electrical Overstress/Electrostatic Discharge (EOS/ESD) Symposium*, 2007; 1–8.
43. B. Atwood, Y. Zhou, D. Clarke, and T. Weyl. Effect of large device capacitance on FICDM peak current. *Proceedings of the Electrical Overstress/Electrostatic Discharge (EOS/ESD) Symposium*, 2007; 273–282.
44. A. Jahanzeb, Y. Lin, S. Marum, J. Schichl, and C. Duvvury. CDM peak current variations and impact upon CDM performance thresholds. *Proceedings of the Electrical Overstress/Electrostatic Discharge (EOS/ESD) Symposium*, 2007; 283–288.
45. A. Gerdemann, E. Rosenbaum, and M. Stockinger. A novel testing approach for full-chip CDM characterization. *Proceedings of the Electrical Overstress/Electrostatic Discharge (EOS/ESD) Symposium*, 2007; 289–296.
46. H. Wolf, H. Gieser, and D. Walter. Investigating the CDM susceptibility of IC's at package and wafer level by capacitively coupled TLP. *Proceedings of the Electrical Overstress/Electrostatic Discharge (EOS/ESD) Symposium*, 2007; 297–302.
47. A. Jahanzeb, C. Duvvury, J. Schichl, J. McGee, S. Marum, P. Koeppens, S. Ward, and Y.Y. Lin. Single pulse CDM testing and relevance to IC reliability. *Proceedings of the Electrical Overstress/Electrostatic Discharge (EOS/ESD) Symposium*, 2008; 94–98.
48. A. Olney, B. Gifford, J. Guravage, and A. Righter. Real world charged board model (CBM) failures. *Proceedings of the Electrical Overstress/Electrostatic Discharge (EOS/ESD) Symposium*, 2003; 34–43.
49. S. Voldman. *ESD: Physics and Devices*. Chichester, England: John Wiley and Sons, Ltd., 2004.
50. S. Voldman. *ESD: Circuits and Devices*. Chichester, England: John Wiley and Sons, Ltd., 2005.
51. S. Voldman. *ESD: Circuits and Devices 2nd Edition*. Chichester, England: John Wiley and Sons, Ltd., 2015.
52. S. Voldman. *ESD: RF Circuits and Technology*. Chichester, England: John Wiley and Sons, Ltd., 2006.
53. S. Voldman. *Latchup*. Chichester, England: John Wiley and Sons, Ltd., 2007.
54. S. Voldman. *ESD: Circuits and Devices*. Beijing, China: Publishing House of Electronic Industry (PHEI), 2008.
55. S. Voldman. *ESD: Failure Mechanisms and Models*. Chichester, England: John Wiley and Sons, Ltd., 2009.
56. S. Voldman. *ESD: Design and Synthesis*. Chichester, England: John Wiley and Sons, Ltd., 2011.
57. S. Voldman. *ESD Basics: From Semiconductor Manufacturing to Product Use*. Chichester, England: John Wiley and Sons, Ltd., 2012.
58. S. Voldman. *Electrical Overstress (EOS): Devices, Circuits, and Systems*. Chichester, England: John Wiley and Sons, Ltd., 2013.
59. S. Voldman. *ESD: Analog Design and Circuits*. Chichester, England: John Wiley and Sons, Ltd., 2014.
60. ANSI/ESD ESD-SP 5.3.2 – 2013. *ESD Association Standard Test Method for the Protection of Electrostatic Discharge Sensitive Items – Electrostatic Discharge Sensitivity Testing – Socketed Device Model (SDM) Testing – Component Level*. Standard Test Method (STM) document, 2013.
61. ESD ESD-TR 5.3.2 – 2013. *ESD Association Technical Report – Socketed Device Model (SDM) Tester – An Applications and Technical Report*, 2013.

62. Hanwa Electronics. http://www.hanwa-ei.co.jp. Hanwa HED-5000 CD test system, 2016.
63. Thermo KeyTek Corporation. http:www.thermofischer.com. Orion™ CDM tester, 2016.
64. D. Abessolo-Bidzon, T. Smedes, and A.J. Huitsing. CDM simulation based on tester, package, and full integrated circuit modeling: case study. *IEEE Transactions on Electron Devices*, Vol. 59, (11), Nov 2012; 2869.
65. V. Shukla, G. Boselli, M. Dissegna, C. Duvvury, R. Sankaralingam, and E. Rosenbaum. Prediction of charged device model peak discharge current for microelectronic components. *IEEE Transactions on Device Materials Reliability*, Vol. 14, (3), Sept 2014.

5

Transmission Line Pulse (TLP) Testing

5.1 History

Today, there is a standard test method for the transmission line pulse (TLP) test methodology [1–3]. The standard test method for TLP test simulation is supported by the electrostatic discharge (ESD) Association (ANSI/ESD – ESD-STM 5.5.1 – 2008) [3]. The standard test method was developed with a cable length that is consistent with the energy under the human body model (HBM) waveform. Today, there are significant research as well as commercial TLP test systems that support this TLP test method [1–26].

TLP systems are designed in different configurations. TLP system configurations include current source, time domain reflectometry (TDR), time domain transmission (TDT), and time domain reflectometry and transmission (TDRT) [1–3]. In all configurations, the source is a transmission line whose characteristic time constant is determined by the length of the transmission line cable. The various TLP configurations influence the system characteristic impedance, the device under test (DUT) location, and the measurement of the transmitted or reflected signals. For this method, the choice of pulse width is determined by the interest to use TLP testing as an equivalent or substitute method to the HBM methodology. The standard practice today is the TLP cable length is chosen as to provide a TLP pulse width of 100 ns with less than 10 ns rise time (Figure 5.1).

TLP testing was initially standardized for equivalency to the HBM event [1–3]. As a result, the pulse width is in the thermal diffusion time regime. Hence, the present testing method is faster than desired for analysis of electrical overstress (EOS) events. EOS events are at the slow end of the thermal diffusion regime and in the steady state regime.

Historically, for electromagnetic pulse (EMP), the quantification of the robustness of components was performed for all pulse widths. The Wunsch–Bell power-to-failure curve was established by varying all pulse widths, and in essence, by evaluating the EOS robustness for all pulse phenomena.

ESD Testing: From Components to Systems, First Edition. Steven H. Voldman.
© 2017 John Wiley & Sons, Ltd. Published 2017 by John Wiley & Sons, Ltd.

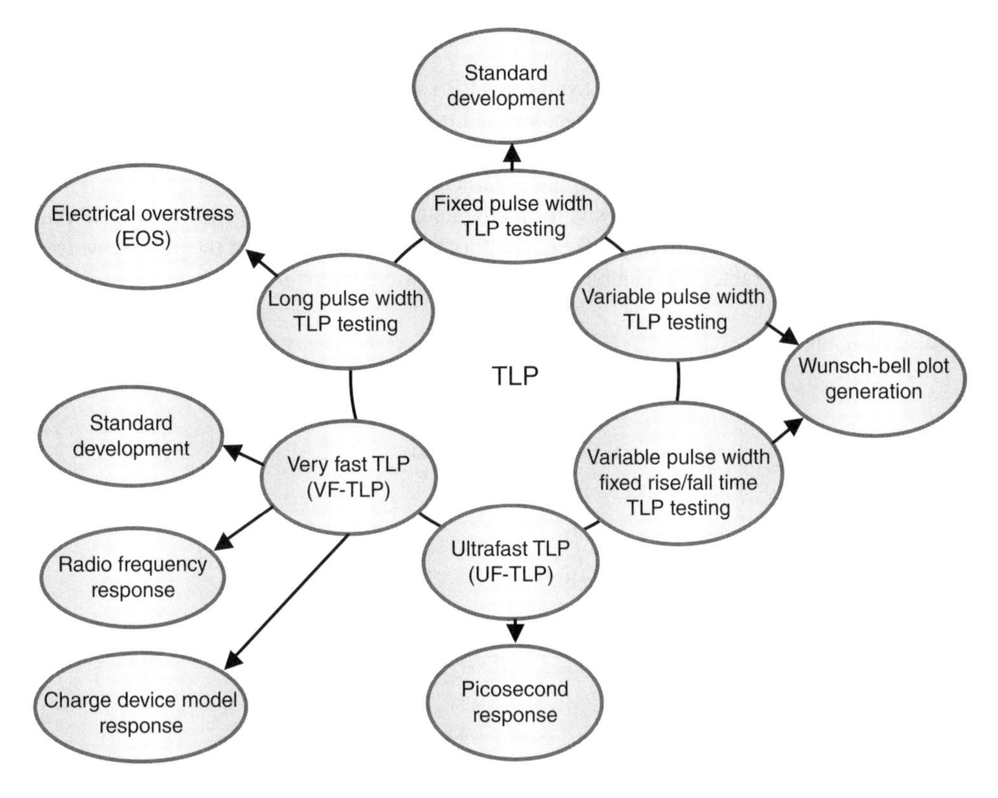

Figure 5.1 Pulse testing

A key note is that this test method and test simulation can be modified for EOS analysis by choosing a longer pulse width. The EOS event simulation can be established using a longer cable length in the test system. High current square pulse systems do exist that allow variable pulse widths that can be programmed from very short pulse to very long pulse events. Today, there is a return of growing interest in "long-duration" TLP techniques to apply to EOS events.

5.2 Scope

The scope of the TLP ESD test is for the testing, evaluation, and classification of components and micro- to nanoelectronic circuitry. The test is to quantify the sensitivity or susceptibility of these components to damage or degradation to the defined TLP test [1–3].

5.3 Purpose

The purpose of the TLP ESD test is for establishment of a test methodology to evaluate the repeatability and reproducibility of components to a defined pulse event in order to classify or compare ESD sensitivity levels of components [2, 3].

5.4 Pulse Waveform

TLP testing provides a square pulse from the TLP source. The TLP test system source can provide different pulse widths, rise times, and fall times. For the semiconductor industry, the TLP standard was developed with a fixed pulse width for comparison of TLP testing results. The TLP results change with different pulse characteristics.

Figure 5.2 shows the TLP pulse waveform characteristics. Figure 5.3 shows the TLP pulse waveform expanded view to highlight the characteristics at the front end of the pulse waveform.

The waveform characteristic can be defined by specific parameters [2, 3]:

- Pulse plateau
- Pulse width

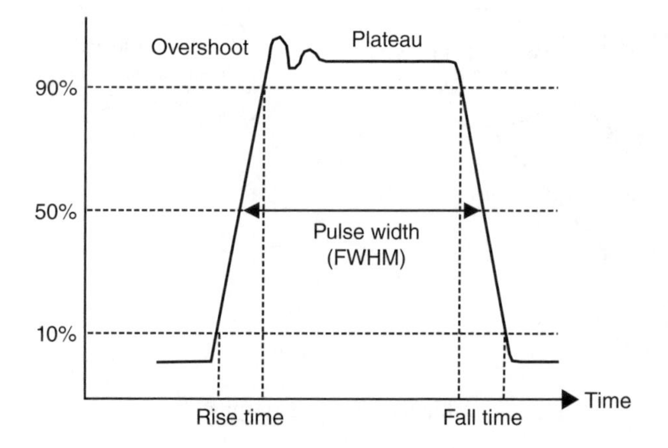

Figure 5.2 Transmission line pulse (TLP) waveform

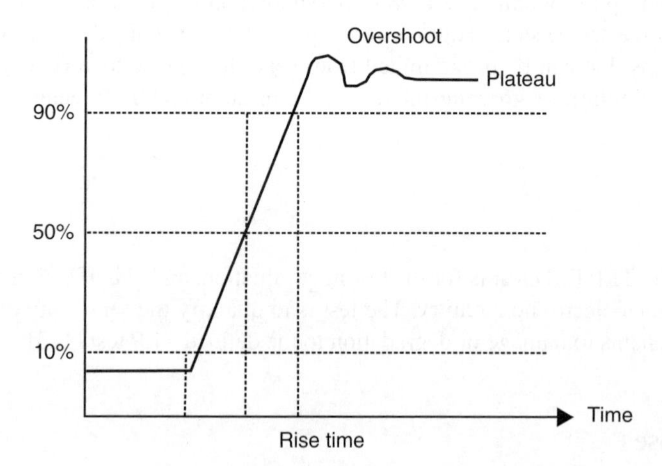

Figure 5.3 Transmission line pulse waveform (expansion)

- Rise time
- Fall time
- Maximum peak current overshoot
- Maximum current ringing duration
- Maximum peak voltage overshoot
- Maximum voltage ringing duration
- Measurement window.

Pulse plateau: The plateau of the pulse is the maximum value in the measurement window. The plateau is a time-averaged value after the overshoot and the oscillation region.

Pulse width: The pulse width is defined as the full width half maximum (FWHM). This is measured at 50% magnitude. For the TLP standard, this is defined as 100 ns [2, 3]. The reason with this pulse width is to make the energy under the TLP waveform equivalent to the energy of the HBM waveform.

Rise time: The rise time is defined as the time it takes to rise from 10% to 90% of the pulse plateau. For the TLP standard, the rise time of 10 ns is used [2, 3].

Fall time: The fall time is defined as the time it takes to fall from 90% to 10% of the pulse plateau. For the TLP standard, the fall time is equal to or greater than the rise time. For the TLP standard, the fall time of 10 ns or greater is used [2, 3].

Maximum peak current overshoot: The maximum peak current overshoot is defined as the magnitude of the overshoot peak current above the current plateau. The TLP standard states that this should be less than 20% of the plateau current. This is defined into a short circuit.

Maximum current ringing duration: The maximum current ringing duration is defined as the length of time between the beginning of the current pulse and the time that the ringing reaches less than 5% of the plateau current. For the TLP standard, the maximum current ringing duration should be less than 25% of the pulse width [2, 3].

Maximum peak voltage overshoot: The maximum peak voltage overshoot is defined as the magnitude of the overshoot peak voltage above the voltage plateau. The TLP standard states that this should be less than 20% of the plateau voltage. This is defined into an open circuit.

Maximum voltage ringing duration: The maximum voltage ringing duration is defined as the length of time between the first overshoot beyond the plateau voltage to the time it reaches 5% of the plateau voltage. For the TLP standard, the maximum voltage ringing duration should be less than 25% of the pulse width [2, 3].

Measurement window: The measurement window is the range of time within the pulse width where the voltage and current can be measured and extracted in the plateau [2, 3]. According to the TLP standard, the measurement window should be greater than 10% of the TLP pulse width [2, 3].

5.5 Equivalent Circuit

For the TLP test system, an equivalent model circuit representation can be defined. The TLP circuit is a transmission line (TL) source. The TLP source does not demonstrate any resistive loss and hence can be represented as a lossless inductor–capacitor (LC) transmission line (Figure 5.4). Figure 5.4 shows two representations of the equivalent circuit.

5.6 Test Equipment

TLP test systems can be used in different test configurations. The TLP test system can be in a configuration defined as a current source, TDR, TDT, and TDRT configuration (Figures 5.5–5.8, respectively) [2, 3].

There are requirements for the equipment independent of the test configuration. The equipment utilized for a TLP test system includes the following:

- Oscilloscope
- Voltage probe
- Current probe
- Transmission line
- High-voltage power supply
- High-voltage switch
- Attenuator
- Rise time filter.

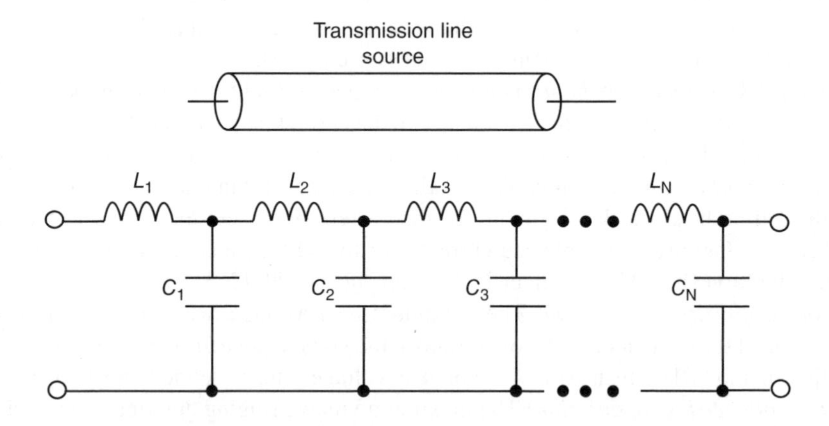

Figure 5.4 Transmission line pulse (TLP) equivalent circuit

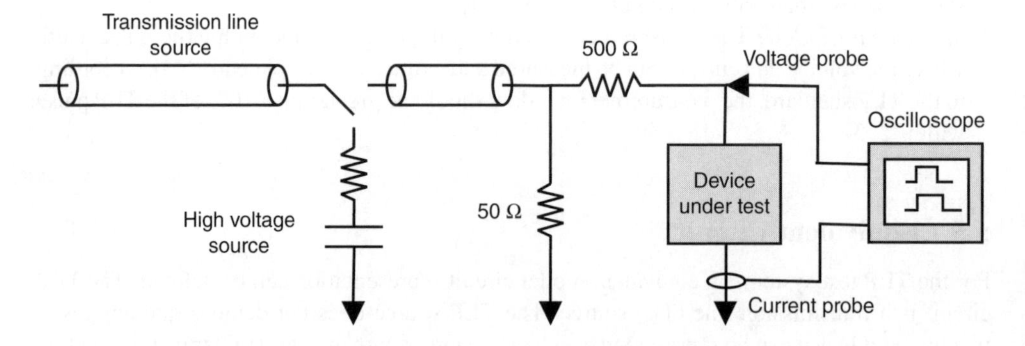

Figure 5.5 Transmission line pulse (TLP) current source configuration

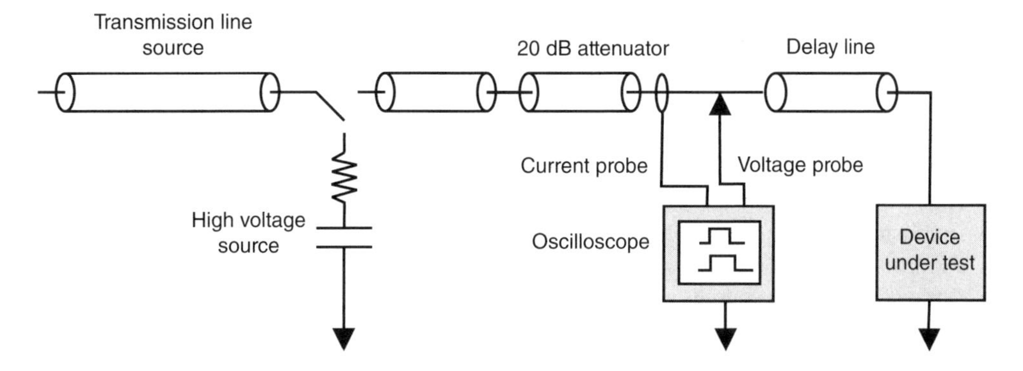

Figure 5.6 Transmission line pulse (TLP) time domain reflectometry (TDR) configuration

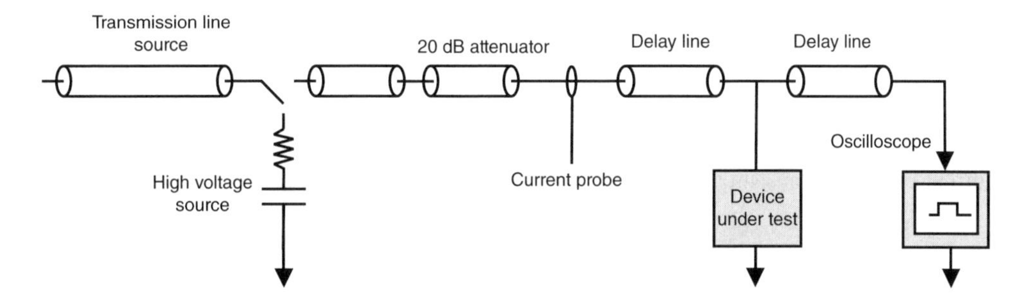

Figure 5.7 Transmission line pulse (TLP) time domain transmission (TDT) configuration

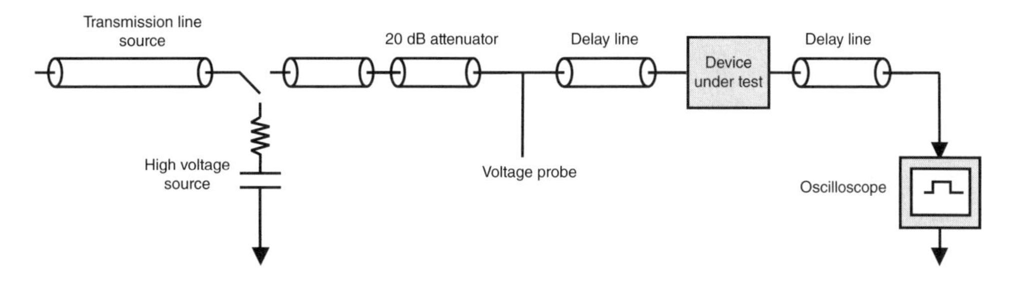

Figure 5.8 Transmission line pulse (TLP) time domain reflection transmission (TDRT) configuration

There are requirements for the equipment in the TLP system. For all the equipment, the equipment must be able to withstand the current and voltage levels without electrical damage. The oscilloscope must have a single shot bandwidth of at least 200 MHz. The voltage and current probes must have a minimum bandwidth of 200 MHz.

5.6.1 Current Source

TLP test system method is known as the current source TLP (Figure 5.5). The current source TLP method comprises the following test equipment [2, 3]:

- High-voltage charging source
- High-voltage charging source series resistor
- A transmission line for matching (50 Ω)
- A transmission line switch
- A transmission line source
- A 56 Ω ground termination
- A 500 Ω series resistor
- A current probe
- A voltage probe
- A two-channel oscilloscope.

In the current source TLP method, the components are arranged as shown in Figure 5.5.

The high-voltage source charges the transmission line to a high voltage through the 10 MΩ resistor element. In this method, there is a 500 Ω resistor element in series with the DUT. There is also a termination resistor of 56 Ω. A two-channel oscilloscope has a voltage probe and a current probe that measures the voltage and current across the DUT [2, 3].

5.6.2 Time Domain Reflection (TDR)

TLP test system method is known as the TDR TLP (Figure 5.6). The TDR–TLP method comprises the following test equipment [2, 3]:

- High-voltage charging source
- High-voltage charging source series resistor
- A transmission line for matching (50 Ω)
- A transmission line switch
- A transmission line source
- A transmission line connection
- A transmission line delay
- An attenuator
- A voltage probe
- A single-channel oscilloscope.

In the TDR–TLP method, the components are arranged as shown in Figure 5.6 [2, 3]. The TDR–TLP system is a 50 Ω system. The high-voltage source charges the transmission line to a high voltage through the 10 MΩ resistor element. An attenuator is placed in series with the DUT. A single-channel oscilloscope is used with a voltage probe that measures the voltage across the DUT. A reference pulse is required if the reflected pulse overlaps the initial TLP pulse

A TDR TLP system can also be formed using a balun (Figure 5.6). The TDR–TLP method with a balun comprises the following test equipment [2, 3]:

- High-voltage charging source
- High-voltage charging source series resistor
- A 50 Ω coaxial charge line
- A single-pole double-throw (SPDT) coaxial switch
- A 20 dB attenuator network
- A coaxial sensor
- A 50 Ω transmission line connection cable
- A balun
- A set of probe needles.

5.6.3 Time Domain Transmission (TDT)

TLP test system method is known as the TDT TLP (Figure 5.7). The TDT–TLP method comprises the following test equipment [2, 3]:

- High-voltage charging source
- High-voltage charging source series resistor (10 MΩ)
- A transmission line switch
- An attenuator
- A transmission line for matching
- A high impedance voltage probe
- A single-channel oscilloscope.

In the TDT–TLP method, the components are arranged as shown in Figure 5.6. The TDR–TLP system is a 50 Ω system. The high-voltage source charges the transmission line to a high voltage through the 10 MΩ resistor element. No attenuator is placed in series with the DUT. A single-channel oscilloscope is used with a voltage probe that measures the voltage across the DUT. A reference pulse is required [2, 3].

5.6.4 Time Domain Reflection and Transmission (TDRT)

TLP test system method is known as the TDRT TLP (Figure 5.8). The TDRT–TLP method comprises the following test equipment [2, 3]:

- High-voltage charging source
- High-voltage charging source series resistor (10 MΩ)
- A transmission line switch
- A transmission line connection
- An transmission line delay
- A transmission line termination
- A high impedance voltage probe
- A two-channel oscilloscope.

In the TDRT–TLP method, the components are arranged as shown in Figure 5.8. The DUT is placed in a series configuration in the center of the system and allows for both reflection and transmission of the signal. The TDRT–TLP system is a 100 Ω system. The high-voltage source charges the transmission line to a high voltage through the 10 MΩ resistor element. No attenuator is placed in series with the DUT. A two-channel oscilloscope is used with a voltage probe that measures the voltage across the DUT. The transmitted current is used for measuring the current through the DUT. A reference pulse is required.

5.6.5 Commercial Transmission Line Pulse (TLP) Systems

In the early development of TLP testing, there was no commercial test equipment. Figure 5.9 shows an example of the first commercial TLP system, the Barth Model 4002 TLP™ [38]. Prior to the Barth TLP Model 4002, many corporations were doing TLP testing internal to their corporation for technology characterization using "home grown" systems. The Barth 4002 TLP+™ Pulse Curve Tracer characterizes the ESD robustness of silicon chip protection circuitry.

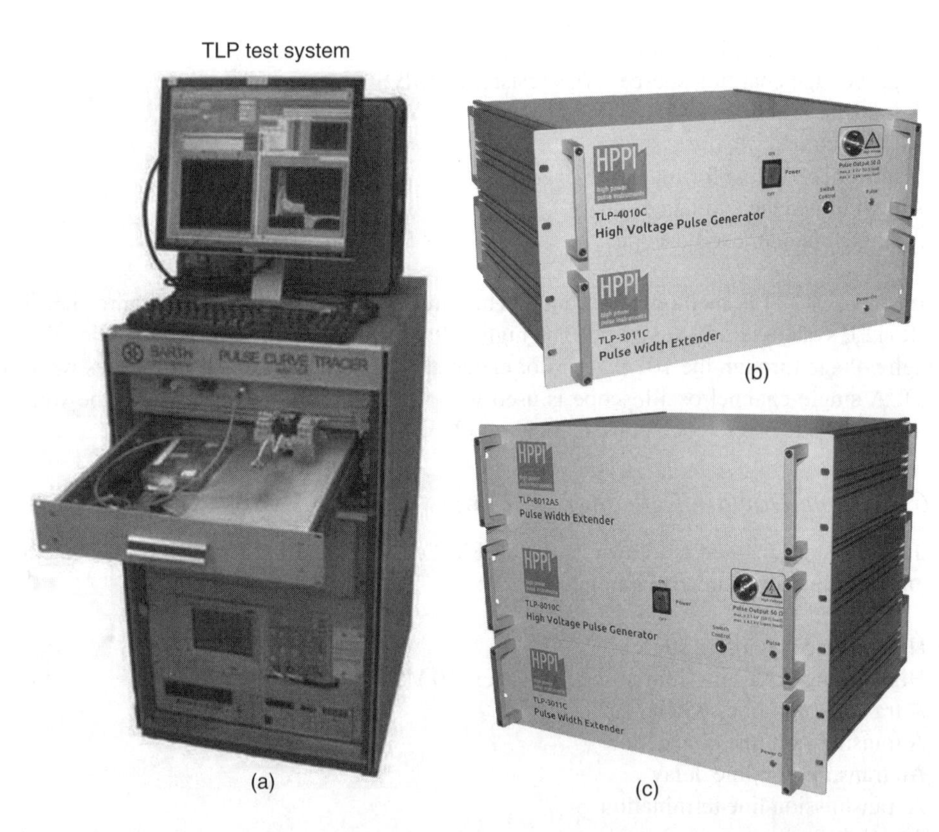

Figure 5.9 (a) Transmission line pulse (TLP) commercial test equipment – Barth system (Reproduced with permission of Barth Electronics Inc.). (b) Transmission line pulse (TLP) commercial test equipment – High power pulse instruments (HPPI) TLP 4010C/3011C (Reproduced with permission of HPPI GmBH). (c) HPPI TLP–8012/8010C/3051C Pulse Width Extender (Reproduced with permission of HPPI GmBH)

The Barth Model 4002 TLP system components comprise the following [38]:

- Barth 40021 control box/pulse generator
- Tektronix 500 MHz two-channel digitizing oscilloscope
- Keithley Pico-ammeter/voltage source
- Stanford Research Systems high-voltage power supply
- Barth 44001A-48 DIP Test Fixture
- A computer
- High-resolution monitor
- Color printer.

Programmed rectangular pulses are applied to the DUT during the testing process. A leakage current measurement is made after each pulse to obtain the leakage evolution current as a function of the TLP pulse voltage.

In this test system, packaged or wafer level testing is possible. A dual wafer probe (Barth Model 45002WP) allows for wafer level testing. The Barth 45002WP Dual Wafer Probe has two separate needles and isolated probe connections that can be independently positioned. To avoid mechanical issues of crossed needles, a constant impedance-reversing switch allows for reversal of the TLP pulse polarity.

The need to retain 50-Ω impedance avoids reflections and measurement errors. A controlled 50-Ω impedance minimizes the influence on the parasitic on the measurement. For both the packaged device socket and the wafer level system, a 50-Ω controlled impedance exists for the packaged socket and the Barth TLP wafer prober. The Barth TLP probe station also has a controlled 50 Ω impedance throughout its connections to the two needle contacts at any two pads. From an inherently low 50 Ω source impedance, a test pulse without ringing, undershoot, or overshoot, or pulse distortion can be formed.

The TLP test system has a manually selectable pulse width of 75–150 ns. The TLP pulse voltage can range from 0 to 250 V, with a TLP pulse current to 5 A for a 50 Ω load, and 10 A for a short circuit. The 10–90% TLP pulse rise time can be varied from 0.2, 2, and 10 ns. The system can provide up to 10 test pulses per minute. The leakage voltage can be varied from 0 to 100 V and can vary from picoamps to 2.5 mA.

Figure 5.9(b) shows an example of a commercial benchtop TLP system that provides the capability to test multiple ESD tests in a common test system. The HPPI TLP test system TLP-4010C/3011C provides both wafer and packaged testing [39]. The test system has two basic configurations. The first configuration is the TLP 4010C high-voltage pulse generator unit, and the second configuration is the TLP 4010C combined with the TLP-3011C pulse width extender. This test system is a 50 Ω impedance system avoiding ringing and allowing for fast rise times without ringing. The TLP-4010C includes printed circuit board (PCB) adapter, current sensors, 18 GHz DUT switch, and cable connections [39].

The TLP 4010C has eight programmable pulse widths from 1 to 100 ns, allowing for evaluation of the power-to-failure over a wider range from the standard TLP test (e.g., 100 ns) to shorter pulse widths. This explores the response of electronics from the HBM to CDM pulse time regime [39].

The TLP 3011C pulse width extender can extend the pulse width well into the microsecond regime. The TLP 3011C allows for extension of the pulse into 1.6 μs [39].

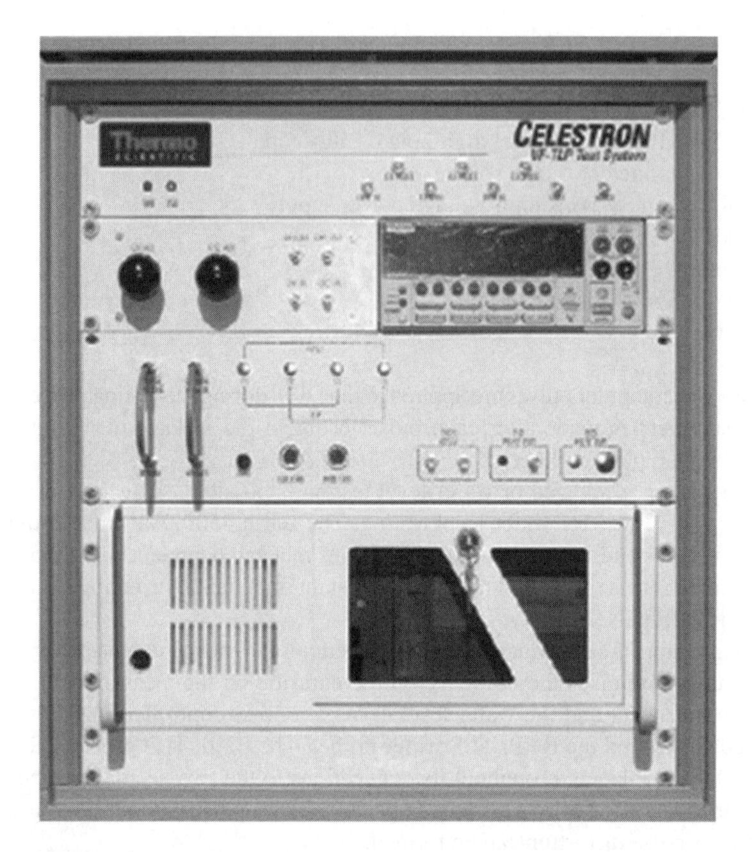

Figure 5.10 Transmission line pulse (TLP) commercial test equipment – Thermo KeyTek Bench Top Celestron™ Flexible Bench Top TLP/VF-TLP System (Reproduced with permission of Thermo Fisher Scientific)

Figure 5.10 shows an example of a commercial benchtop TLP system that provides the capability to test multiple ESD tests in a common test system. The Thermo Scientific™ Celestron™ Flexible Bench Top TLP/VF-TLP System can provide both the TLP event and the very fast transmission line pulse (VF-TLP) event [40]. In addition, this test system can provide testing of HBM and machine model (MM) at wafer level, or packaged level.

The test system can be configured in TDR-O, TDR-S, TDT, TDRT, and Hi-Z TDRT test modes. The system can provide a standard TLP pulse width of 100 ns, as well as a VF-TLP pulse duration of 10 ns [40].

During the testing procedure, the operator of the system can select a range of test stress voltages, pulse polarity, as well as leakage and curve tracing values. The operator is free to modify the measurement test window within the TLP pulse before or after the testing completion. This allows for reduction of the error in the testing measurement in extraction of the current and voltage values.

5.7 Test Sequence and Procedure

For the TLP testing, there is a test sequence and a procedure to provide the TLP pulsed current–voltage characteristic. The TLP test procedure uses a series of pulses with increasing pulse amplitude (Figure 5.11). The steps in the procedure are as follows:

- Select a pulse of a given pulse width, rise time, fall time, amplitude, and polarity.
- Select a step size increment.
- Test a first component.
- Select a minimum time between successive steps (e.g., greater than 0.1 s).
- Define a failure criteria.

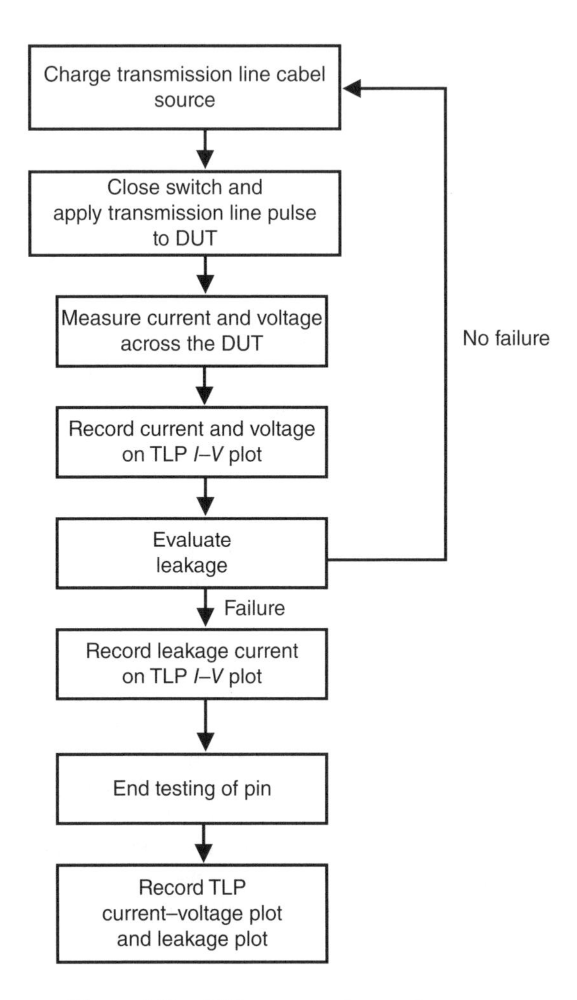

Figure 5.11 TLP test procedure and sequence

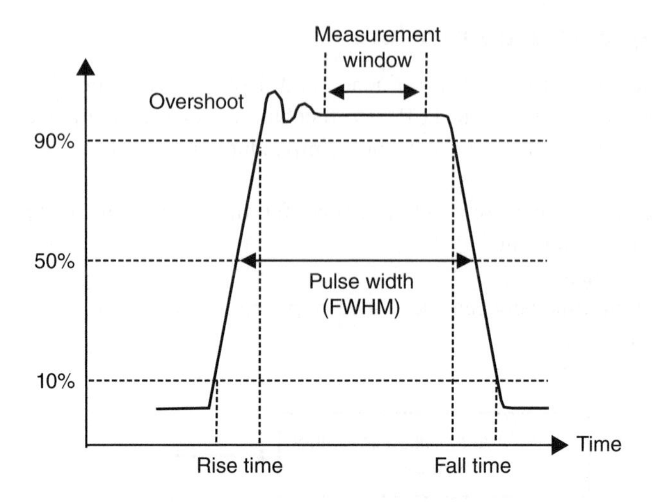

Figure 5.12 TLP measurement window and the current overshoot

- Apply a stress pulse of fixed width to the DUT.
- Measure and record stress pulse voltage and current, representing one *I–V* point on the DUT of the *I–V* characteristic,
- Perform a poststress evaluation of the DUT.
- If no "failure" occurs, the process is repeated with the next pulse applied at a higher magnitude; if failure occurs, the testing is stopped.

5.7.1 TLP Pulse Analysis

For the TLP testing methodology, TLP pulse event analysis is required. The TLP methodology must define a measurement window to extract current and voltage from the DUT.

5.7.2 Measurement Window

The TLP methodology must define a measurement window to extract current and voltage from the DUT. A measurement window is chosen within the pulse event that avoids the rise time, fall time, and the region of current and voltage overshoot, oscillation, and ringing (Figure 5.12).

The measurement is taken in a quiescent region so that an accurate value for the voltage and current is obtained. This is achieved by sampling a number of points and averaging them. The measurement time window is greater than 10% of the pulse width to allow for averaging of many data points. The measurement window is taken in the time outside of the overshoot region.

5.7.3 Measurement Analysis – TDR Voltage Waveform

In this section, the measurement analysis of a pulse waveform in a TDR TLP system is discussed. Figure 5.13 illustrates a voltage characteristic for a measurement where the DUT is

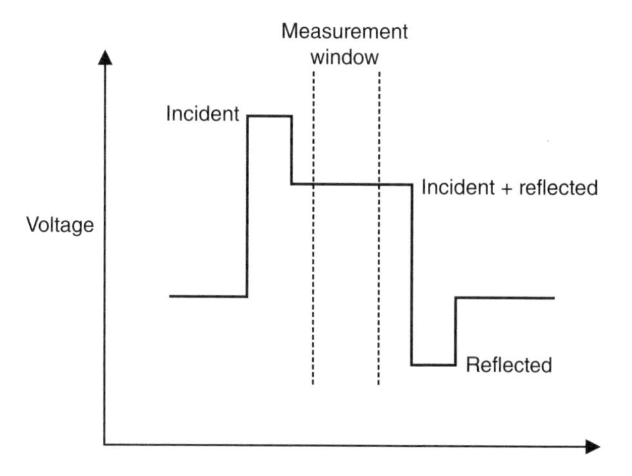

Figure 5.13 TLP analysis – TDR voltage waveform

under 50 Ω. In the plot, the incident pulse is shown, as well as the reflected pulse. In the center region, the incident and reflected pulses overlap in a subtractive fashion. The measurement window is shown in the region where the incident and reflected pulses overlap.

5.7.4 Measurement Analysis – Time Domain Reflection (TDR) Current Waveform

In this section, the measurement analysis of a current pulse waveform in a TDR TLP system is discussed. Figure 5.14 illustrates a current characteristic for a measurement where the DUT is under 50 Ω. In the plot, the incident pulse is shown, as well as the reflected pulse. In the

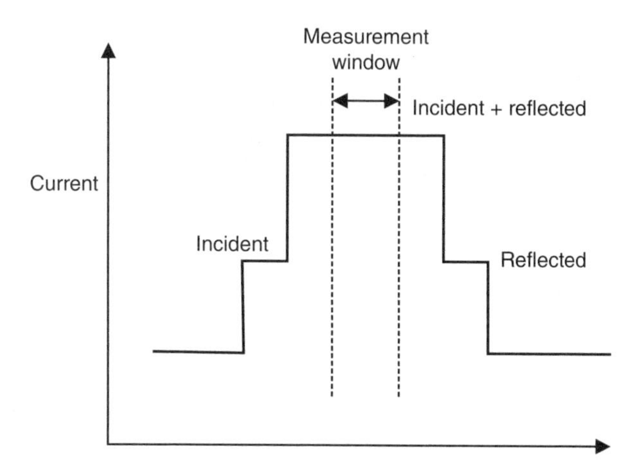

Figure 5.14 TLP analysis – TDR current waveform

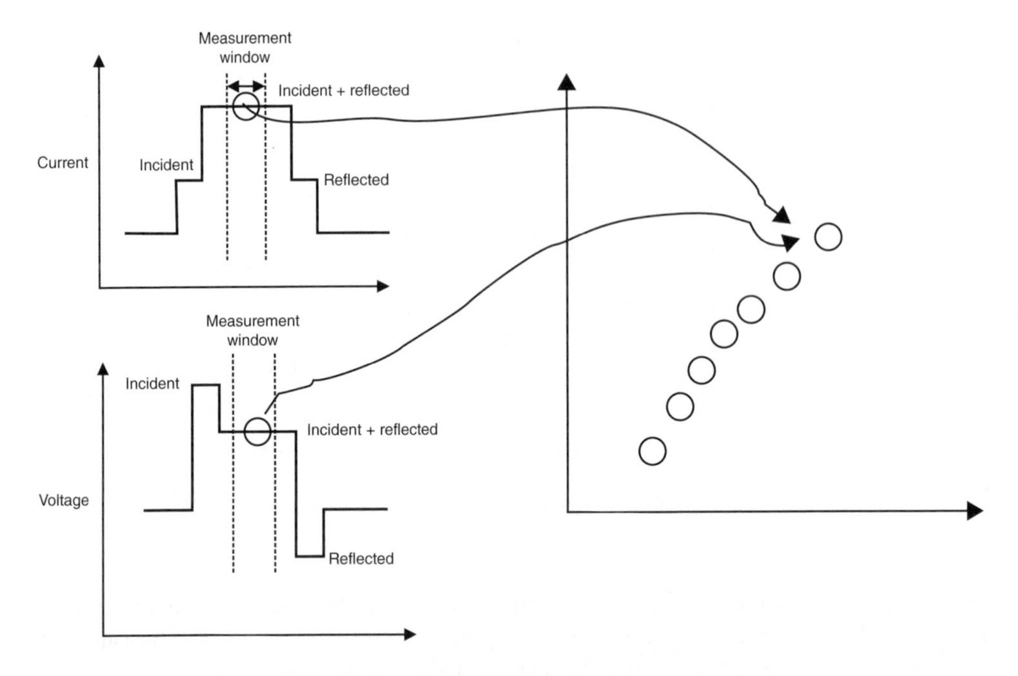

Figure 5.15 TLP analysis – TDR results for current and voltage

center region, the incident and reflected pulses overlap in an additive fashion. The measurement window is shown in the region where the incident and reflected pulses overlap.

5.7.5 Measurement Analysis – Time Domain Reflection (TDR) Current–Voltage Characteristic

In this section, the measurement analysis of a pulse waveform in a TDR TLP system are superimposed on the leakage characteristic plot is shown. Figure 5.15 illustrates a voltage and current versus time characteristic for a measurement as well as the current–voltage (*I*–*V*) characteristic. The measurement for the voltage and the current is achieved by evaluating the average within the measurement window. These averaged values are then recorded in the *I*–*V* plot. In addition to the (*I*, *V*) point placed on the *I*–*V* plot, the leakage current poststress is also recorded.

5.8 TLP Pulsed *I*–*V* Characteristic

From TLP testing, a pulsed current–voltage (*I*–*V*) characteristic can be constructed. Each data point in the pulsed *I*–*V* characteristic is the current and voltage across the DUT. Each data point represents a pulse in the test sequence. The voltage on the transmission line (TL) source is increased for each step in the *I*–*V* characteristic.

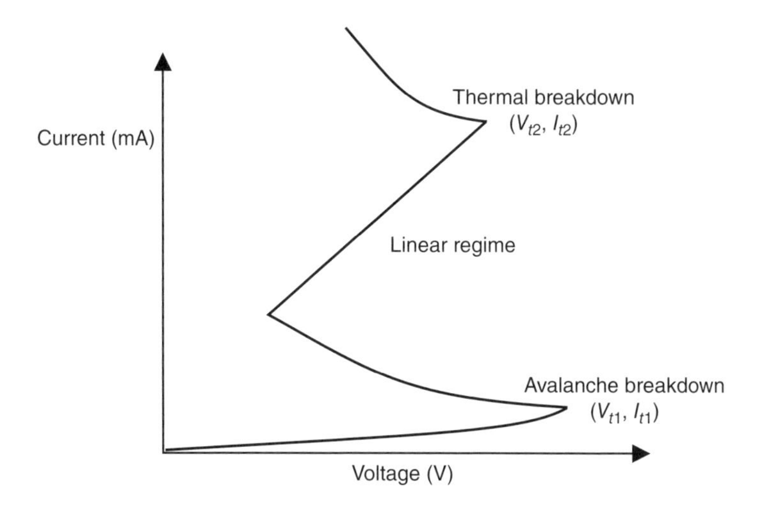

Figure 5.16 Construction of transmission line pulse (TLP) *I–V* characteristic

Figure 5.16 illustrates the current and voltage from a pulse event and how it is projected onto the current–voltage characteristic. Each current and voltage point is an average of the measured pulse event where the information is from an average of measurements from the measurement window.

5.8.1 TLP I–V Characteristic Key Parameters

The TLP *I–V* characteristics contain key metrics. The TLP *I–V* characteristic comprises the following key parameters:

- Avalanche voltage
- Trigger point
- Holding point
- Turn-on voltage
- On-resistance
- Clamping voltage
- Second trigger point
- Soft failure current
- Failure current.

5.8.2 TLP Power Versus Time

In the early development of TLP testing, the focus was not on the current and voltage *I–V* characteristics but on power. The primary reason was the researchists were interested in the power-to-failure. For the power analysis, the techniques were to evaluate the incident, transmitted, and reflected power as a function of time.

The first system developed by J.J. Whalen and H. Domingos was a pulse system in a TDRT configuration [4]. In this configuration, the measured parameters were the incident, transmitted, and reflected power as a function of time. From this method, the absorbed power in the DUT was the incident power minus the reflected power and the transmitted power. Waveforms of the incident power, the reflected power, and the transmitted power as a function of time were recorded. In this methodology, the time-to-failure was observable from transitions in the power waveforms.

The energy-to-failure can be defined as

$$E_{\mathrm{A}} = E_{\mathrm{I}} - E_{\mathrm{R}} - E_{\mathrm{T}}$$

where E_{I} is the incident energy, E_{R} is the reflected energy, and E_{T} is the transmitted energy.

5.8.3 *TLP Power Versus Time – Measurement Analysis*

A second test system, published in 1979, was simplified to include a single pulse unit, a square pulse matching network, a voltage probe, a current probe, and a dual-channel oscilloscope [4]. Note that this system is essentially the first TLP system published in the EOS/ESD Symposium proceedings [4].

In this system, the voltage across the device, current across the device, and the time-to-failure were used to calculate the energy at failure. The absorbed energy versus time-to-failure can be quantified.

The energy-to-failure, absorbed energy E_{A} can be defined as

$$E_{\mathrm{A}} = V \times I \times T_{\mathrm{F}}$$

where V is the voltage across the DUT before failure, I is the current across the DUT before failure, and T_{F} is the time duration before failure.

Figure 5.17 shows the power versus time plot for a TDRT TLP system. The power versus time can be observed for the incident power, the reflected power, and the transmitted power.

5.8.4 *TLP Power-to-Failure Versus Pulse Width Plot*

For construction of the Wunsch–Bell power-to-failure curve, a series of successive tests are made for different pulse widths. The Wunsch–Bell power-to-failure characteristics are formed by sweeping all possible pulse widths capable of the test system [27–37]. During this test sequence, the rise time and fall time are made fixed. For each pulse width, a TLP pulsed I–V plot (for a given pulse width) is formed. From the test, the failure point (I–V current and voltage) is extracted. This failure point is the product of the current-to-failure (I_{f}) and the voltage-to-failure (V_{f}); this can be represented as the power-to-failure (P_{f}). The data is plotted as the power-to-failure (P_{f}) versus the logarithm of the pulse width. Figure 5.18 shows the test sequence.

For construction of the Wunsch–Bell power-to-failure curve, a series of test are made for different pulse widths. The Wunsch–Bell power-to-failure characteristics are formed by sweeping all possible pulse widths capable of the test system (Figure 5.19).

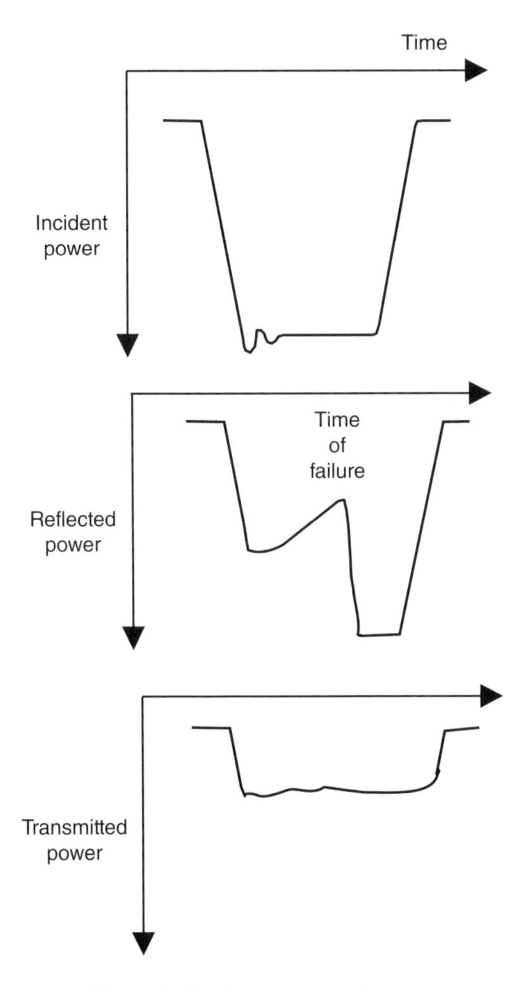

Figure 5.17 Power versus time plot

5.9 Alternate Methods

Alternate methods will occur in the future. With increasing interest in EOS events, there is growth in the long pulse events. Long duration transmission line pulse (LD-TLP) events (e.g., also referred to as long pulse TLP (LPTLP) provide an understanding of the response of devices and circuits closer to the steady state regime [36].

5.9.1 Long Duration TLP (LD-TLP)

LD-TLP event waveforms are rectangular pulses with a long pulse width. The distinction between the TLP standard and the long pulse TLP events is the length of the pulse width. Figure 5.20 shows an example of the LD-TLP waveform. The defined LD-TLP pulse width is 500 ns.

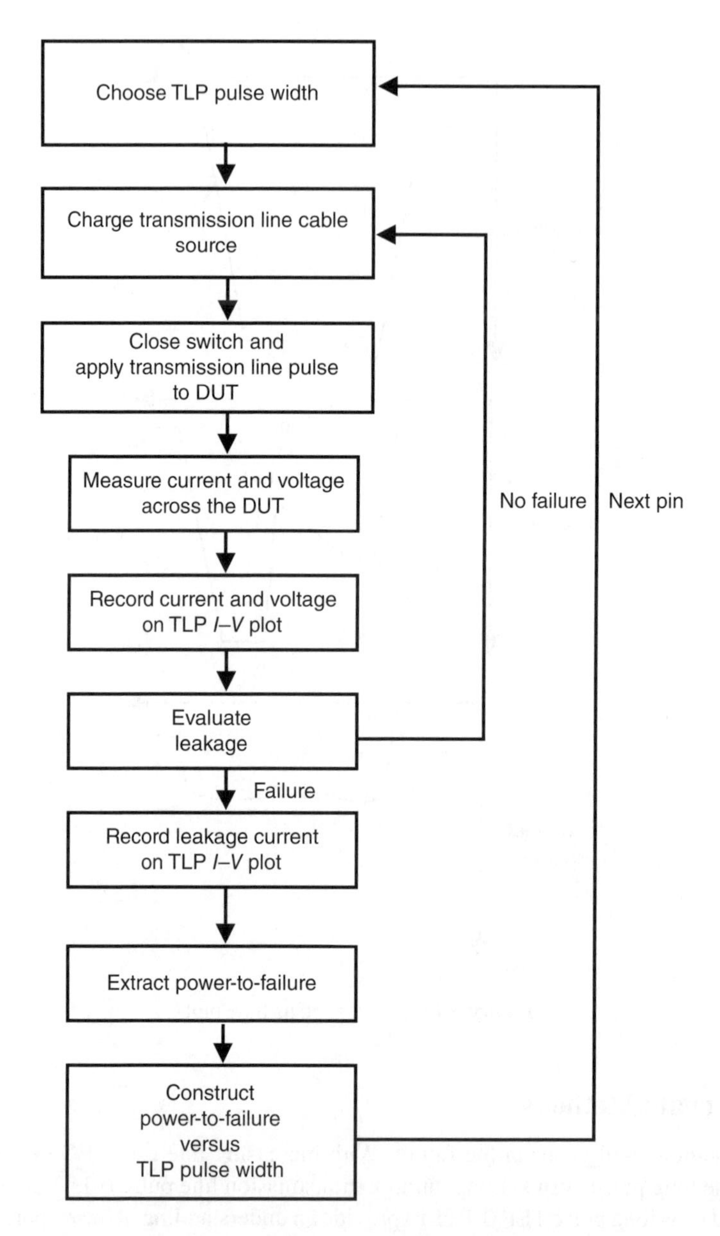

Figure 5.18 Test sequence for Wunsch–Bell curve construction

5.9.2 Long Duration TLP Time Domain

The long duration TLP (LD-TLP) can be plotted on a Wunsch–Bell power-to-failure plot [36]. Figure 5.21 is an example of the Wunsch–Bell plot highlighting the long-pulse TLP time domain. The LD-TLP pulse width is near the end of the thermal diffusion regime and/or steady state regime.

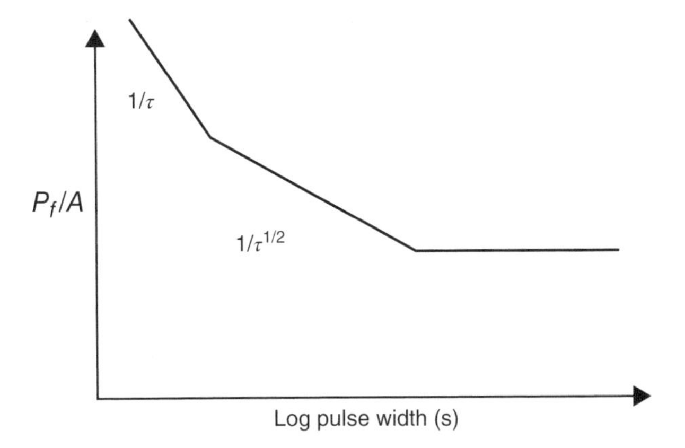

Figure 5.19 Wunsch–Bell curves: power-to-failure versus pulse width

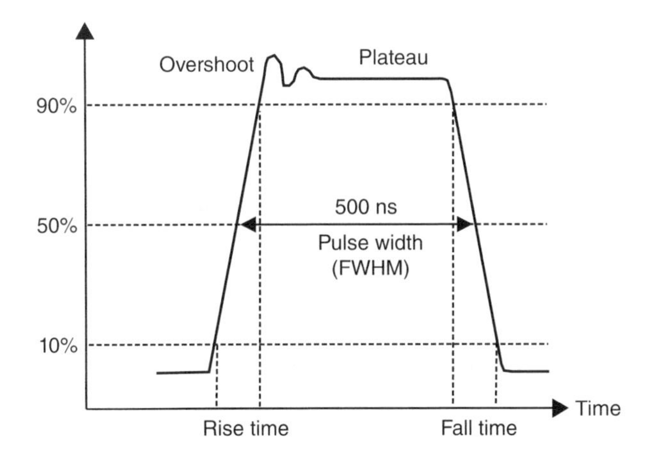

Figure 5.20 Long-pulse TLP waveform

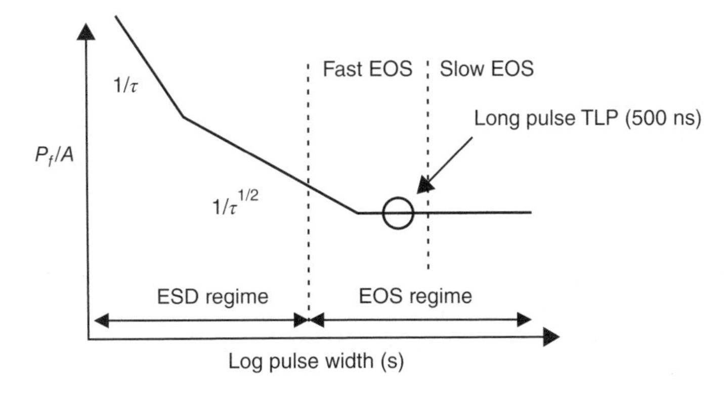

Figure 5.21 Wunsch–Bell curve highlighting long duration TLP time domain

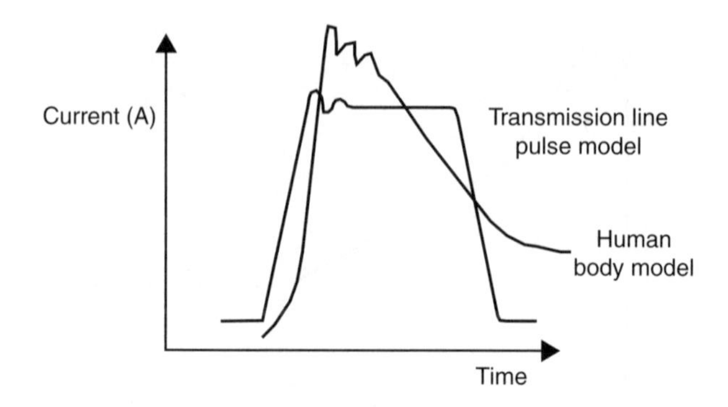

Figure 5.22 Transmission line pulse (TLP) and HBM pulse waveform comparison

5.10 TLP-to-HBM Ratio

5.10.1 Comparison of Transmission Line Pulse (TLP) and Human Body Model (HBM) Pulse Width

The TLP can be compared to the HBM. The motivation is the interest of eliminating HBM testing and replace it with TLP testing only. A metric can be established between the TLP current-to-failure and the HBM failure voltage level. Figure 5.22 shows a comparison of the HBM voltage level and TLP current-to-failure [41]. For example, this can be expressed as [41]:

$$V_{\text{HBM}}(\text{V}) = 1680 I_{\text{TLP}}(\text{A}).$$

5.11 Summary and Closing Comments

In this chapter, one of the most important tests in the history of ESD testing, the TLP model is discussed. This method had a rapid growth in popularity in the late 1990s due to its powerful value in semiconductor device and circuit development. The ability to see the current and voltage characteristics across the DUT was of significant value to ESD engineers, circuit designers, to failure analysis teams who need to develop, understand, or debug the response of devices and circuits. This was a large step forward in increasing the technical nature of ESD development.

In Chapter 6, another important test in the history of ESD testing, the high-speed short-pulse version of the TLP model, known as the VF-TLP test, is discussed.

Problems

1. An HBM test result quotes a number as a result, typically an ESD HBM level. A TLP test also describes some metrics from the TLP results. How does it differ? Where is the voltage or current measured in the case of an HBM result? Where is it measured for a TLP test?

2. What information does a TLP test provide that an HBM test does not provide?
3. Explain the differences between an HBM test and TLP procedure.
4. Explain the difference between the test sequence of an HBM test and a TLP test.
5. An HBM test is a multipin test, whereas the TLP test is a two-pin test. Explain how the tests are the same, and how they are different.
6. Show an example of a TLP *I–V* characteristic of a diode. Explain each region of a TLP *I–V* characteristic.
7. Show an example of a TLP *I–V* characteristic of an n-channel MOSFET. Explain each region of a TLP *I–V* characteristic.
8. Show an example of a TLP *I–V* characteristic of a p-channel MOSFET. Explain each region of a TLP *I–V* characteristic. How does the p-channel MOSFET TLP *I–V* characteristic differ from the n-channel MOSFET TLP *I–V* characteristic?
9. Show an example of a TLP *I–V* characteristic of a bipolar transistor. Explain each region of a TLP *I–V* characteristic. How does it differ from an n-channel MOSFET?
10. Show an example of a TLP *I–V* characteristic of a silicon controlled rectifier. Explain each region of a TLP *I–V* characteristic. How does it differ from an n-channel MOSFET? How does it differ from a bipolar transistor?
11. Show an example of a TLP *I–V* characteristic of an n-well ballast resistor. Explain each region of a TLP *I–V* characteristic. How does it differ from an n-channel MOSFET? How does it differ from a bipolar transistor?

References

1. S. Voldman, R. Ashton, J. Barth, D. Bennett, J. Bernier, M. Chaine, J. Daughton, E. Grund, M. Farris, H. Gieser, L.G. Henry , M. Hopkins, H. Hyatt, M.I. Natarajan, P. Juliano, T.J. Maloney , B. McCaffrey, L. Ting, and E. Worley. Standardization of the transmission line pulse (TLP) methodology for electrostatic discharge (ESD). *Proceedings of the Electrical Overstress/Electrostatic Discharge (EOS/ESD) Symposium*, 2003; 372–381.
2. ANSI/ESD Association. ESD-SP 5.5.1 – 2004. *ESD Association Standard Practice for the Protection of Electrostatic Discharge Sensitive Items – Electrostatic Discharge Sensitivity Testing – Transmission Line Pulse (TLP) Testing Component Level.* Standard Practice (SP) document, 2004.
3. ANSI/ESD Association. ESD-STM 5.5.1 – 2008. *ESD Association Standard Test Method for the Protection of Electrostatic Discharge Sensitive Items – Electrostatic Discharge Sensitivity Testing – Transmission Line Pulse (TLP) Testing Component Level.* Standard Test Method (STM) document, 2008.
4. J.J. Whalen and H. Domingos. Square pulse and RF pulse overstressing of UHF transistors. *Proceedings of the Electrical Overstress/ Electrostatic Discharge (EOS/ESD) Symposium*, 1979; 140–146.
5. J.J. Whalen, M.C. Calcatera, and M. Thorn. Microwave nanosecond pulse burnout properties of GaAs MESFETs. *Proceedings of the IEEE MTT-S International Microwave Symposium*, 1979; 443–445.
6. T. Maloney and N. Khurana. Transmission line pulsing techniques for circuit modeling of ESD phenomena. *Proceedings of the Electrical Overstress/Electrostatic Discharge (EOS/ESD) Symposium*, 1985; 49–54.
7. R.A. Ashton. Modified transmission line pulse system and transistor test structures for the study of ESD. *Proceedings of the International Conference on Microelectronic Test Structures (ICMTS)*, 1995; 127–132.
8. W. Stadler, X. Guggenmos, P. Egger, H. Gieser, and C. Musshoff. Does the ESD failure current obtained by transmission line pulsing always correlate to human body model tests? *Proceedings of the Electrical Overstress/ Electrostatic Discharge (EOS/ESD) Symposium*, 1997; 366–372.
9. G. Notermans, P. de Jong, and F. Kuper. Pitfalls when correlating TLP, HBM and MM testing. *Proceedings of the Electrical Overstress/Electrostatic Discharge (EOS/ESD) Symposium*, 1998; 170–176.
10. J. Barth. *Calibrating TLP Systems. TLP Application Note, No. 2.* Boulder City, Nevada: Barth Electronics Inc., 1999 (www.barthelectronics.com).

11. J. Barth. *TLP to HBM Rise Time Correlation. TLP Application Note, No. 3*. Boulder City, Nevada: Barth Electronics Inc., 1999 (www.barthelectronics.com).
12. J. Barth, J. Richner, K. Verhaege, and L.G. Henry. TLP calibration, correlation, standards, and new techniques. *Proceedings of the Electrical Overstress/Electrostatic Discharge (EOS/ESD) Symposium*, 2000; 85–97.
13. H. Hyatt, J. Harris, A. Alanzo, and P. Bellew. TLP measurements for verification of ESD protection device response. *Proceedings of the Electrical Overstress/Electrostatic Discharge (EOS/ESD) Symposium*, 2000; 111–121.
14. J. Lee, M. Hoque, J. Liou, G. Croft, W. Young, and J. Bernier. A method for determining a transmission line pulse shape that produces equivalent results to human body model testing methods. *Proceedings of the Electrical Overstress/Electrostatic Discharge (EOS/ESD) Symposium*, 2000; 97–105.
15. L.G. Henry, J. Barth, K. Verhaege, and J. Richner. Transmission line pulse testing of the ESD protection structures of IC's: a failure analysts perspective. *Proceedings of the International Symposium for Testing and Failure Analysis (ITSFA)*, 2000; 203–213.
16. L.G. Henry, J. Barth, K. Verhaege, and J. Richner. Transmission line pulse ESD testing of IC's: a new beginning. *Compliance Engineering*, Mar/Apr 2001; 46–53.
17. J. Barth. Pulse circuit. U.S. Patent No. 6,429,674, August 6, 2002.
18. E. Grund. Transmission line pulse measurement system for measuring the response of a device under test. U.S. Patent No. 7,545,152, June 9, 2009.
19. K. Kato, and Y. Fukuda. ESD evaluation by TLP method on advanced semiconductor devices. *Proceedings of the Electrical Overstress/Electrostatic Discharge (EOS/ESD) Symposium*, 2001; 419–425.
20. T. Smedes, R.M.D.A. Velghe, R.S. Ruth, and A.J. Huitsing. The application of transmission line pulse testing for the ESD analysis of integrated circuits. *Proceedings of the Electrical Overstress/Electrostatic Discharge (EOS/ESD) Symposium*, 2001; 426–434.
21. J. Barth, and J. Richter. Correlation considerations: real HBM to TLP and HBM testers. *Proceedings of the Electrical Overstress/Electrostatic Discharge (EOS/ESD) Symposium*, 2001; 453–460.
22. B. Keppens, V. De Heyn, N. Mahadeva, G. Groeseneken, Contributions to standardization of transmission line pulse testing methodology. *Proceedings of the Electrical Overstress/Electrostatic Discharge (EOS/ESD) Symposium,* 2001; 461–468.
23. ESD Association. ESD-SP 5.5.1. *ESD Association Standard Practice for the Protection of Electrostatic Discharge Sensitive Items – Electrostatic Discharge Sensitivity Testing Transmission Line Pulse (TLP) Testing Component Level*. Standard Practice document, 2003.
24. J. Barth. *Evaluation of the TLP System by the Use of Known SOLZ Elements to Determine the Pulse Measurement Range, Accuracy, and Resolution. TLP Application Note*. Boulder City, Nevada: *Barth Electronics* Inc., 2003 (www.barthelectronics.com).
25. S. Hynoven, S. Joshi, and E. Rosenbaum. Combined TLP/RF testing system for detection of ESD failures in RF circuits. *Proceedings of the Electrical Overstress/Electrostatic Discharge (EOS/ESD) Symposium*, 2003; 346–353.
26. S. Joshi, and E. Rosenbaum. Transmission line pulsed waveform shaping with microwave filters. *Proceedings of the Electrical Overstress/Electrostatic Discharge (EOS/ESD) Symposium*, 2003; 364–371.
27. S. Voldman. *ESD: Physics and Devices*. Chichester, England: John Wiley and Sons, Ltd., 2004.
28. S. Voldman. *ESD: Circuits and Devices*. Chichester, England: John Wiley and Sons, Ltd., 2005.
29. S. Voldman. *ESD: Circuits and Devices 2nd Edition*. Chichester, England: John Wiley and Sons, Ltd., 2015.
30. S. Voldman. *ESD: RF Circuits and Technology*. Chichester, England: John Wiley and Sons, Ltd., 2006.
31. S. Voldman. *Latchup*. Chichester, England: John Wiley and Sons, Ltd., 2007.
32. S. Voldman. *ESD: Circuits and Devices*. Beijing, China: Publishing House of Electronic Industry (PHEI), 2008.
33. S. Voldman. *ESD: Failure Mechanisms and Models*. Chichester, England: John Wiley and Sons, Ltd., 2009.
34. S. Voldman. *ESD: Design and Synthesis*. Chichester, England: John Wiley and Sons, Ltd., 2011.
35. S. Voldman. *ESD Basics: From Semiconductor Manufacturing to Product Use*. Chichester, England: John Wiley and Sons, Ltd., 2012.
36. S. Voldman. *Electrical Overstress (EOS): Devices, Circuits, and Systems*. Chichester, England: John Wiley and Sons, Ltd., 2013.
37. S. Voldman. *ESD: Analog Design and Circuits*. Chichester, England: John Wiley and Sons, Ltd., 2014.

38. Barth Electronics. http://www.barthelectronics.com. Barth Model 4002 TLP™, 2016.
39. High Power Pulse Instruments GmBH. http://www.hppi.de. HPPI TLP test system TLP-4010C/3011C, 2016.
40. Thermo Fischer Scientific. http://www.thermofisher.com. Thermo Scientific™ Celestron™ Flexible Bench Top TLP/VF-TLP System, 2016.
41. G. Noterman, O. Quittard, A. Heringa, et al., Designing HV active clamp for HBM robustness. *Proceedings of the Electrical Overstress/Electrostatic Discharge (EOS/ESD) Symposium*, 2007; 47–52.

6

Very Fast Transmission Line Pulse (VF-TLP) Testing

6.1 History

Very fast transmission line pulse (VF-TLP) became popular to quantify the electrostatic discharge (ESD) robustness for short pulse event. This method is in the "adiabatic regime" of the Wunsch–Bell power-to-failure curve, where in contrast, the electrical overstress (EOS) events of interest are in the "steady-state" time-independent region of the Wunsch–Bell power-to-failure curve.

After completion of the standard and standard practice for the transmission line pulse (TLP) test [1–24], the new methodology for the VF-TLP was initiated [5–9].

Today, there is a standard test method for the VF-TLP test methodology [5–9]. The standard practice for VF-TLP test simulation is supported by the ESD Association (ANSI/ESD – ESD-SP 5.5.2 – 2007) [6].

Figure 6.1 illustrates the evolution of the VF-TLP method. The method will be extended to applications to radio frequency (RF) technology, to ultrafast transmission line pulse (UF-TLP), to methods combining TLP and VF-TLP for Wunsch–Bell plot development.

6.2 Scope

The scope of the VF-TLP ESD test is for the testing, evaluation, and classification of components and micro- to nanoelectronic circuitry. The test is to quantify the sensitivity or susceptibility of these components to damage or degradation to the defined VF-TLP test [6, 9].

6.3 Purpose

The purpose of the VF-TLP ESD test is for establishment of a test methodology to evaluate the repeatability and reproducibility of components to a defined pulse event in order to classify or compare ESD sensitivity levels of components [9].

ESD Testing: From Components to Systems, First Edition. Steven H. Voldman.
© 2017 John Wiley & Sons, Ltd. Published 2017 by John Wiley & Sons, Ltd.

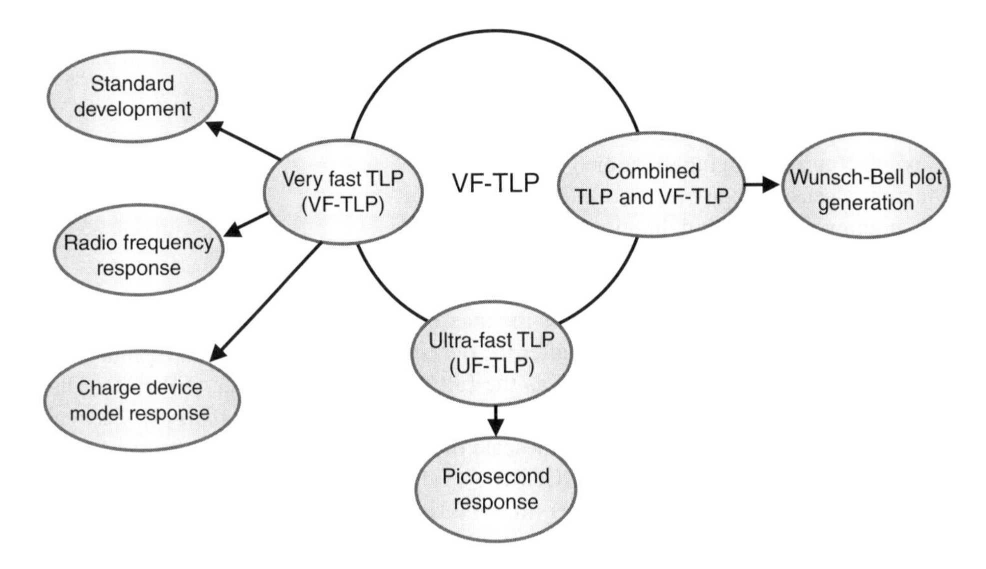

Figure 6.1 Very fast TLP testing overview

6.4 Pulse Waveform

TLP testing provides a square pulse from the TLP source. The VF-TLP test system source can provide different pulse widths, rise times, and fall times. For the semiconductor industry, the VF-TLP standard was developed with a fixed pulse width for comparison of TLP testing results. The TLP results change with different pulse characteristics.

Figure 6.2 shows the VF-TLP pulse waveform characteristics. The waveform characteristic can be defined by specific parameters [9]:

- Pulse plateau

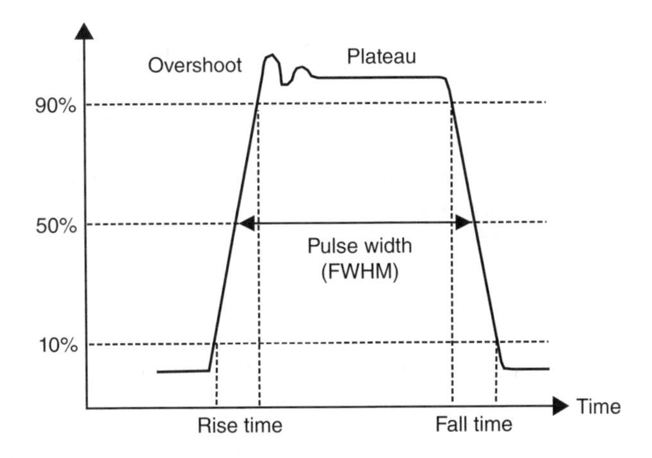

Figure 6.2 Very fast transmission line pulse (VF-TLP) pulse waveform

- Pulse width
- Rise time
- Fall time
- Maximum peak current overshoot
- Maximum current ringing duration
- Maximum peak voltage overshoot
- Maximum voltage ringing duration
- Measurement window.

Pulse plateau: The plateau of the pulse is the maximum value in the measurement window. The plateau is a time-averaged value after the overshoot and the oscillation region.

Pulse width: The pulse width is defined as the full width half maximum (FWHM). This is measured at 50% magnitude. For the VF-TLP standard, this defined as 10 ns [9]. The reason with this pulse width is to make the timescale of the VF-TLP waveform equivalent to the timescale of a charged device model (CDM).

Rise time: The rise time is defined as the time it takes to rise from the 10–90% of the pulse plateau. For the VF-TLP standard, the rise time of 1 ns is used [9].

Fall time: The fall time is defined as the time it takes to fall from the 90–10% of the pulse plateau. For the VF-TLP standard, the fall time is equal to or greater than the rise time. For the VF-TLP standard, the fall time of 1 ns or greater is used [9].

Maximum peak current overshoot: The maximum peak current overshoot is defined as the magnitude of the overshoot peak current above the current plateau. The VF-TLP standard states that this should be less than 20% of the plateau current. This is defined into a short circuit.

Maximum current ringing duration: The maximum current ringing duration is defined as the length of time between the beginning of the current pulse and the time that the ringing reaches less than 5% of the plateau current. For the VF-TLP standard, the maximum current ringing duration should be less than 25% of the pulse width [9].

Maximum peak voltage overshoot: The maximum peak voltage overshoot is defined as the magnitude of the overshoot peak voltage above the voltage plateau. The VF-TLP standard states that this should be less than 20% of the plateau voltage. This is defined into an open circuit.

Maximum voltage ringing duration: The maximum voltage ringing duration is defined as the length of time between the first overshoot beyond the plateau voltage to the time it reaches 5% of the plateau voltage. For the VF-TLP standard, the maximum voltage ringing duration should be less than 25% of the pulse width [9].

Measurement window: The measurement window is the range of time within the pulse width where the voltage and current can be measured and extracted in the plateau [9]. According to the VF-TLP standard, the measurement window should be greater than 10% of the VF-TLP pulse width [9].

6.4.1 Comparison of VF-TLP Versus TLP Waveform

Comparing the VF-TLP test source to the TLP source, both are transmission line sources. The distinction between the two sources is the VF-TLP source is significantly shorter than the TLP source. Figure 6.3 shows both the TLP and VF-TLP pulse waveforms.

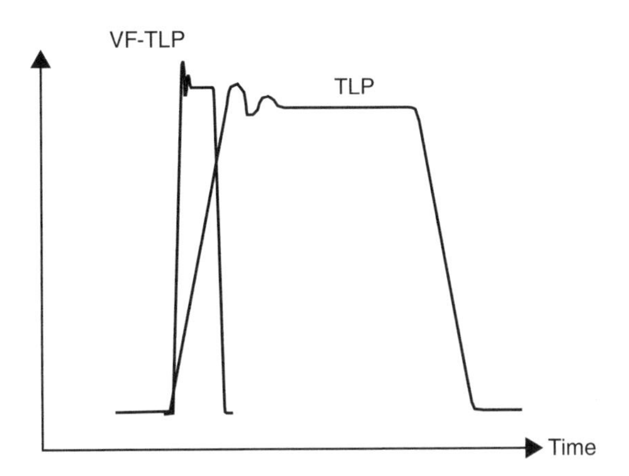

Figure 6.3 Very fast transmission line pulse (VF-TLP) and TLP waveform comparison

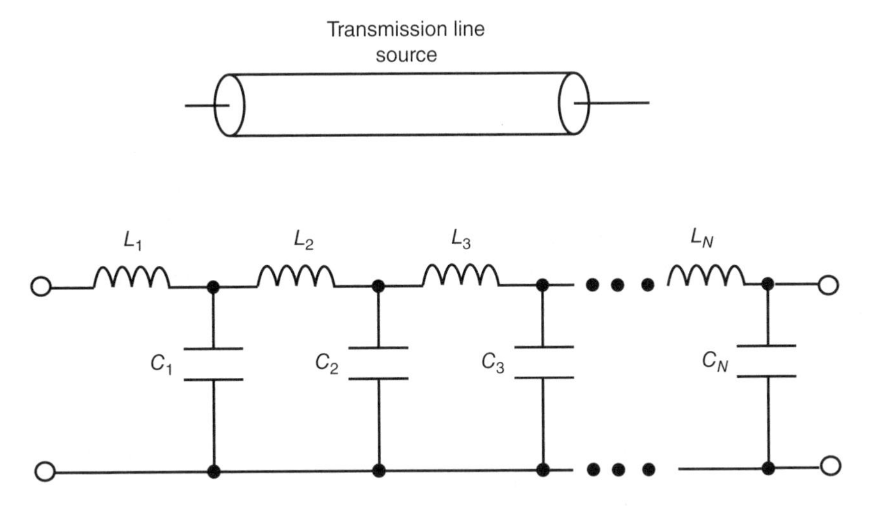

Figure 6.4 Very fast transmission line pulse (VF-TLP) equivalent circuit

6.5 Equivalent Circuit

For the VF-TLP test system, an equivalent model circuit representation can be defined. The VF-TLP circuit is a transmission line source. The VF-TLP source is short and does not demonstrate any resistive loss. Hence, it can be represented as a lossless inductor–capacitor (LC) transmission line (Figure 6.4).

6.6 Test Equipment Configuration

VF-TLP test systems can be placed in different configurations, as was true for the TLP testing. In the following section, the acceptable test configurations are discussed.

6.6.1 *Current Source*

VF-TLP test system method known as the current source method is a high impedance test method. Due to the high impedance, this method is not suitable for VF-TLP methods.

6.6.2 *Time Domain Reflection (TDR)*

VF-TLP test system method is known as the time domain reflection (TDR) VF-TLP (Figure 6.5). The TDR–VF-TLP method comprises the following test equipment [9]:

- High-voltage charging source
- High-voltage charging source series resistor
- A transmission line for matching (50 Ω)
- A transmission line switch
- A transmission line source
- A transmission line connection
- A transmission line delay
- An attenuator
- A voltage probe
- A single-channel oscilloscope.

In the TDR–VF-TLP method, the components are arranged as shown in Figure 6.5. The TDR–VF-TLP system is a 50 Ω system. The high-voltage source charges the transmission line to a high voltage through the 10 MΩ resistor element. An attenuator is placed in series with the device under test (DUT). A single-channel oscilloscope is used with a voltage probe that measures the voltage across the DUT. A reference pulse is required if the reflected pulse overlaps the initial VF-TLP pulse [9].

6.6.3 *Time Domain Transmission (TDT)*

VF-TLP test systems can be configured to look at the transmitted signal. This method is known as the time domain transmission (TDT) VF-TLP. The TDT VF-TLP method comprises the following test equipment [9]:

- High-voltage charging source

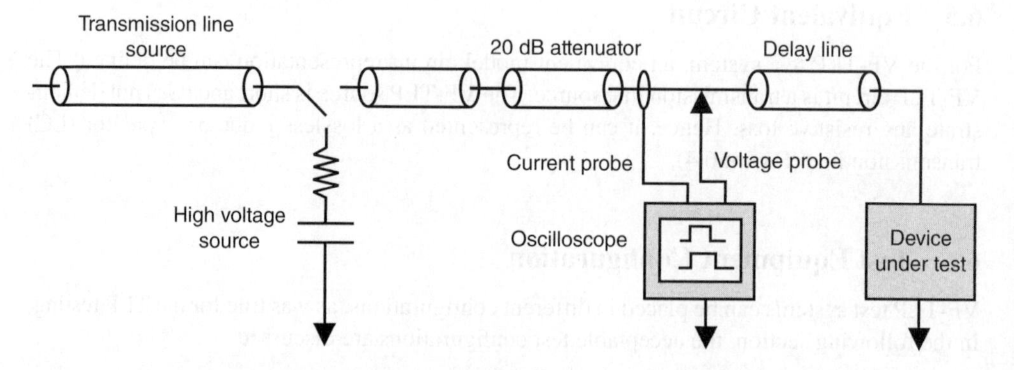

Figure 6.5 Very fast transmission line pulse (VF-TLP) TDR configuration

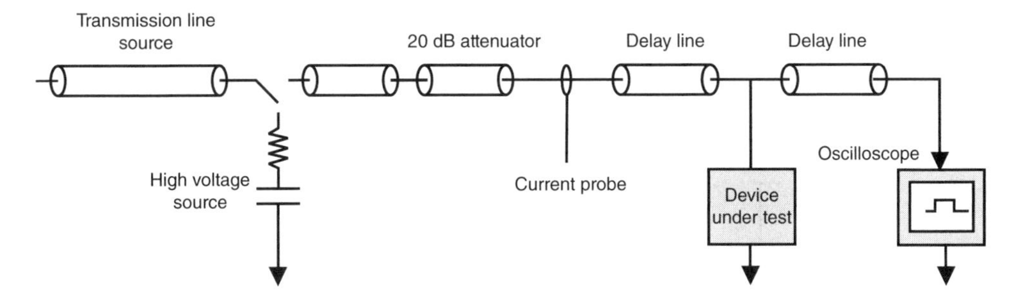

Figure 6.6 Very fast transmission line pulse (VF-TLP) time domain transmission (TDT) configuration

- High-voltage charging source series resistor (10 MΩ)
- A transmission line switch
- An attenuator
- A transmission line for matching
- A high impedance voltage probe
- A single-channel oscilloscope.

In the TDT VF-TLP method, the components are arranged as shown in Figure 6.6. The TDR VF-TLP system is a 50 Ω system. The high-voltage source charges the transmission line to a high voltage through the 10 MΩ resistor element. No attenuator is placed in series with the DUT. A single-channel oscilloscope is used with a voltage probe that measures the voltage across the DUT. A reference pulse is required.

6.6.4 Time Domain Reflection and Transmission (TDRT)

VF-TLP test systems can also be configured to measure the reflected and the transmitted signal. This method is known as the time domain reflection and transmission (TDRT) VF-TLP. The TDRT VF-TLP method comprises the following test equipment [9]:

- High-voltage charging source
- High-voltage charging source series resistor (10 MΩ)
- A transmission line switch
- A transmission line connection
- An transmission line delay
- A transmission line termination
- A high impedance voltage probe
- A two-channel oscilloscope.

In the TDRT VF-TLP method, the components are arranged as shown in Figure 6.7. The DUT is placed in a series configuration in the center of the system and allows for both reflection and transmission of the signal. The TDRT VF-TLP system is a 100 Ω system. The high-voltage source charges the transmission line to a high voltage through the 10 MΩ resistor element. No attenuator is placed in series with the DUT. A two-channel oscilloscope is used with a voltage probe that measures the voltage across the DUT. The transmitted current is used for measuring the current through the DUT. A reference pulse is required.

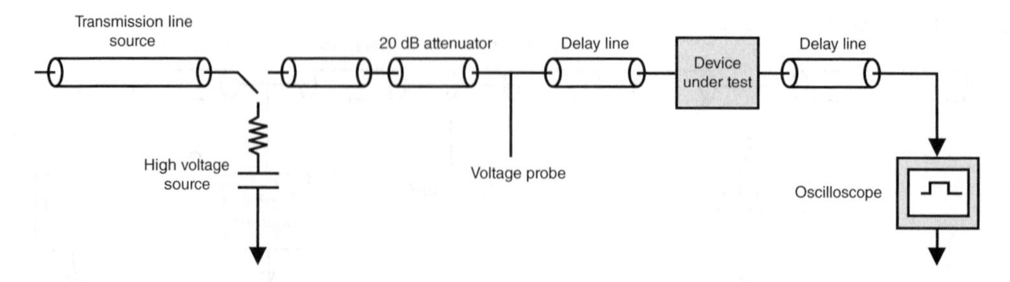

Figure 6.7 Very fast transmission line pulse (VF-TLP) time domain transmission (TDRT) configuration

6.6.5 Early VF-TLP Systems

In the 1970s, VF-TLP systems were used for understanding the robustness of semiconductors. Figure 6.8 shows a VF-TLP pulse system for evaluation of the EOS of an RF component [10]. For evaluation of the power-to-failure, a system was established that could evaluate the incident, reflected, and transmitted power through a DUT. An ultrahigh-frequency source signal was transmitted through a 0–30 dB attenuator. The signal was passed through a first directional coupler for the incident power, and a second directional coupler for the reflected power evaluation. A switch followed by a load is placed after the directional couplers to allow for isolation of UHF RF source from the DUT. In the TDRT VF-TLP system, the DUT is placed in series with the incident and the transmitted signals. A directional coupler is also utilized to evaluate the transmitted power through the DUT. The three directional coupler elements are followed by crystal detectors and 50 Ω terminations, whose signals are transferred to oscilloscopes and camera sources; these diagnostics capture the incident power, reflected power, and transmitted power waveforms in time [10, 11].

In this test system, the incident, reflected, and transmitted power waveforms were captured. J.J. Whalen and H. Domingos showed that the time of failure is observable in the reflected power waveform prior to the end of the applied pulse waveform [10]. In the case of the incident power pulse, a constant power pulse is observable. But, in the reflected power waveform, variation can be observed in the peak power as a function of time. It was noted that a transition occurs in the reflected power waveform, indicating a change in the power absorption and a reduction in the reflected power, followed by an abrupt increase in the reflected power characteristic. This was not observed in power waveforms of RF devices that did not fail at lower power levels.

The absorbed energy of the DUT can be calculated by the difference between the incident energy and the sum of the reflected and transmitted energy.

$$E_A = E_I - E_R - E_T$$

where the incident energy is

$$E_I = \int P_I(t)dt$$

and the reflected energy is

$$E_R = \int P_R(t)dt$$

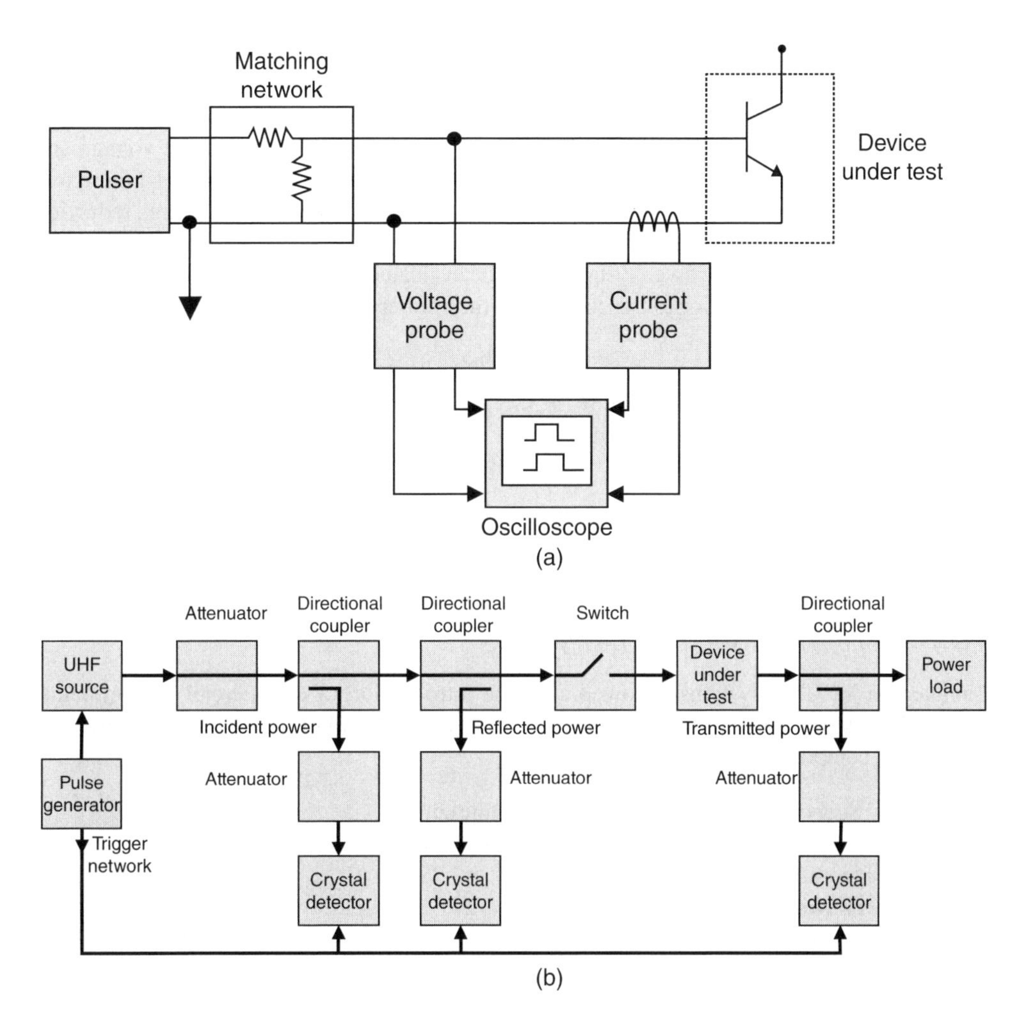

Figure 6.8 (a) Very fast transmission line pulse (VF-TLP) test system with matching network. (b) Very fast transmission line pulse (VF-TLP) test system (Whalen and Domingos)

and transmitted energy is

$$E_T = \int P_T(t)dt$$

In the case of component failure, the integration is terminated at the time-of-failure, t_f.

From this test methodology, the energy-to-failure from a pulse source can be determined.

J.J. Whalen and H. Domingos also performed a square pulse test that could vary the polarity of the pulse waveform to either positive pulses or negative pulses [10]. Figure 6.8(b) shows the square pulse system utilized. As in a TLP test system, this system utilized a square pulse waveform and was able to capture the voltage and current characteristics across the DUT. In this test system, the test system integrated a pulse source with a manual trigger, a resistor matching network, and the RF DUT. To capture the voltage characteristic across the RF DUT, a voltage probe was placed across the device, whose output was transferred to an oscilloscope. In addition, a current probe was placed over the ground return line, observing the current that flowed through the RF DUT. The output of the current probe was transferred to an oscilloscope to observe the current waveform. From the voltage and current waveforms, the voltage and current across the RF DUT were evaluated. From these characteristics, the time-of-failure was also observable by the change in the forward bias value across the RF transistor (e.g., reduction of V_{BE}).

From the test system, the power-to-failure can be evaluated as the product of the current and voltage across the device prior to the collapse of the emitter-base voltage,

$$P_f = I_B V_{BE}$$

whereas the absorbed-energy-to-failure is the product of the power-to-failure and time-of-failure,

$$E_A = P_f t_f = \{I_B V_{BE}\} t_f$$

From the test system, the voltage, current, and the time-of-failure are obtainable from the waveforms [10].

6.6.6 *Commercial VF-TLP Test Systems*

Commercial VF-TLP systems followed after the introduction of commercial TLP systems. One of the first VF-TLP systems was the Barth Model 4012 VF-TLP+™ (Figure 6.9) [25]. The system components comprise the following:

- Tektronix Scope Model DPO 70604, 6 GHz Digitizing
- High Reliability Barth Electronics Control Box/Pulse Generator
- Keithley Pico-ammeter/Voltage Source
- Main power switch/power distribution panel
- Test control computer (Dell Optiplex Minitower; Dell 19″ flat panel; Windows XP; Keyboard; Mouse)
- High-speed business class color printer
- LabView® Runtime Control and Analysis Software.

With the fast pulse width, wide bandwidth pulse current and voltage sensors are required to capture an accurate signal. To avoid signal losses, cable connections and interconnects are optimized.

The Barth Model 4012 is capable of different pulse widths, from 1 to 10 ns, as well as a variable rise time of 100, 200, and 400 ps. The test pulse voltage range is from 0 to 500 V into a 50 Ω load. The test pulse current can be changed from 0 to 10 A at 50 Ω load. The sensors rise time is 30 ps (DUT voltage sensor) and 35 ps (DUT current sensor). The leakage current range is from 1 pA to 2.5 mA [25].

The HPPI TLP-8010C is a versatile test system that can test VF-TLP, TLP, and HMM standard [26]. Figure 6.9(b) shows the HPPI TLP-8010C. For VF-TLP mode of operation, fast pulses of 1 ns can be initiated. With the addition of the pulse width extender, pulse widths of 5 and 10 ns can be generated. Note that this system can also generate much longer pulse widths. From the value of the longer pulse widths, Wunsch–Bell curves can be generated from the VF-TLP time regime, TLP regime, and longer.

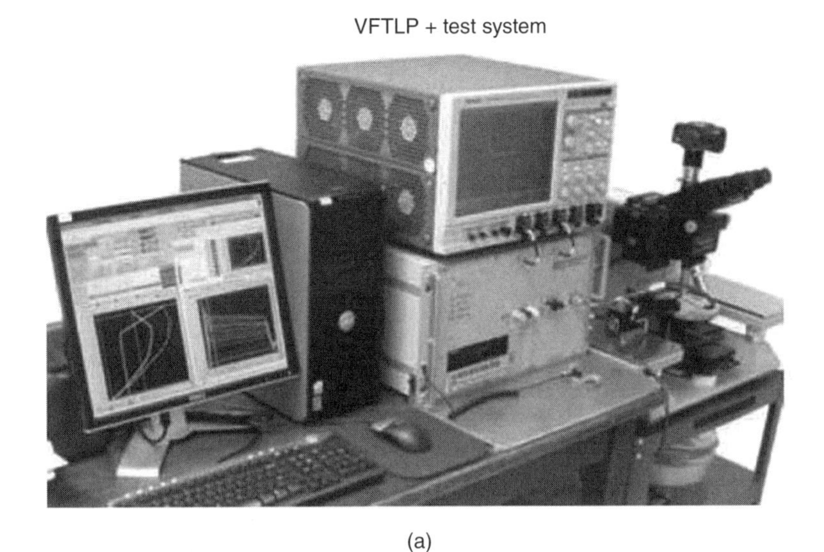

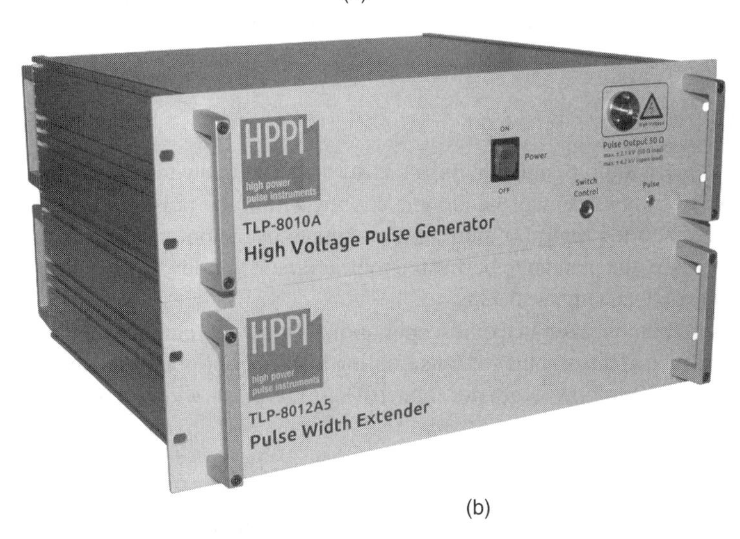

Figure 6.9 (a) Very fast transmission line pulse (VF-TLP) commercial test equipment – Barth Model 4012 VF-TLP+™ (Reproduced with permission of Barth Electronics Inc). (b) Very fast transmission line pulse (VF-TLP) commercial test equipment HPPI TLP-8010C (Reproduced with permission of HPPI GmBH)

6.7 Test Sequence and Procedure

For the VF-TLP testing, there is a test sequence and a procedure to provide the VF-TLP pulsed current–voltage characteristics. The VF-TLP test procedure uses a series of pulses with increasing pulse amplitude. The steps in the procedure are as follows:

- Select a pulse of a given pulse width, rise time, fall time, amplitude, and polarity.

- Select a step size increment.
- Test a first component.
- Select a minimum time between successive steps (e.g., greater than 0.1 s).
- Define a failure criterion.
- Apply a stress pulse of fixed width to the DUT.
- Measure and record stress pulse voltage and current, representing one *I–V* point on the DUT of the *I–V* characteristics.
- Perform a poststress evaluation of the DUT.
- If no "failure" occurs, the process is repeated with the next pulse applied at a higher magnitude; if failure occurs, the testing is stopped.

Figure 6.10 illustrates a flowchart of the test sequence for VF-TLP testing [9].

6.7.1 VF-TLP Pulse Analysis

For the VF-TLP testing methodology, TLP pulse event analysis is required. The TLP methodology must define a measurement window to extract current and voltage from the DUT.

6.7.2 Measurement Window

The VF-TLP methodology must define a measurement window to extract current and voltage from the DUT. A measurement window is chosen within the pulse event that avoids the rise time, fall time, and the region of current and voltage overshoot, oscillation, and ringing. For the VF-TLP pulse, the plateau to define the measurement window is significantly smaller than the TLP pulse plateau (Figure 6.11).

The measurement is taken in a quiescent region so that an accurate value for the voltage and current is obtained. This is achieved by sampling a number of points and averaging them. The measurement time window is greater than 10% of the pulse width to allow for averaging of many data points.

6.7.3 Measurement Analysis – VF-TLP Voltage Waveform

In this section, the measurement analysis of a pulse waveform in a TDR TLP system is shown. Figure 6.14 illustrates a voltage characteristic for a measurement where the DUT is under 50 Ω. In the plot, the incident pulse is shown, as well as the reflected pulse. In a VF-TLP event, there is no overlap region of the incident and reflected pulses. The measurement window is shown in the region where the incident and reflected pulses overlap (Figure 6.12).

6.7.4 Measurement Analysis – Time Domain Reflectometry (TDR) Current Waveform

In this section, the measurement analysis of a pulse waveform in a TDR VF-TLP system is illustrated. Figure 6.13 illustrates a current versus time characteristic for a measurement where

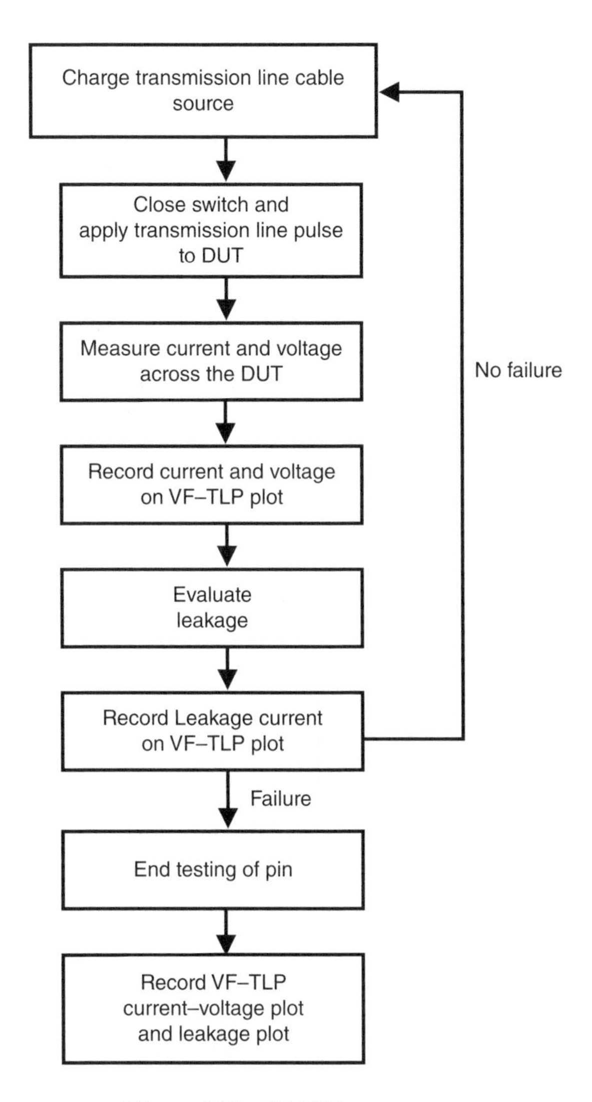

Figure 6.10 VF-TLP test sequence

the DUT is under 50 Ω. In the plot, the incident pulse is shown, as well as the reflected pulse. The incident and the reflected pulse waveforms do not overlap.

6.7.5 Measurement Analysis – Time Domain Transmission (TDR) Current–Voltage Characteristics

In this section, the measurement analysis of a pulse waveform in a TDR VF-TLP system is illustrated. Figure 6.14 illustrates a voltage and current versus time characteristic for a measurement as well as the current–voltage (*I–V*) characteristics. The measurement for the voltage

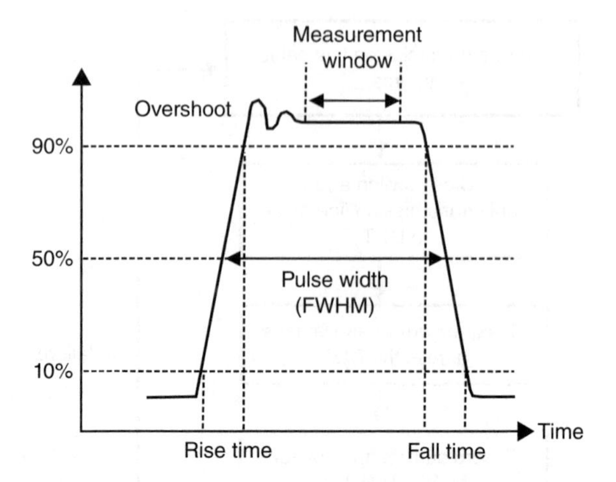

Figure 6.11 VF-TLP measurement window

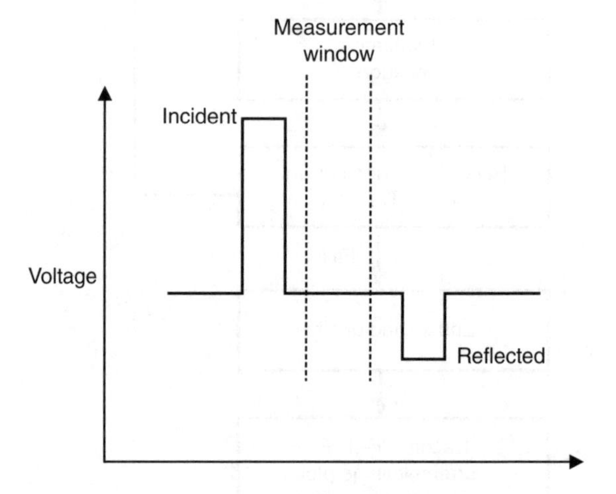

Figure 6.12 VF-TLP analysis – TDR voltage waveform

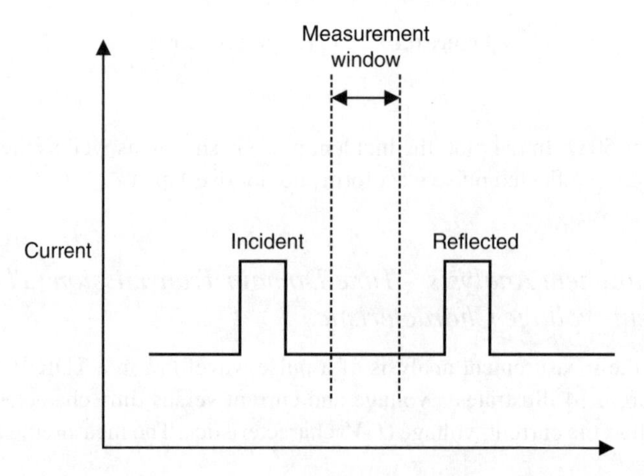

Figure 6.13 VF-TLP analysis – TDR current waveform

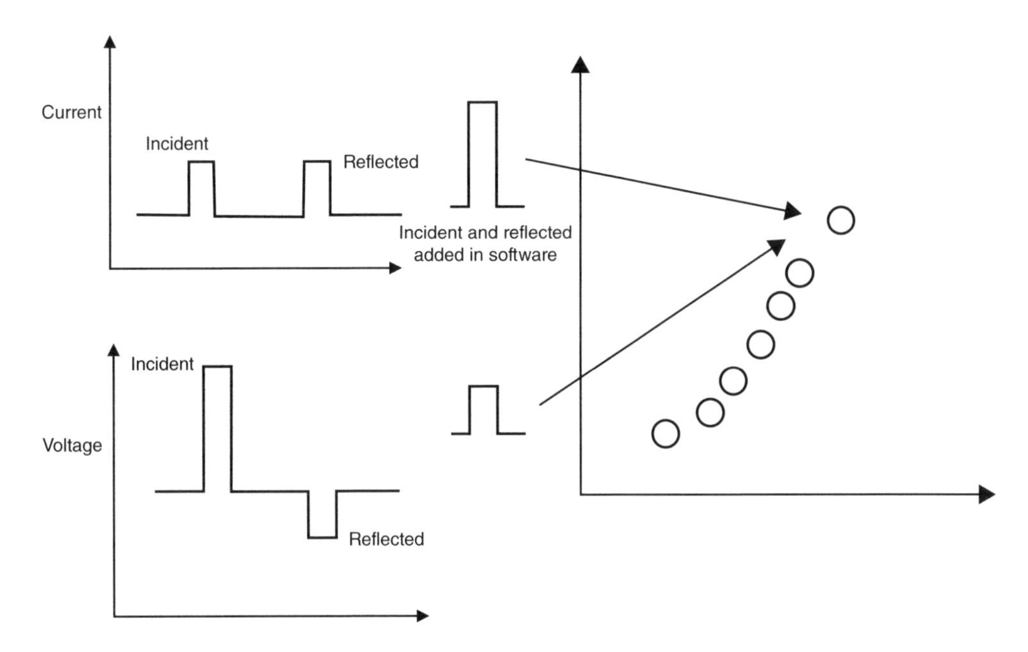

Figure 6.14 Very fast transmission line pulse (TLP) *I–V* characteristic analysis construction

and current is achieved by evaluating the average within the measurement window. These averaged values are then recorded in the *I–V* plot. In addition to the (*I*, *V*) point placed on the *I–V* plot, the leakage current poststress is also recorded.

6.8 VF-TLP Pulsed *I–V* Characteristics

From VF-TLP testing, a pulsed current–voltage (*I–V*) characteristic can be constructed. Each data point in the pulsed *I–V* characteristic is the current and voltage across the DUT. Each data point represents a pulse in the test sequence. The voltage on the transmission line (TL) source is increased for each step in the *I–V* characteristic.

Figure 6.15 illustrates the current and voltage from a pulse event, and how it is projected onto the current–voltage characteristics. Each current and voltage point is an average of the measured pulse event where the information is from an average of measurements from the measurement window.

6.8.1 VF-TLP Pulsed I–V Characteristic Key Parameters

The VF-TLP *I–V* characteristics contain key metrics. Similar to TLP testing, the VF-TLP *I–V* characteristics comprise the following key parameters:

- Avalanche voltage
- Trigger point
- Holding point

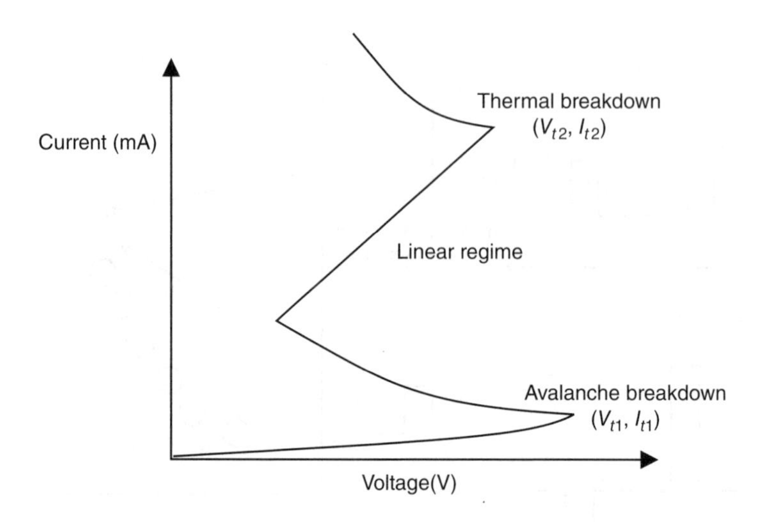

Figure 6.15 Very fast transmission line pulse (TLP) *I–V* characteristic analysis construction

- Turn-on voltage
- On-resistance
- Clamping voltage
- Second trigger point
- Soft failure current
- Failure current.

6.8.2 VF-TLP Power Versus Time Plot

In the early development of VF-TLP testing, the focus was not on the current and voltage *I–V* characteristics but on power. The primary reason was the researchers were interested in the power-to-failure, not the *I–V* characteristics. For the power analysis, the technique was to evaluate the incident, transmitted, and reflected power as a function of time. The first system developed by J.J. Whalen and H. Domingos was a pulse system in a TDRT configuration [10]. In this configuration, the measured parameters were the incident, transmitted, and reflected power as a function of time. From this method, the absorbed power in the DUT was the incident power minus the reflected power and the transmitted power. Waveforms of the incident power, the reflected power, and the transmitted power as a function of time were recorded. In this methodology, the time-to-failure was observable from transitions in the power waveforms.

The energy-to-failure can be defined as

$$E_A = E_I - E_R - E_T$$

where E_I is the incident energy, E_R is the reflected energy, and E_T is the transmitted energy. In the plot, the incident, reflected and transmitted power versus time is shown. In addition, the time-of-failure is also shown (Figure 6.16).

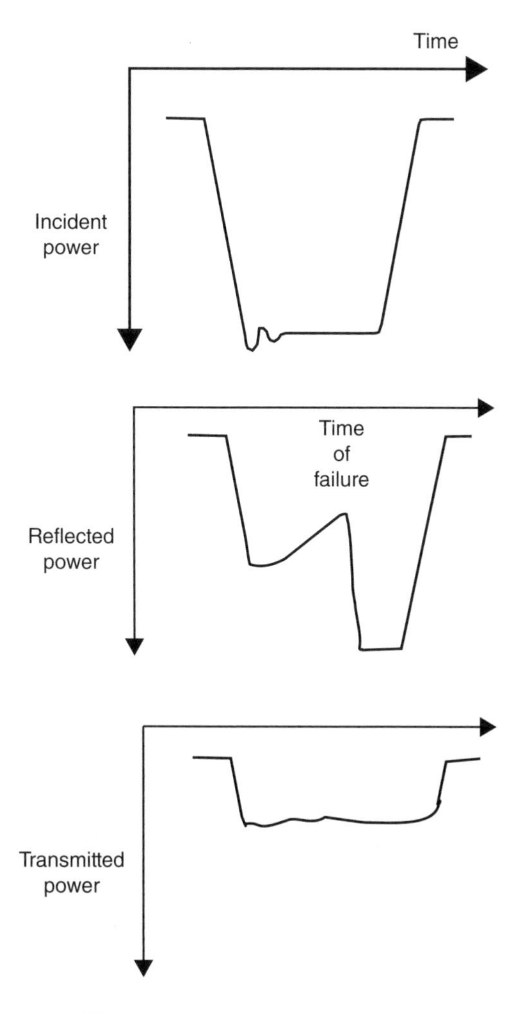

Figure 6.16 Power versus time plot

6.8.3 VF-TLP Power Versus Time – Measurement Analysis

A second test system, published in 1979, was simplified to include a single pulse unit, a square pulse matching network, a voltage probe, a current probe, and a dual-channel oscilloscope [10].

6.8.4 VF-TLP Power-to-Failure Versus Pulse Width Plot

Figure 6.17 illustrates the power-of-failure as a function of the pulse width. In the plot, the point of the VF-TLP event is highlighted. The VF-TLP event will be in the adiabatic regime of the Wunsch–Bell plot.

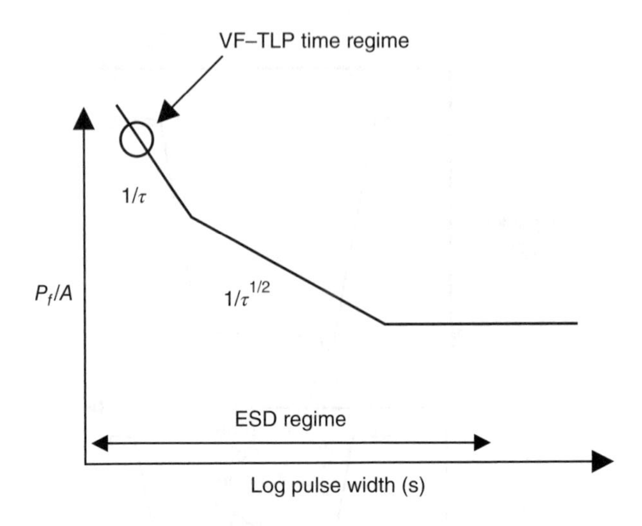

Figure 6.17 Wunsch–Bell curves: power-to-failure versus pulse width

6.8.5 VF-TLP and TLP Power-to-Failure Plot

One of the advantages of using both the TLP method and the VF-TLP method is that a two-point plot can be constructed for estimating the power-to-failure. For construction of the Wunsch–Bell power-to-failure curve, a series of successive tests are made for different pulse widths. The Wunsch–Bell power-to-failure characteristics are formed by sweeping all possible pulse widths capable of the test system. But, nowadays, many test systems do not have the ability to change the pulse width. Using a TLP system, with one fixed pulse width, and a second VF-TLP system, with a faster pulse width, a two-point curve can be created for rough estimation of the robustness in the time regime of interest.

For each pulse width, both the TLP and the VF-TLP systems, a pulsed *I–V* plot is formed. From the test, the failure point (*I–V* current and voltage) is extracted. This failure point is the product of the current-to-failure (I_f) and the voltage-to-failure (V_f); this can be represented as the power-to-failure (P_f). The data is plotted as the power-to-failure (P_f) versus the logarithm of the pulse width as shown in Figure 6.18. This can be used a projection of the anticipated results for a fast pulse and slower pulse.

6.9 Alternate Test Methods

Today, there are alternate test systems and test methods for exploring high-speed pulse testing. In the following section, pulse testing of RF components and UF-TLP is shown.

6.9.1 Radio Frequency (RF) VF-TLP Systems

For high-speed testing and RF components, different criteria are used to determine the failure of a device or component [17]. For RF devices and circuits, different RF parameters must be

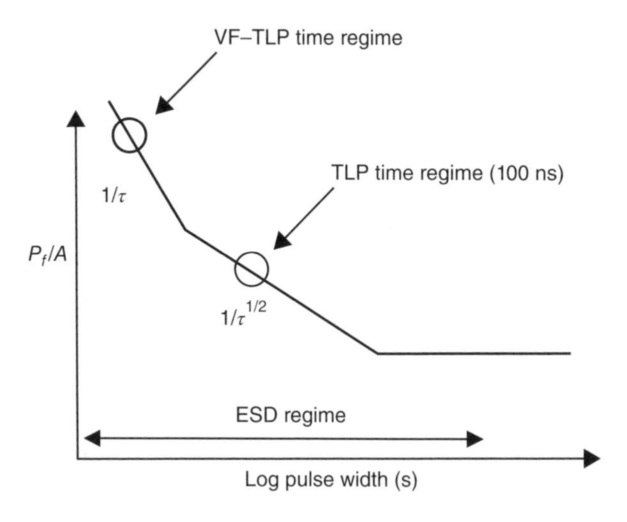

Figure 6.18 Wunsch–Bell curve with VF-TLP and TLP result

evaluated to determine the device failure. In CMOS technology for digital circuits, DC leakage current is used to evaluate the device failure.

S. Hynoven and E. Rosenbaum developed an alternate TLP system to evaluate parameters [27]. Figure 6.19(a) and (b) show a test system used for evaluation of RF parameters as the criteria for failure. This will be of increased importance for evaluation of RF components. Evaluation of the RF parameters of devices, components, and systems will need to have this capability in the future [17].

6.9.2 Ultrafast Transmission Line Pulse (UF-TLP)

Today's VF-TLP system evaluates the sensitivity of components in the nanosecond time regime. In the future, it will be important to understand the robustness of devices and components in the subnanosecond to picosecond time frame.

Figure 6.20 is an example of an UF-TLP system that produces pulses in the tens of picoseconds [28]. The UF-TLP system was constructed using single-pole double-throw (SPDT) switches [28]. The "transmission line" was constructed by disassembling one of the SPDT and adding a small element to store the charge. Figure 6.21 shows an example of the pulse widths produced from this system.

In Figure 6.22, the UF-TLP system was used to evaluate a tunneling magnetoresistor (TMR) head used in the magnetic recording industry [29].

6.10 Summary and Closing Comments

In this chapter, the VF-TLP model was discussed. This method had an increasing popularity due to its powerful value in semiconductor device and circuit development to evaluate high-speed interactions in the GHz regime. The ability to see the current and voltage

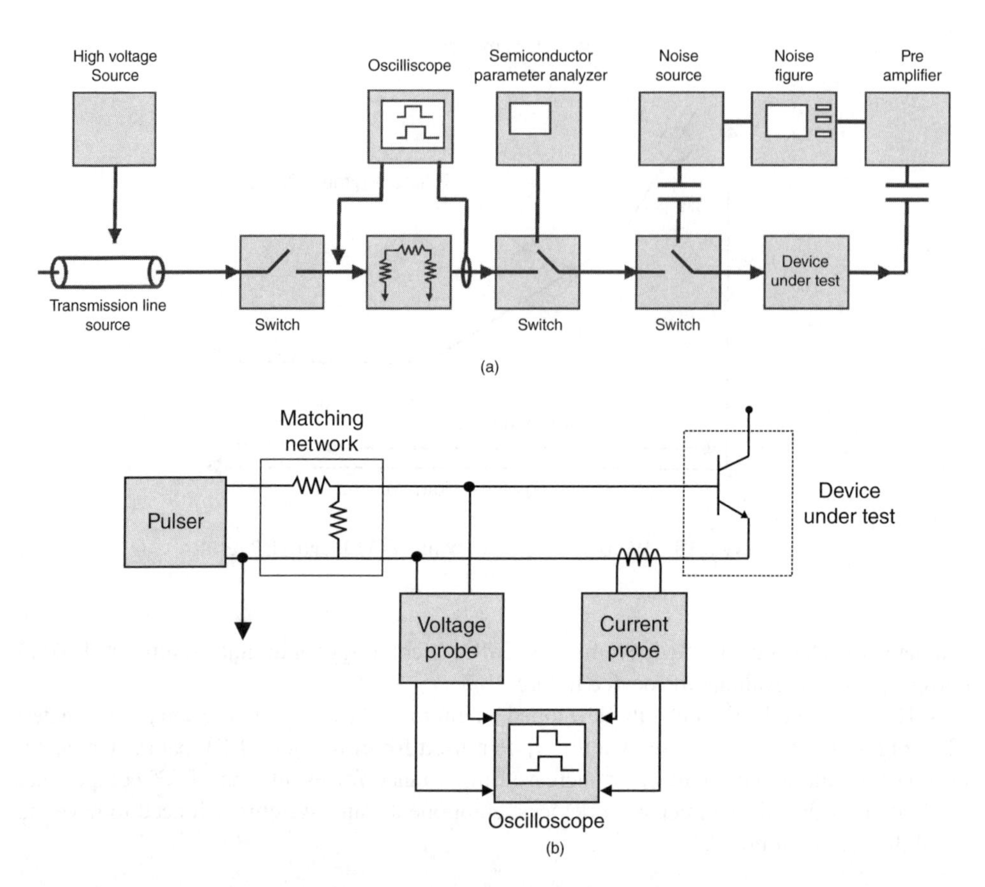

Figure 6.19 (a) Very fast transmission line pulse (VF-TLP) test system. (b) Very fast transmission line pulse (VF-TLP) test system for RF parameter analysis with matching network

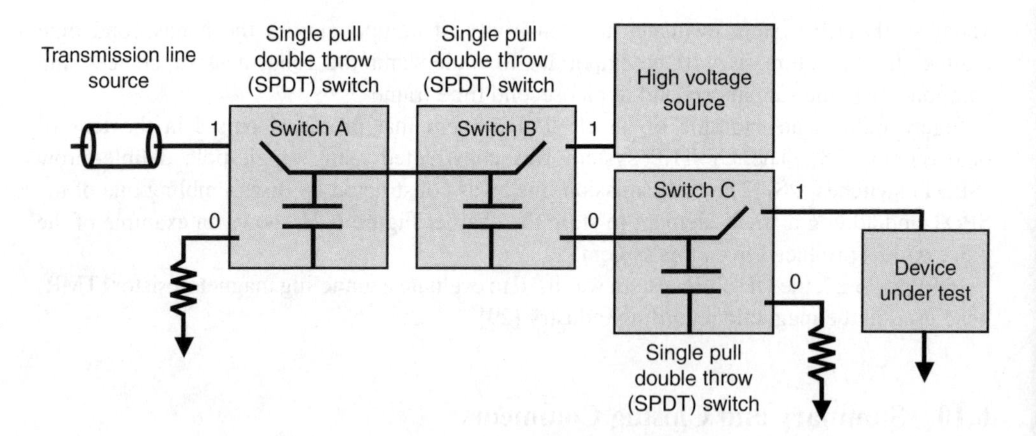

Figure 6.20 Ultrafast transmission line pulse (UF-TLP) test system

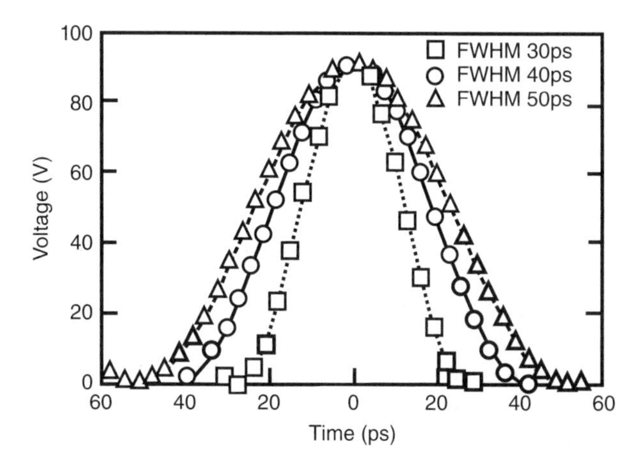

Figure 6.21 Ultrafast transmission line pulse (UF-TLP) pulse waveform

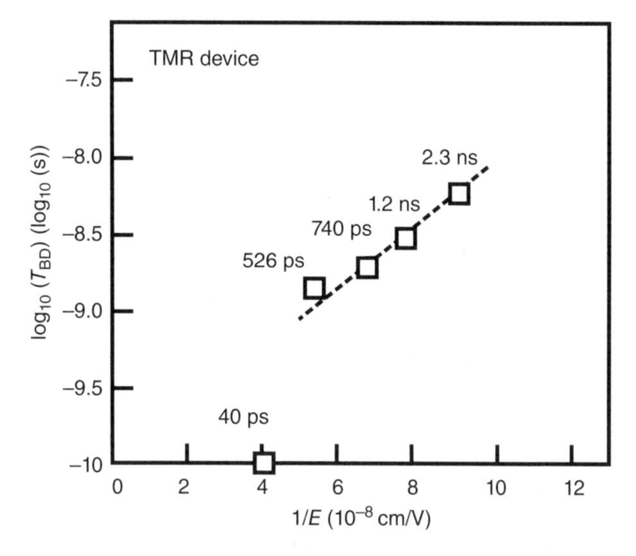

Figure 6.22 Ultrafast transmission line pulse (UF-TLP) results for a tunneling magnetoresistor (TMR) structure

characteristics across the DUT in the GHz time regime was of significant value to ESD engineers, circuit designers, and to failure analysis teams who need to develop, understand, or debug the response of high-speed digital logic and RF devices and circuits. This was a large step forward in increasing the technical nature of high-speed phenomena in ESD development.

Chapter 7 addresses system level concerns in semiconductor development known as IEC 61000-4-2. This test introduces a fast transient followed by a slower HBM-like waveform. With the growing interest in system level EOS, the interest in this test has emerged recently in semiconductor chip suppliers to test equipment.

Problems

1. How does the VF-TLP results compare to a TLP result? How does it differ?
2. What information does a VF-TLP test provide that a TLP test does not provide?
3. What is the frequency of the VF-TLP pulse? How does this compare to the application frequencies of DRAMs, SRAMs, ASICs, microprocessors, or RF products?
4. For semiconductor materials, is the VF-TLP pulse in the adiabatic, thermal diffusion, or steady state time regime? Does heat have time to diffuse in the semiconductor devices?
5. Combining VF-TLP and TLP measurements, show how segments of the Wunsch–Bell power-to-failure curve can be formed. What information can be obtained from this?
6. An RF cell phone product has a 5 GHz application frequency. An RF oscillating power source is applied to the product. How can we use a VF-TLP test system to quantify the power-to-failure of a cell phone product?
7. A commercial RF tester can cost millions of dollars. The test head of the RF test system must be resistant to fast high current events. Semiconductor RF chips are inserted into the socket of the test head for RF analysis. Assume the person and the chip are precharged prior to inserting the chip into the test system. Explain the event and how the VF-TLP test system can be used to quantify the robustness of the RF test system.
8. A charged device model (CDM) pulse is initiated by charging the substrate, power grid, or package. Explain the differences between VF-TLP and CDM events. Show the current path of the VF-TLP event. Show the current path of the CDM event.

References

1. S. Voldman, R. Ashton, J. Barth, D. Bennett, J. Bernier, M. Chaine, J. Daughton, E. Grund, M. Farris, H. Gieser, L.G. Henry , M. Hopkins, H. Hyatt, M.I. Natarajan, P. Juliano, T.J. Maloney , B. McCaffrey, L. Ting, and E. Worley. Standardization of the transmission line pulse (TLP) methodology for electrostatic discharge (ESD). *Proceedings of the Electrical Overstress/Electrostatic Discharge (EOS/ESD) Symposium*, 2003; 372–381.
2. ESD Association. ESD-SP 5.5.1. *ESD Association Standard Practice for the Protection of Electrostatic Discharge Sensitive Items – Electrostatic Discharge Sensitivity Testing Transmission Line Pulse (TLP) Testing Component Level.* Standard Practice document, 2003.
3. ANSI/ESD Association. ESD-SP 5.5.1 – 2004. *ESD Association Standard Practice for the Protection of Electrostatic Discharge Sensitive Items – Electrostatic Discharge Sensitivity Testing – Transmission Line Pulse (TLP) Testing Component Level.* Standard Practice (SP) document, 2004.
4. ANSI/ESD Association. ESD-STM 5.5.1 – 2008. *ESD Association Standard Test Method for the Protection of Electrostatic Discharge Sensitive Items – Electrostatic Discharge Sensitivity Testing – Transmission Line Pulse (TLP) Testing Component Level.* Standard Test Method (STM) document, 2008.
5. ESD Association. ESD-SP 5.5.2. *ESD Association Standard Practice for the Protection of Electrostatic Discharge Sensitive Items – Electrostatic Discharge Sensitivity Testing Very Fast Transmission Line Pulse (VF-TLP) Testing Component Level.* Standard Practice (SP) document, 2007.
6. ANSI/ESD Association. ESD-SP 5.5.2 – 2007. *ESD Association Standard Practice for the Protection of Electrostatic Discharge Sensitive Items – Electrostatic Discharge Sensitivity Testing – Very Fast Transmission Line Pulse (VF-TLP) Testing Component Level.* Standard Practice (SP) document, 2007.
7. S. Voldman. Standardization of the VF-TLP methodology for ESD evaluation of semiconductor components. Invited Talk, *Proceedings of the Taiwan Electrostatic Discharge Conference (T-ESDC)*, 2007, 13–20.
8. K. Muhonen, R. Ashton, J. Barth, M. Chaine, H. Gieser, E. Grund, L.G. Henry, T. Meuse, N. Peachey, T. Prass, W. Stadler, and S. Voldman. VF-TLP round robin study, analysis, and results. *Proceedings of the Electrical Overstress/ Electrostatic Discharge (EOS/ESD) Symposium*, 2008; 40–49.
9. ESD Association. ESD-STM 5.5.2. *ESD Association Standard Test Method for the Protection of Electrostatic Discharge Sensitive Items – Electrostatic Discharge Sensitivity Testing Very Fast Transmission Line Pulse (VF-TLP) Testing Component Level.* Standard Test Method (STM) document, 2009.

10. J.J. Whalen and H. Domingos. Square pulse and RF pulse overstressing of UHF transistors. *Proceedings of the Electrical Overstress/Electrostatic Discharge (EOS/ESD) Symposium*, 1979; 140–146.
11. J.J. Whalen, M.C. Calcatera, and M. Thorn. Microwave nanosecond pulse burnout properties of GaAs MESFETs. *Proceedings of the IEEE MTT-S International Microwave Symposium*, 1979; 443–445.
12. T. Maloney and N. Khurana. Transmission line pulsing techniques for circuit modeling of ESD phenomena. *Proceedings of the Electrical Overstress/Electrostatic Discharge (EOS/ESD) Symposium*, 1985; 49–54.
13. H. Gieser and M. Haunschild. Very fast transmission line pulsing of integrated structures and the charged device model. *IEEE Transactions on Components, Packaging and Manufacturing Technology, Part C*, Vol. 21, (4), Oct 1998; 278–285.
14. S. Voldman. *ESD: Physics and Devices*. Chichester, England: John Wiley and Sons, Ltd., 2004.
15. S. Voldman. *ESD: Circuits and Devices*. Chichester, England: John Wiley and Sons, Ltd., 2005.
16. S. Voldman. *ESD: Circuits and Devices 2nd Edition*. Chichester, England: John Wiley and Sons, Ltd., 2015.
17. S. Voldman. *ESD: RF Circuits and Technology*. Chichester, England: John Wiley and Sons, Ltd., 2006.
18. S. Voldman. *Latchup*. Chichester, England: John Wiley and Sons, Ltd., 2007.
19. S. Voldman. *ESD: Circuits and Devices*. Beijing, China: Publishing House of Electronic Industry (PHEI), 2008.
20. S. Voldman. *ESD: Failure Mechanisms and Models*. Chichester, England: John Wiley and Sons, Ltd., 2009.
21. S. Voldman. *ESD: Design and Synthesis*. Chichester, England: John Wiley and Sons, Ltd., 2011.
22. S. Voldman. *ESD Basics: From Semiconductor Manufacturing to Product Use*. Chichester, England: John Wiley and Sons, Ltd., 2012.
23. S. Voldman. *Electrical Overstress (EOS): Devices, Circuits, and Systems*. Chichester, England: John Wiley and Sons, Ltd., 2013.
24. S. Voldman. *ESD: Analog Design and Circuits*. Chichester, England: John Wiley and Sons, Ltd., 2014.
25. Barth Electronics. http://www.barthelectronics.com. Barth Model 4012 VF-TLP+™, 2016.
26. High Power Pulse Instruments GmBH. http://www.hppi.de. HPPI TLP test system HPPI TLP-8010C, 2016.
27. S. Hynoven, S. Joshi, and E. Rosenbaum. Combined TLP/RF test systems for detection of ESD failures in RF circuits. *Proceedings of the Electrical Overstress/Electrostatic Discharge (EOS/ESD) Symposium*, 2003; 346–353.
28. T.W. Chen, R.W. Dutton, C. Ito et al., Gate oxide reliability characterization in the 100 ps regime with an ultra-fast transmission line pulsing system. *Proceedings of the Electrical Overstress/Electrostatic Discharge (EOS/ESD) Symposium*, 2007; 102–106.
29. T.W. Chen, A. Wallash, and R.W. Dutton. Ultra-fast transmission line pulse testing of tunneling and giant magneto-resistor recording heads. *Proceedings of the Electrical Overstress/Electrostatic Discharge (EOS/ESD) Symposium*, 2008; 258–261.

7

IEC 61000-4-2

7.1 History

Electrostatic discharge (ESD) events can impact electronic and electrical systems in assembly, shipping, installation, and postinstallation. Systems, subsystems, and peripherals can be involved in ESD events that can include destructive and nondestructive events. Environmental conditions, such as low humidity, can influence electronic systems. In addition, installation and assembly conditions also can influence the sensitivity, susceptibility, and vulnerability to ESD events. ESD immunity requirements and test methods for electrical and electronic equipments subjected to ESDs from operators and personnel are necessary in today's electronic systems.

7.2 Scope

The scope of the IEC 61000-4-2 ESD test is for the testing, evaluation, and classification of systems [1]. The test is to quantify the sensitivity or susceptibility of these components to damage or degradation to the defined IEC 61000-4-2 test [1–32]. This includes ESD immunity requirements and test methods for electrical and electronic equipments subjected to ESDs from operators and personnel. The IEC 61000-4-2 test defines test level ranges that relate to environmental and installation conditions.

7.3 Purpose

The purpose of the IEC 61000-4-2 ESD test is for establishment of a test methodology to evaluate the repeatability and reproducibility of systems to a defined pulse event in order to classify or compare ESD sensitivity levels of systems [1]. The purpose of the IEC 61000-4-2 test is to establish a common and reproducible basis for evaluation of electronic and electrical systems to ESD events. The IEC 61000-4-2 test defines the following (Figure 7.1) [1]:

- Pulse waveform
- Range of test levels
- Test equipment

ESD Testing: From Components to Systems, First Edition. Steven H. Voldman.
© 2017 John Wiley & Sons, Ltd. Published 2017 by John Wiley & Sons, Ltd.

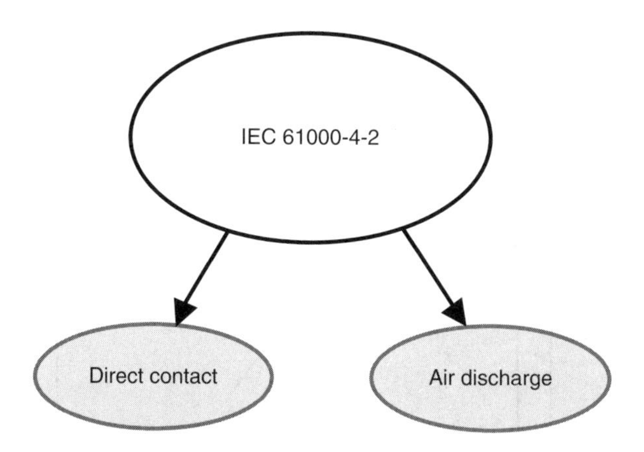

Figure 7.1 IEC 61000-4-2

- Test setup and configuration
- Test procedure
- Test calibration.

7.3.1 Air Discharge

The IEC 61000-4-2 ESD test addresses air discharge events [1, 2]. The event involves the current arc discharge event between a metal tip and the system under test. In this case, the energy of the electromagnetic (EM) fields introduces electromagnetic interference (EMI) that can couple to the system. The test quantifies the sensitivity or susceptibility of the system to disturbs, upsets, or failure.

7.3.2 Direct Contact Discharge

The IEC 61000-4-2 ESD test also addresses direct contact discharge events [1–6]. In this case, direct contact is made with the system prior to initiation of the pulse event. In this case, the current of the IEC 61000-4-2 event is applied to any location in the system. The test quantifies the sensitivity or susceptibility of the system to disturbs, upsets, or failure.

7.4 Pulse Waveform

The IEC 61000-4-2 ESD is a well-defined pulse waveform. The waveform is defined by the ESD gun specifically designed to provide this pulse event to mimic the ESD event from a human body or metal object with a built-up static charge [1].

Figure 7.2 illustrates an IEC 61000-4-2 event [1]. The event consists of a fast high current event followed by a second slower event. The fast event has a rise time of 0.7–1 ns. The slower event has a lower current with a peak at 30 ns and a decay time of 60 ns. The energy accumulation capacity is associated with a 150 pF capacitor. The discharge resistance is a 330 Ω resistor.

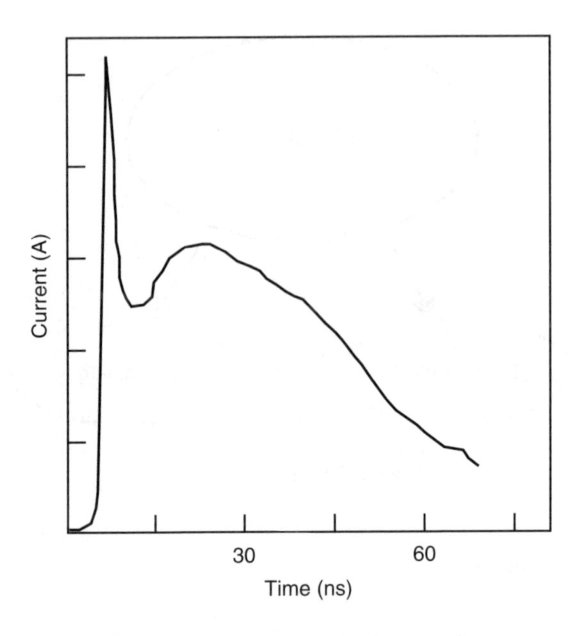

Figure 7.2 IEC 61000-4-2 pulse waveform

This provides an 8 kV event for direct contact resistance and 15 kV event for air discharge. The polarity of the pulse event addresses positive and negative events.

7.4.1 Pulse Waveform Equation

For the IEC 61000-4-2 pulse waveform, an analytical formulation was established by Heidler [31, 32]. The Heidler formulation can be expressed as two exponential terms as follows:

$$i(t) = i_1/k_1 A(t, \tau_1)\exp\{-t/\tau_2\} + i_2/k_2 B(t, \tau_3)\exp\{-t/\tau_4\}$$

where

$$k_1 = \exp\{-(\tau_1/\tau_2)(n\tau_2/\tau_1)^{1/n}\}$$
$$k_2 = \exp\{-(\tau_3/\tau_4)(n\tau_4/\tau_3)^{1/n}\}$$

and additionally,

$$A(t, \tau_1) = (t/\tau_1)/\{1 + (t/\tau_1)^n\}$$

and

$$B(t, \tau_3) = (t/\tau_3)/\{1 + (t/\tau_3)^n\}$$

In this formulation, the variable i_1 and i_2 are currents in amperes, τ_1, τ_2, τ_3, and τ_4 are time constants in nanoseconds, and n is a constant [31, 32]. An example of parameters, for a 2 kV positive pulse, $i_1 = 3.6$ A, and $i_2 = 3.74$ A, $\tau_1 = 0.29$, $\tau_2 = 18.44$, $\tau_3 = 34.34$, and $\tau_4 = 31.02$ (in nanoseconds), and $n = 3.26$ [31, 32].

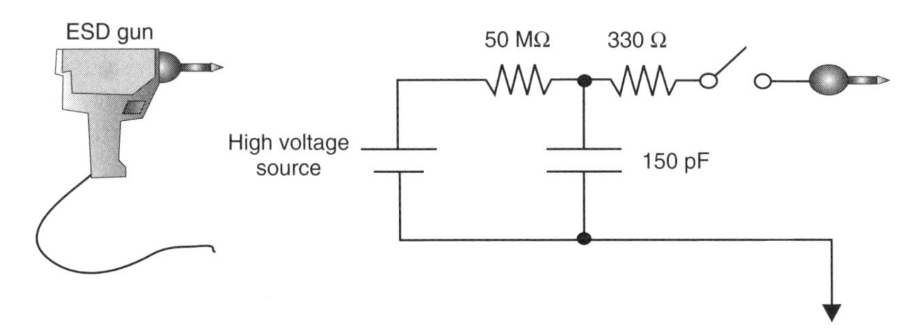

Figure 7.3 IEC 61000-4-2 ESD gun equivalent circuit (Reproduced with permission of Thermo Fisher Scientific)

7.5 Equivalent Circuit

Figure 7.3 shows the equivalent circuit schematic for the IEC 61000-4-2 [1]. The circuit schematic shows a 150 pF capacitor and a 330 Ω resistor. The charging resistance is a 50–100 MΩ source. This provides an 8 kV event for direct contact resistance and 15 kV event for air discharge. The polarity of the pulse event addresses both positive and negative events.

7.6 Test Equipment

An example of a commercial test simulator is shown in Figure 7.4(a). This ESD simulator is a handheld ESD gun with a metal tip allowing for easy access to test systems. Many different vendors have produced ESD guns for evaluation of manufacturing environments.

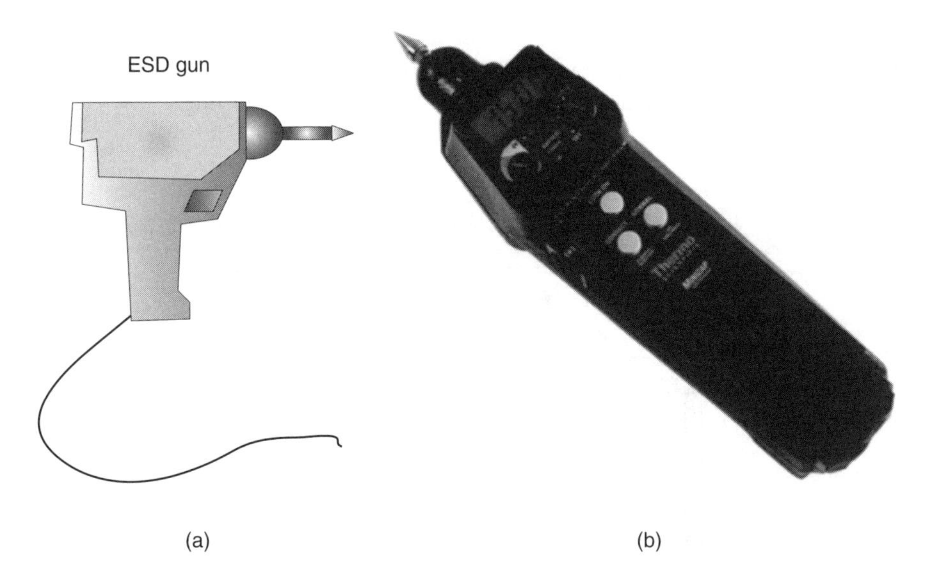

Figure 7.4 (a) IEC 61000-4-2 commercial test equipment – ESD gun. (b) IEC 61000-4-2 MiniZap ESD simulator – Thermo KeyTek (Reproduced with permission of Thermo Fisher Scientific)

An example of a commercial test simulator is shown in Figure 7.4(b). This ESD simulator is a handheld ESD gun with a metal tip allowing for easy access to test systems, subsystems, and assemblies. Figure 7.4(b) is the Thermo Scientific™ MiniZap-15 ESD Simulator [33]. The MiniZap-15 Simulator supports testing to the international standard IEC/EN 61000-4-2 and ANSI C63.16. The tester is suitable for 8 kV contact mode discharge and 16 kV air discharge testing. This simulator generates "real-world" ESD pulses, which are repeatable. It has both E-field and H-field diagnostics and suitable for vertical coupling plane (VCP) and horizontal coupling plane (HCP) for indirect ESD testing. The MiniZap tester does not use a voltage multiplier. As a result, it will not introduce testing errors and uncertainties. Additional ESD test standards can be done with the simulator, by changing of the plug-in tips.

7.6.1 Test Configuration

In the IEC 61000-4-2 test specification, the table configuration and grounding condition are specified [1]. This is significantly different from the human body model (HBM) component ESD test specification.

Using a table top for testing, a horizontal metal coupling plane is used below an insulation sheet (Figure 7.5). The HCP is connected to a ground reference plane on the floor using a $2 \times 470\,\text{k}\Omega$ bleed resistor cable. This is acceptable when the equipment under test (EUT) can be placed on the table top. Given that the EUT does not fit on the table, and the system is placed on an insulation pallet whose dimension is greater than 0.1 m high.

For indirect contact discharge events, a VCP should be placed 10 cm from the EUT. The VCP should also be connected to the ground reference using the $2 \times 470\,\text{k}\Omega$ bleed resistor cable [1].

7.6.2 ESD Guns

To perform the IEC 61000-4-2, there are ESD guns to apply the ESD event to the component or system. In the following sections, modules exist for system level testing.

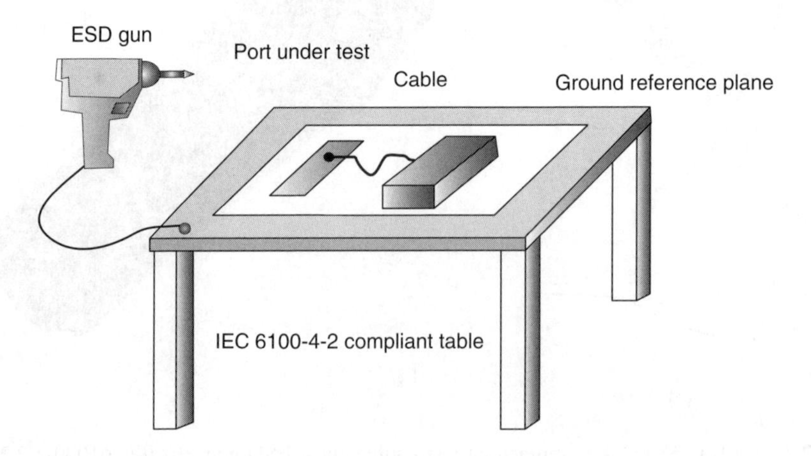

Figure 7.5 IEC 61000-4-2 test equipment – table configuration

7.6.3 *ESD Guns – Standard Versus Discharge Module*

For the IEC 61000-4-2 standard, there is a module for the gun that is specified. But there are many discharge modules, equivalent circuit schematic for the IEC 61000-4-2. The IEC 61000-4-2 standard defines a 150 pF capacitor and a 330 Ω resistor [1]. The charging resistance is a 50–100 MΩ source. This provides an 8 kV event for direct contact resistance and 15 kV event for air discharge. There are additional discharge modules that can be substituted into test systems to provide the equivalent test.

7.6.4 *Human Body Model Versus IEC 61000-4-2*

In Figure 7.6, the HBM and the IEC 61000-4-2 waveform are illustrated. There are a few distinct differences between the two waveforms. The HBM waveform is a double exponential waveform of single polarity. The HBM waveform includes an RC network from the 1500 Ω resistor and 100 pF capacitor. For the IEC 61000-4-2 standard, there is also an RC network contained within the ESD gun, but with different resistor and capacitor values. A second key feature that is present in the IEC 61000-4-2 waveform is the rapid peak at the beginning of the waveform. The fast peak of an IEC 61000-4-2 waveform is associated with the charge present at the tip of the ESD gun [8, 12, 13].

7.7 Test Sequence and Procedure

The test procedure and test sequence for the IEC 61000-4-2 event is a simple test procedure (Figure 7.7). First, the environment for the room must be between 15 and 35 °C. The room relative humidity (RH) must be between an RH level of 30–60% [1].

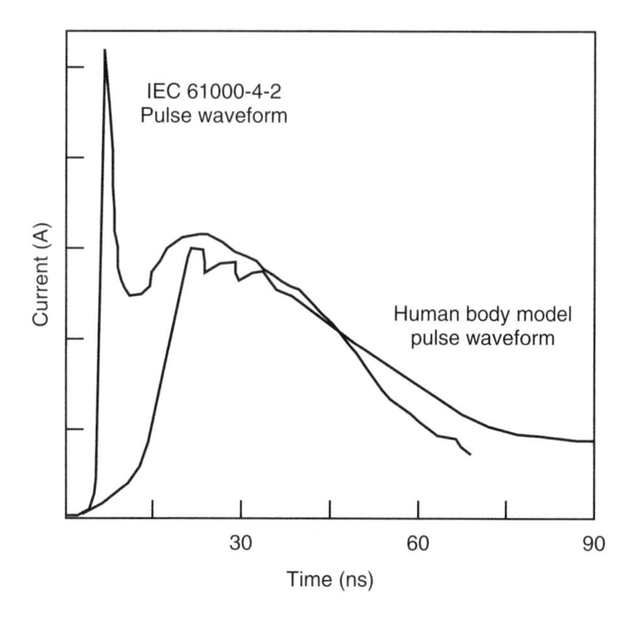

Figure 7.6 IEC 61000-4-2 versus HBM waveform comparison

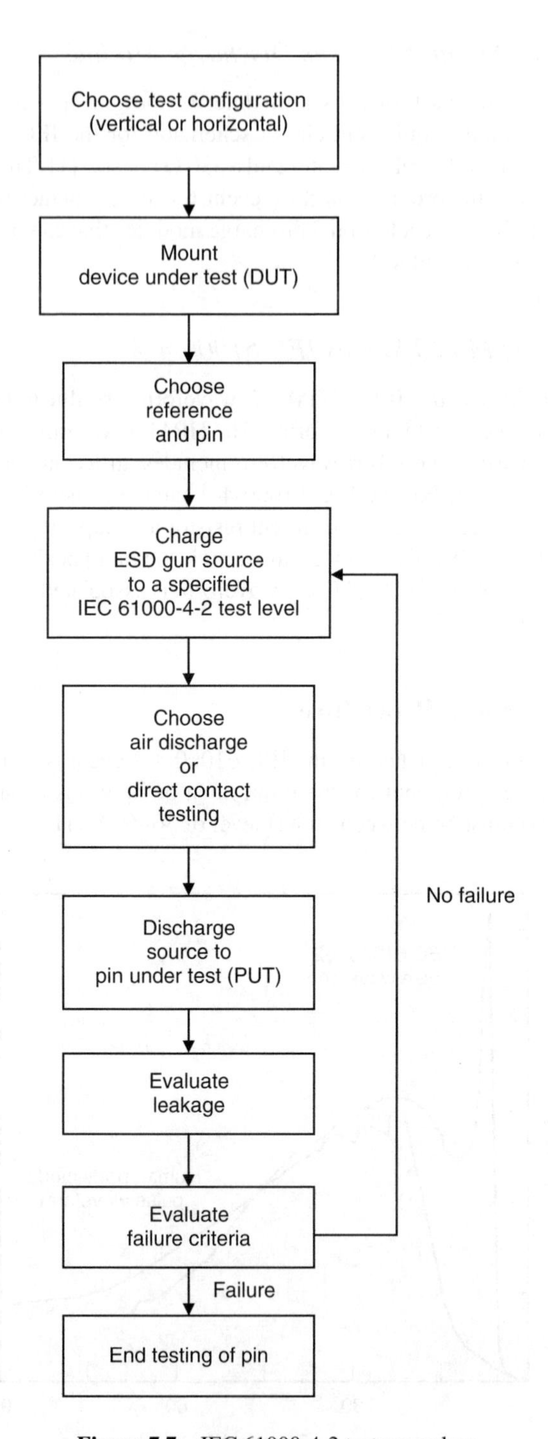

Figure 7.7 IEC 61000-4-2 test procedure

For direct discharge events, the ESD gun is to be placed to the location of the system under test. The ESD gun is then discharged at 1 s intervals. The test is to be performed for both polarities and repeated at least 10 times. The results of the test are then recorded.

For indirect discharge events, the ESD gun is to be discharged to the horizontal and VCPs instead of directly to the EUT. The ESD gun is then discharged at 1 s intervals. The test is to be performed for both polarities and repeated at least 10 times. The results of the test are then recorded.

The recording of the test results can be classified based on the response. The response of the test can be defined as follows [1]:

- Normal operation within the operational specification of the system
- Loss of operation, loss of function, or temporary degradation that is recoverable by system self-recovery or "self-healing" function
- Loss of operation, loss of function, or temporary degradation that is nonrecoverable by system self-recovery or "self-healing" function requiring operator intervention (e.g., power down, power up, or system reset)
- Loss of operation, loss of function, or nonrecoverable degradation that is nonrecoverable and damage to the system, components, software, or data.

7.8 Failure Mechanisms

Failure mechanisms on a system level can be destructive and nondestructive events (Figure 7.8). The following responses of the IEC 61000-4-2 test can be defined as failure mechanisms [1]:

- Loss of operation, loss of function, or temporary degradation that is recoverable by system self-recovery or "self-healing" function

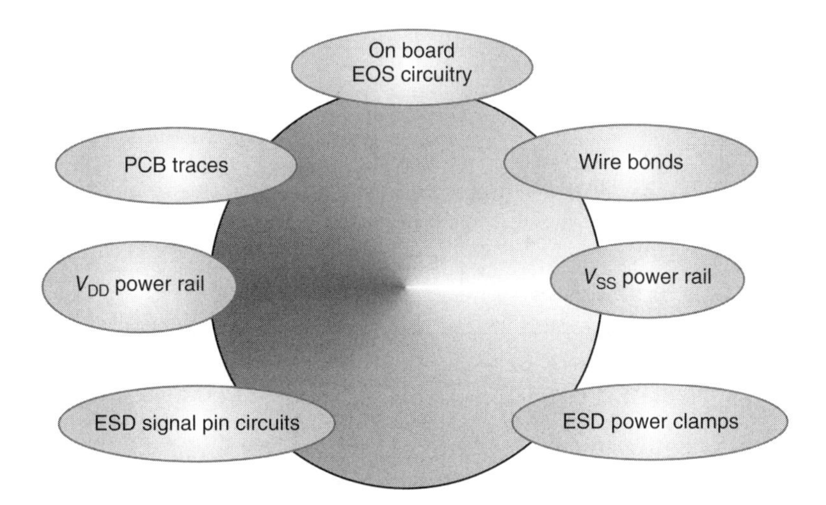

Figure 7.8 IEC 61000-4-2 induced failure mechanism

- Loss of operation, loss of function, or temporary degradation that is nonrecoverable by system self-recovery or "self-healing" function requiring operator intervention (e.g., power down, power up, or system reset)
- Loss of operation, loss of function, or nonrecoverable degradation that is nonrecoverable and damage to the system, components, software, or data
- Permanent damage to the components and system.

7.9 IEC 61000-4-2 ESD Current Paths

Figure 7.9 is an example of an IEC 61000-4-2 ESD event applied to the ground rail V_{SS}. The IEC 61000-4-2 current flows through the V_{SS} ground rail. There are at least two paths for the current flow. One path is through the ESD network to the signal pad. A second current path is through the ESD power clamp. The current through the ESD power clamp, as well as the ESD network, can flow to the V_{DD} power supply. The amount of current that flows to each is dependent on the resistance of the elements in the circuit. Current can also continue into the semiconductor chip through the ground rail [39, 44, 46, 48].

Figure 7.10 is an example of an IEC 61000-4-2 ESD event applied to the power rail V_{DD}. The IEC 61000-4-2 current flows through the V_{DD} power rail. There are at least two paths for the current flow. One path is through the ESD network to the signal pad. A second current path is into the V_{DD} power bus and through the ESD power clamp. The current through the ESD power clamp, as well as the ESD network, can flow to the V_{SS} ground rail and into the V_{DD} power supply. The amount of current that flows to each is dependent on the resistance of the bus resistance and the ESD power clamp resistance. Current can also continue into the semiconductor chip through the power supply rail or the ground rail. The IEC 61000-4-2 current flowing into the I/O and core can lead to ESD failures [39, 44, 46].

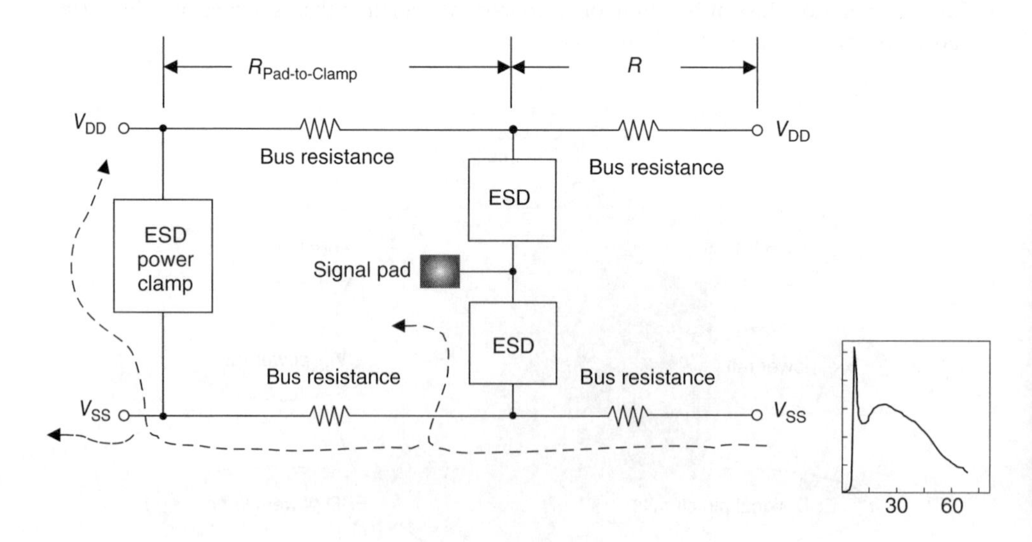

Figure 7.9 IEC 61000-4-2 ESD event current paths – V_{SS} positive polarity

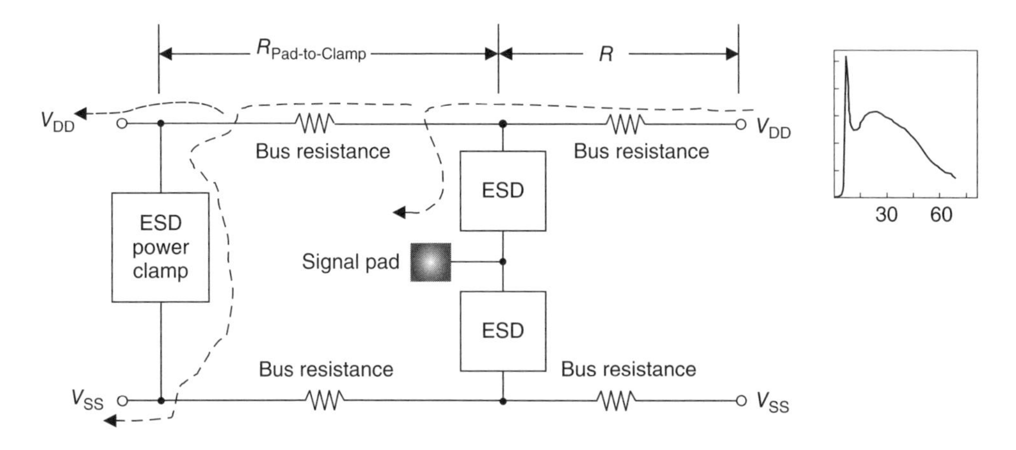

Figure 7.10 IEC 61000-4-2 ESD event current paths

7.10 ESD Protection Circuitry Solutions

One of the issues with the IEC 61000-4-2 ESD event is that ESD circuitry designed for HBM and MM events are not responsive to this ESD pulse event [37–47]. As a result, some of the circuitry must be modified due to the faster event. Figure 7.11 shows an example of a solution to address the fast pulse portion of the IEC 61000-4-2 event. The standard circuit for the HBM and MM events used an inverter drive stage of three successive inverter stages to drive the gate of the output MOSFET between V_{DD} and V_{SS} power rails. Each complementary metal oxide semiconductor (CMOS) inverter stage has a switching delay time. In order to make the circuit more responsive, the three CMOS stages were reduced to a single CMOS inverter stage [39, 44].

An ESD protection solution to avoid failures from the IEC 61000-4-2 ESD event is to establish separate IEC dedicated power rails. The high current pulse can lead to damage of the internal circuitry, as well as the power buses. If the resistance of the internal power bus is

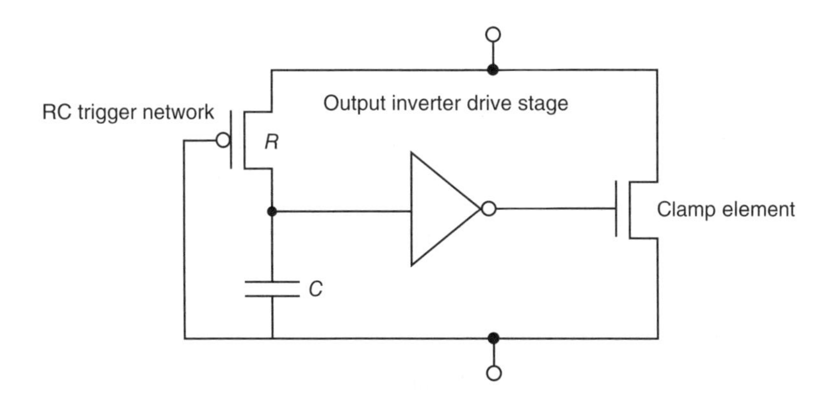

Figure 7.11 IEC 61000-4-2 ESD protection power clamp

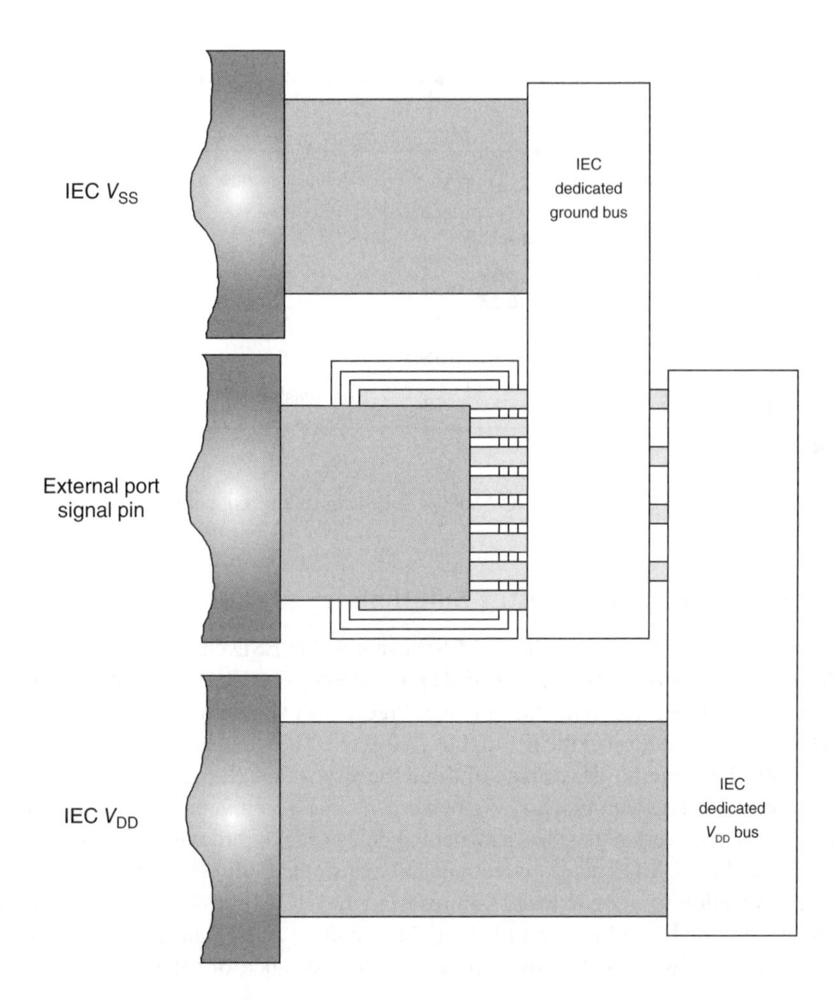

Figure 7.12 IEC 61000-4-2 ESD protection circuits

significantly higher than the IEC dedicated bus, then the residual current flowing into the semiconductor chip internal circuits will be minimized. Figure 7.12 illustrates an example layout of an IEC bond pad, IEC wide interconnects, and IEC dedicated power buses [39, 44].

Figure 7.13 illustrates a circuit schematic of the IEC bond pad, IEC wide interconnects, and IEC dedicated power buses. The schematic highlights the distinction between the IEC bus resistance, the standard bus resistance, and the bond-pad-to-power-clamp resistance.

7.11 Alternative Test Methods

Alternative test requirements and methods for automotive, medical, avionics, and military equipment exist based on other system level standards [34–36]. Table 7.1 shows a listing of some alternate standard discharge modules. In many cases, the IEC 61000-4-2 waveform are used but with other requirements. In these alternative test standards, the resistor and capacitor

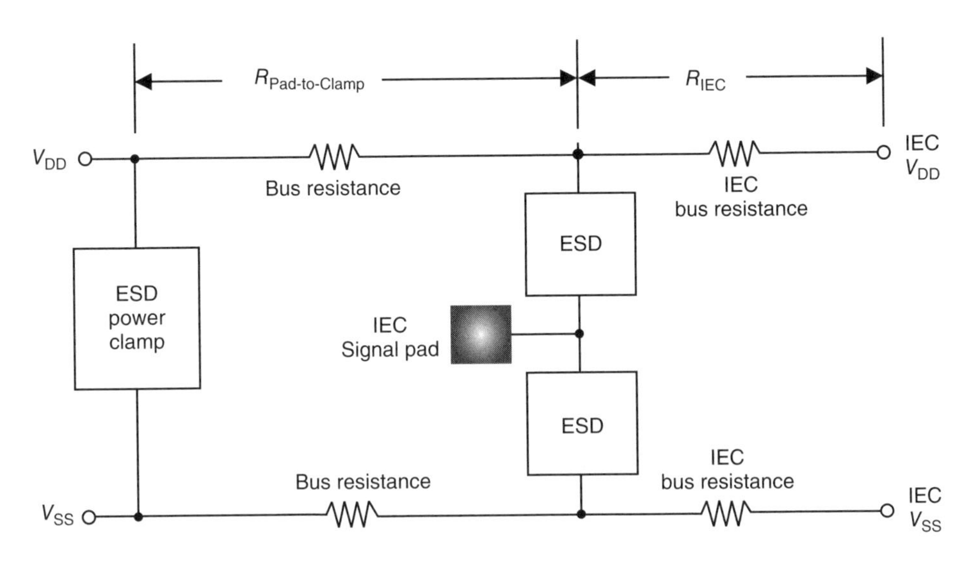

Figure 7.13 IEC 61000-4-2 ESD event – power grid definition

Table 7.1 Alternative standards and associated discharge modules

Standard	Discharge module (capacitor/resistor)
IEC 61000-4-2	150 pF/330 Ω
ISO 10605	150 pF/330 Ω
	330 pF/330 Ω
	150 pF/2000 Ω
	330 pF/2000 Ω
IEC 60601-1-2	150 pF/330 Ω
RTCA DO-160 Section 25	150 pF /330 Ω
SAE J1113	150 pF / 2000 Ω
	330 pF/2000 Ω
ANSI C63.16	150 pF/330 Ω
	150 pF/15 Ω
	150 pF/75 Ω
MIL-STD-883	100 pF/1500 Ω
MIL-STD-464	500 pF/500 Ω

elements of the discharge module vary. To test these other standards, different ESD guns with different modules are required.

7.11.1 Automotive ESD Standards

The automotive industry uses the ISO 10605:2008 standard [34]. This standard uses for different RC networks for the ESD gun source as shown in Table 7.1. In these tests, the initial peak remains the same, but the waveform differs due to the different RC decay.

In all cases, the test setup is the same. A distinction between the IEC 61000-4-2 test and the ISO 10605 test is the ISO test requires an insulating support above the HCP of 2–3 mm, and be able to support a 25 kV event without electrical breakdown. Both standards have a 2–15 kV contact mode discharge, but the ISO 10605 standard requires an air discharge simulator event that can test to a 25 kV test level. Also note that the IEC test requires a repetition rate of 20 discharges per second, whereas the ISO standard only requires 10 discharges per second.

The automotive standard also requires a different failure criterion with a different function performance status classification (FPSC). The FPSC includes the following classifications [34]:

- Status I: The function performs as designed, during and after test.
- Status II: The function does not perform as designed during the test but returns automatically to normal operation after the test.
- Status III: The function does not perform as designed during the test and does not return automatically to normal operation after the test without intervention.
- Status IV: The function does not perform as designed during the test and cannot be returned to proper operation without more extensive intervention. The function shall not have sustained any permanent damage as a result of the testing.

7.11.2 Medical ESD Standards

In the medical industry, the standard that is used is IEC 60601-1-2 Edition 4.0 [35]. The ESD testing is equivalent to IEC 61000-4-2. For the medical industry, some unique features are added:

- Guidance for immunity test level for special environments
- Harmonization with risk concepts for safety and essential performance
- Specification of immunity testing and test levels associated with medical equipment ports and medical electrical systems.

The IEC requirement for medical equipment requires both contact and air discharge events. The ESD test levels for air discharge are 2, 4, and 8 kV. The ESD test levels for contact discharge are 2, 4, and 6 kV.

The failure criterion for medical equipment are different from standard electronic equipment. These include the following [35]:

- False alarms
- Cessation or interruption of any operation
- Changes in operational mode
- Changes in programmable parameters
- Reset to factory defaults
- Initiation of any unintended operation
- Noise on a waveform that is indistinguishable from physiological produced signal
- Distortion of an image that is indistinguishable from physiological produced signal
- Failure of components
- Failure of automatic diagnosis.

7.11.3 Avionic ESD Standard

The avionic standard for ESD testing is RTCA DO-160 Section 25 [36]. This standard uses the IEC 61000-4-2 test waveform but only addresses the air discharge event. The test requires 10 positive discharge events and 10 negative discharge events at a level of 15 kV.

7.11.4 Military-Related ESD Standard

There are additional military standards for ESD of parts, assemblies, and equipment. These include the following:

- *MIL-HDBK-263B*: ESD Control Handbook for Protection of Electrical and Electronic Parts, Assemblies, and Equipment
- *MIL-STD-1686*: ESD Control Program for Protection of Electrical and Electronic Parts, Assemblies, and Equipment
- *MIL-STD-785*: Reliability Program for Systems and Equipment Development and Production.

7.12 Summary and Closing Comments

This chapter addressed system level concerns in semiconductor development known as IEC 61000-4-2. This test introduces a fast transient followed by a slower HBM-like waveform. With the growing interest in system level electrical overstress (EOS), the interest in this test has emerged recently in semiconductor chip suppliers to test equipment. The chapter addressed the purpose, the scope, to the pulse waveform. ESD protection concepts for the IEC 61000-4-2 were also discussed.

Chapter 8 addresses a new semiconductor chip level test to address IEC 61000-4-2 pulse events into external ports of a semiconductor chip. This test is known as the human metal model (HMM). This test also introduces a fast transient followed by a slower HBM-like waveform that is only applied to specific ports exposed on a system level. With the growing interest in system level EOS, the interest in this test has emerged recently due to the request of system level developer for the semiconductor chip suppliers.

Problems

1. The IEC 61000-4-2 can be an air discharge event. What parameter variables influence the pulse waveform when it is an air discharge event?
2. The IEC 61000-4-2 discharge for an air discharge initiated an E- and H-field outside of the electronic system. Estimate the energy in the E-field and the H-field assuming a given current pulse of current I.
3. An ESD gun can be used to initiate a discharge to a semiconductor DRAM pin. What factors influence the discharge event?
4. An ESD gun can be used to initiate a discharge to a cell phone that contains external port for charging of the cell phone. What factors influence the discharge event?

5. An ESD gun can be used to initiate a discharge to a laptop. What are the potential sources of system failures? Assume the discharge is to the laptop USB ports. What factors influence the discharge event?
6. An ESD gun is used to zap an antennae of a cell phone. The cell phone is electrically connected to a gallium arsenide (GaAs) bipolar transistor power amplifier. Show the ESD event. What ESD protection circuitry can be used to avoid failure of the GaAs cell phone power amplifier?
7. Compare the IEC 61000-4-2 for a direct contact versus air discharge event. What is the influence on the fast current "peak"? How do the two waveforms differ?

References

1. International Electro-technical Commission (IEC). *IEC 61000-4-2: Electromagnetic Compatibility (EMC) – Part 4-2: Testing and Measurement Techniques – Electrostatic Discharge Immunity Test*. IEC International Standard, 2007.
2. D. Pommerenke, and M. Aidam. To what extent do contact mode and indirect ESD test methods reproduce reality? *Proceedings of the Electrical Overstress / Electrostatic Discharge (EOS/ESD) Symposium*, 1995; 101–109.
3. D. Lin, D. Pommeranke, J. Barth, L.G. Henry, H. Hyatt, M. Hopkins, G. Senko, and D. Smith. Metrology and methodology of system level ESD testing. *Proceedings of the Electrical Overstress/Electrostatic Discharge (EOS/ESD) Symposium*, 1998; 29–39.
4. E. Grund, K. Muhonen, and N. Peachey. Delivering IEC 61000-4-2 current pulses through transmission lines at 100 and 330 ohm system impedances. *Proceedings of the Electrical Overstress/Electrostatic Discharge (EOS/ESD) Symposium*, 2008; 132–141.
5. R. Chundru, D. Pommerenke, K. Wang, T. Van Doren, F.P. Centola, J.S. Huang. Characterization of human metal ESD reference discharge event and correlation of generator parameters to failure levels – Part I: reference event. *IEEE Transactions on Electromagnetic Compatibility*, Vol. 46, (4), Nov 2004; 498–504.
6. K. Wang, D. Pommerenke, R. Chundru, T. Van Doren, F.P. Centola, and J.S. Huang. Characterization of human metal model ESD reference discharge event and correlation of generator parameters to failure levels – Part II: correlation of generator parameters to failure levels. *IEEE Transactions on Electromagnetic Compatibility (EMC)*, Vol. 46, (4), Nov 2004; 505–511.
7. ESD Association. ESD-SP 5.6 – 2008. *ESD Association Standard Practice for the Protection of Electrostatic Discharge Sensitive Items – Electrostatic Discharge Sensitivity Testing – Human Metal Model (HMM) Testing Component Level*. Standard Practice (SP) document, 2008.
8. J. Barth, D. Dale, K. Hall, H. Hyatt, D. McCarthy, J. Nuebel, and D. Smith. Measurement of ESD HBM events, simulator radiation, and other characteristics toward creating a more repeatable simulation or simulator should simulate. *Proceedings of the Electrical Overstress/Electrostatic Discharge (EOS/ESD) Symposium*, 1996; 211–222.
9. W. Stadler, S. Bargstadt-Franke, T. Brodbeck, R. Gaertner, M. Goroll, H. Gossner, N. Jensen, and C. Muller. From the ESD robustness of products to system ESD robustness. *Proceedings of the Electrical Overstress/Electrostatic Discharge (EOS/ESD) Symposium*, 2004; 67–74.
10. ESD Association. *Electrostatic Discharge Sensitivity Testing – Human Metal Model (HMM)*, ESD Standard Practice Document, 2009.
11. R. Ashton. System level ESD testing – the waveforms. *Conformity Magazine*, Vol. 12, (10), Oct 2007; 28–34.
12. R. Ashton. Reliability of IEC 61000-4-2 ESD Testing on Components, *EE Times*, August 10, 2008.
13. On Semiconductor. Human Body Model (HBM) vs. IEC 61000-4-2. www.onsemi.com, 2010.
14. N. Peachey. ESD Open Forum – ESD stressing of devices and components. *Conformity Magazine*, Vol. 13, (6), 2008; 22–23.
15. D. Robinson-Hahn, et al. Evaluating IC components utilizing IEC 61000-4-2. *Proceedings of the International Electrostatic Discharge (ESD) Workshop*, 2008; 274–283.

16. S. Bertonnaud, C. Duvvury, and A. Jahanzeb. IEC system level ESD challenges and effective protection strategy for USB interface. *Proceedings of the Electrical Overstress/Electrostatic Discharge (EOS/ESD) Symposium*, 2012.
17. R. Ashton, and L. Lescouzeres. Characterization of off chip protection devices. *Proceedings of the Electrical Overstress/Electrostatic Discharge (EOS/ESD) Symposium*, 2008; 21–29.
18. J. Koo, Q. Cai, K. Wang, J. Maas, T. Takahasi, A. Martwick, and D. Pommerenke. Correlation between EUT failure levels and ESD generator parameters. *IEEE Transactions on Electromagnetic Compatibility*, Vol. 50, (4) Nov 2008; 794–801.
19. K. Muhonen, J. Dunihoo, E. Grund, N. Peachey and A. Brankov. Failure detection with HMM waveforms. *Proceedings of the Electrical Overstress/Electrostatic Discharge (EOS/ESD) Symposium*, 2009; 396–404.
20. K. Muhonen, N. Peachey, and A. Testin. Human metal model (HMM) testing challenges to using ESD guns. *Proceedings of the Electrical Overstress/Electrostatic Discharge (EOS/ESD) Symposium*, 2009; 387–395.
21. S. Marum, C. Duvvury, J. Park, A. Chadwick, and A. Jahanzeb. Protecting circuits from transient voltage suppressor's residual pulse during IEC 61000-4-2 stress. *Proceedings of the Electrical Overstress/Electrostatic Discharge (EOS/ESD) Symposium*, 2009; 377–386.
22. M. Scholz, G. Vandersteen, D. Linten, S. Thijs, G. Groeseneken, M. Sawada, T. Nakaei, T. Hasebe, D. LaFonteese, V. Vashchenko, and P. Hopper. On-wafer human metal model measurements for system-level ESD analysis. *Proceedings of the Electrical Overstress/Electrostatic Discharge (EOS/ESD) Symposium*, 2009; 405–413.
23. T. Vheriakoski, J. Hilberg, and L. Sillanpaa. ESD event receiver for system level testing. *Proceedings of the Electrical Overstress/Electrostatic Discharge (EOS/ESD) Symposium*, 2009; 414–426.
24. ESD Association. HMM – System level ESD pulses to components: Application and Interpretation. Workshop C.4, *Proceedings of the Electrical Overstress/Electrostatic Discharge (EOS/ESD) Symposium*, 2009; 436.
25. E. Grund. Test circuits and current pulse generator for simulating an electrostatic discharge. U.S. Patent No. 8,278,936, October 2, 2012.
26. D.J. Pommerenke. System and method for testing the electromagnetic susceptibility of an electronic display unit. U.S. Patent No. 7,397,266, July 8, 2008.
27. D.J. Pommerenke, W. Huang, P. Shao and J. Xiao. Resonance scanning system and method for test equipment for electromagnetic resonance. U.S. Patent No. 8,143,903, March 27, 2012.
28. K.J. Min, D.J. Pommerenke, G. Muchaidze, and B. Chickhradze. System and method for electrostatic discharge testing of devices under test. U.S. Patent No. 8,823,383, September 2, 2014.
29. D.J. Pommerenke, D. Dickey, J.J. DeBlanc, V.T. Tam, and K.K. Tang. Method and apparatus for controlling electromagnetic radiation emissions from electronic device enclosures. U.S. Patent No. 6,620,999, September 16, 2003.
30. D.J. Pommerenke, D. Dickey, J.J. DeBlanc, and V. Tsang. Method and apparatus for controlling electromagnetic radiation emissions from electronic device enclosures. U.S. Patent No. 6,825,411, November 30, 2004.
31. H. Heidler. *Analytische blitzstromfunktion zur LEMP-Berechnung*. Munich, Germany: International Conference on Lightning Protection (ICLP), 1995.
32. F.E. Asimakopoulou, G.P. Fotis, I.F. Gonos, and I.A. Stathopulos. Parameter determination of Heidler's equation for the ESD current. *15th International Symposium on High Voltage Engineering.* August 27–31, 2007, T2-208, 1–4.
33. Thermo Fischer Scientific. http://www.thermofisher.com. Thermo Scientific™ MiniZap-15 ESD Simulator, 2016.
34. ISO Standard 10605. Road Vehicles – Test Methods for Electrical Disturbances from Electrostatic Discharge. International Organization of Standardization, 2008.
35. IEC. IEC Medical Electrical Equipment – Part 1-2: General Requirements for Basic Safety and Essential Performance – Collateral Standard: Electromagnetic Disturbances – Requirement and Tests, 2007.
36. RTCA, Inc. Environmental Conditions and Test Procedures for Airborne Equipment, DO-160F, 2007.
37. S. Voldman. *ESD: Physics and Devices*. Chichester, England: John Wiley and Sons, Ltd., 2004.
38. S. Voldman. *ESD: Circuits and Devices*. Chichester, England: John Wiley and Sons, Ltd., 2005.
39. S. Voldman. *ESD: Circuits and Devices 2nd Edition*. Chichester, England: John Wiley and Sons, Ltd., 2015.
40. S. Voldman. *ESD: RF Circuits and Technology*. Chichester, England: John Wiley and Sons, Ltd., 2006.
41. S. Voldman. *Latchup*. Chichester, England: John Wiley and Sons, Ltd., 2007.
42. S. Voldman. *ESD: Circuits and Devices*. Beijing, China: Publishing House of Electronic Industry (PHEI), 2008.

43. S. Voldman. *ESD: Failure Mechanisms and Models*. Chichester, England: John Wiley and Sons, Ltd., 2009.
44. S. Voldman. *ESD: Design and Synthesis*. Chichester, England: John Wiley and Sons, Ltd., 2011.
45. S. Voldman. *ESD Basics: From Semiconductor Manufacturing to Product Use*. Chichester, England: John Wiley and Sons, Ltd., 2012.
46. S. Voldman. *Electrical Overstress (EOS): Devices, Circuits, and Systems*. Chichester, England: John Wiley and Sons, Ltd., 2013.
47. S. Voldman. *ESD: Analog Design and Circuits*. Chichester, England: John Wiley and Sons, Ltd., 2014.
48. C. Duvvury, and H. Gossner. *System Level ESD Co-Design*. Chichester, England: John Wiley and Sons, Ltd., 2015.

8

Human Metal Model (HMM)

8.1 History

In the past, electrostatic discharge (ESD) testing was performed on semiconductor components. As stated in the last chapter, today, there is more interest in the testing of components in powered states and in electrical systems. System manufacturers have begun requiring system level testing to be done on semiconductor components, prior to final assembly and product acceptance. These system level tests are performed with an ESD gun, and without direct contact; these air discharge events produce an ESD event as well as generate EMI. In a true system, the system itself provides shielding from EMI emissions. Hence, an ESD test is of interest that has the following characteristics [1–3]:

- An IEC 61000-4-2 current waveform [4].
- No air discharge (contact discharge).
- Semiconductor component is powered during ESD testing.
- Only addresses pins and ports exposed to the external system.

The human metal model (HMM) addresses these characteristics [1–3]. This standard was released in the time frame of 2008 and 2009 [1–3]. Research in the HMM (and IEC 61000-4-2) began in the 1980s continued to recent times [1–24]. The HMM event is a recent ESD model that has increasing interest as a result of cell phone and small components with exposed ports, where field failures were evident. The HMM uses an "IEC-like" pulse waveform. The discharge from the source and the device under test (DUT) is a direct contact to avoid EMI spurious signals. The test is performed when the system is powered, and only the external ports that are exposed to the outside world are of interest.

Figure 8.1 shows the cases for the HMM event. The HMM event is a direct contact event, which can be applied by an ESD gun source or a pulse source.

8.2 Scope

The scope of the HMM ESD test is for the testing, evaluation, and classification of components to system-like ESD events. The test is to quantify the sensitivity or susceptibility of these components to damage or degradation to the defined IEC 61000-4-2 test [1–4].

ESD Testing: From Components to Systems, First Edition. Steven H. Voldman.
© 2017 John Wiley & Sons, Ltd. Published 2017 by John Wiley & Sons, Ltd.

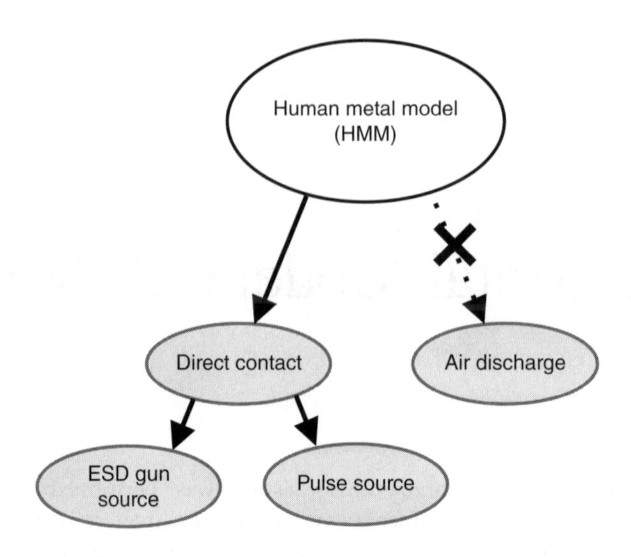

Figure 8.1 Human metal model (HMM) testing overview

8.3 Purpose

The purpose of the HMM ESD test is for establishment of a test methodology to evaluate the repeatability and reproducibility of systems to a defined pulse event in order to classify or compare ESD sensitivity levels of components [1–3].

8.4 Pulse Waveform

Figure 8.2 shows an example of the HMM waveform. Note that this waveform is identical to the IEC 61000-4-2 waveform. A key difference is the HMM test only addresses direct contact discharge. The event consists of a fast high current event followed by a second slower event. The fast event has a rise time of 0.7 to 1 ns. The slower event has a lower current with a peak at 30 ns and a decay time of 60 ns. The energy accumulation capacity is associated with a 150 pF capacitor. The discharge resistance is a 330 Ω resistor. This provides an 8 kV event for direct contact resistance and 15 kV event for air discharge. The polarity of the pulse event addresses positive and negative events [1–3].

8.4.1 Pulse Waveform Equation

For the IEC 61000-4-2 pulse waveform, an analytical formulation was established by Heidler [25, 26]. The Heidler formulation can be expressed as two exponential terms as follows:

$$i(t) = i_1/k_1 A(t, \tau_1) \exp\{-t/\tau_2\} + i_2/k_2 B(t, \tau_3) \exp\{-t/\tau_4\}$$

where

$$k_1 = \exp\{-(\tau_1/\tau_2)(n\tau_2/\tau_1)^{1/n}\}$$

$$k_2 = \exp\{-(\tau_3/\tau_4)(n\tau_4/\tau_3)^{1/n}\}$$

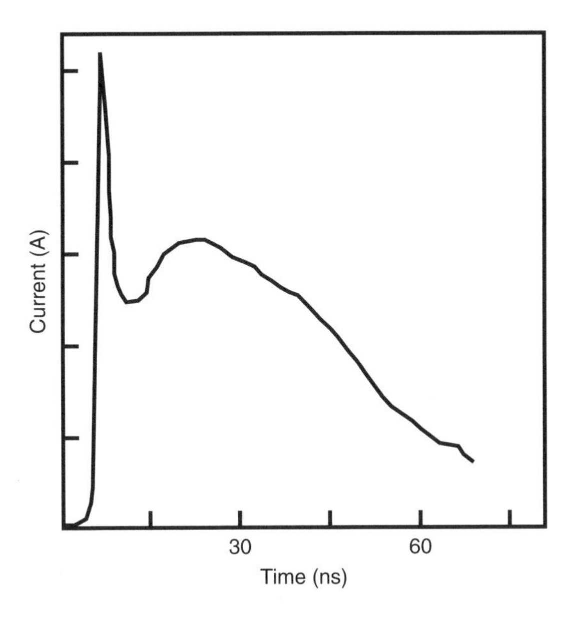

Figure 8.2 Human metal model (HMM) pulse waveform

and additionally,

$$A(t,\tau_1) = (t/\tau_1)/\{1 + (t/\tau_1)^n\}$$

and

$$B(t,\tau_3) = (t/\tau_3)/\{1 + (t/\tau_3)^n\}$$

In this formulation, the variables i_1, and i_2 are currents in amperes, τ_1, τ_2, τ_3, and τ_4 are time constants in nanoseconds, and n is a constant [25, 26]. An example of parameters, for a 2 kV positive pulse, $i_1 = 3.6$ A, and $i_2 = 3.74$ A, $\tau_1 = 0.29$, $\tau_2 = 18.44$, $\tau_3 = 34.34$, and $\tau_4 = 31.02$ (in nanoseconds), and $n = 3.26$ [25, 26].

8.5 Equivalent Circuit

Figure 8.3 shows the equivalent circuit schematic for an ESD gun. The circuit schematic shows a 150 pF capacitor and a 330 Ω resistor. The charging resistance is a 50–100 MΩ source. This provides an 8 kV event for direct contact resistance and 15 kV event for air discharge. The polarity of the pulse event addresses both positive and negative events.

8.6 Test Equipment

Commercial test systems now exist for device level testing to produce a waveform as specified in the IEC 61000-4-2 standard. Figure 8.4 shows a photograph of the Barth Model 4702 HMM+™ test system [27]. The system was designed to eliminate the common testing issues with IEC gun testing. ESD gun testing has reproducibility and repeatability issues due to the ground cable and electromagnetic noise issues.

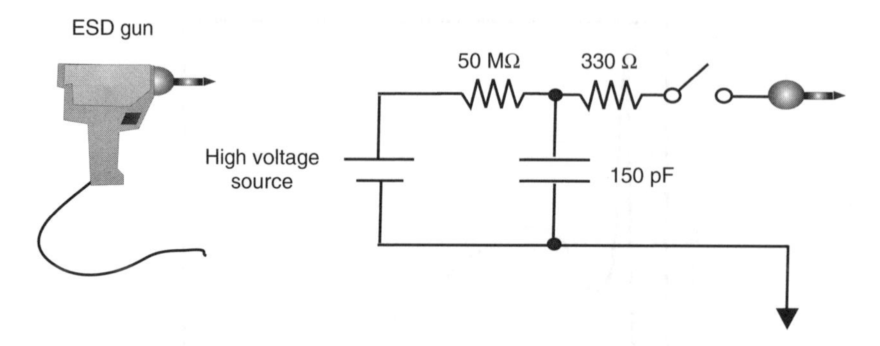

Figure 8.3 Human metal model (HMM) ESD gun equivalent circuit

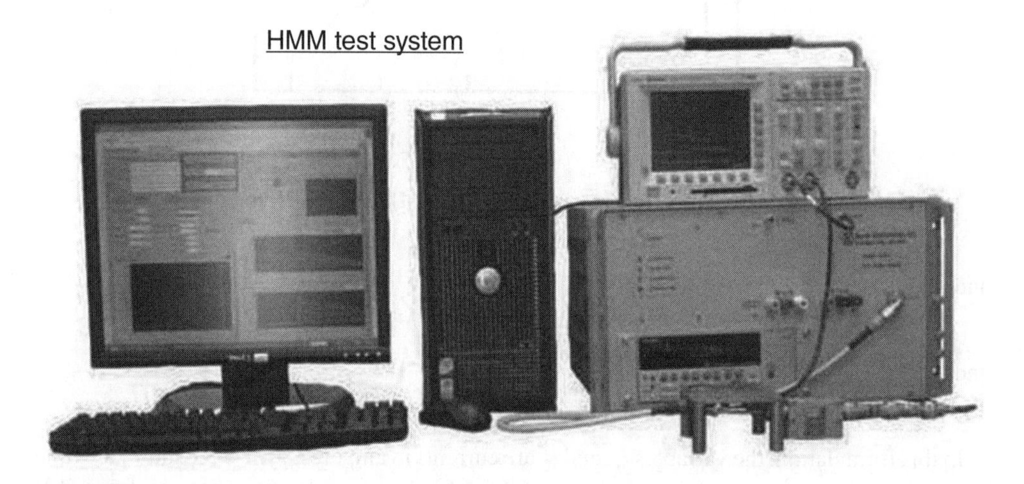

Figure 8.4 Commercial HMM test system – Barth 4702 HMM+™ (Reproduced with permission of Barth Electronics Inc.)

The Barth Model 4702 HMM tester is connected to the DUT with a single 50 Ω coaxial cable; this cable delivers the IEC-HMM pulse and provides connection for leakage measurements between IEC stress pulse events.

The specifications for the waveform are a pulse rise time of 0.7–1.0 ns. The peak current is 3.75 A/kV, a 30 ns current at 2.0 A/kV, and a 60 ns current at 1.0 A/kV [27]. The voltage range is from 500 V to 27 kV, with a maximum current of 100 A.

8.7 Test Configuration

The HMM testing requirements are for two different configurations. The primary reason for the specified orientation is associated with the electromagnetic field and capacitance. These will be shown in the following sections.

8.7.1 Horizontal Configuration

Figure 8.5(a) shows the test system for a horizontal configuration. The method applies an IEC pulse to the DUT without any air discharge. Shown in the figure is an ESD gun. Measurements of 0.5 m × 0.5 m are required for the horizontal surface. Ground clamps are specified as well as the discharge points for application of the HMM pulse. Using a current source, variations in the ESD gun waveform and pulse variation can be removed [1–3].

8.7.2 Vertical Configuration

Figure 8.5(b) shows the test system for a vertical configuration. The method applies an IEC pulse to the DUT without any air discharge. Shown in the figure is an ESD gun. Measurements

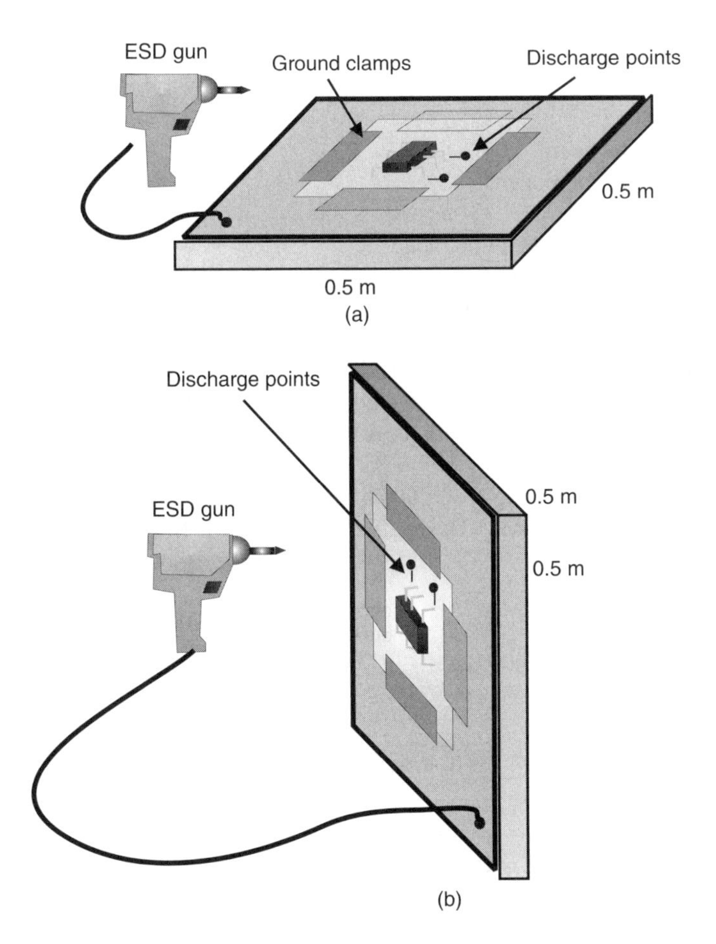

Figure 8.5 (a) Human metal model (HMM) – horizontal test configuration. (b) Human metal model (HMM) test equipment – vertical table test configuration

of 0.5 m × 0.5 m board are required for the vertical surface. Ground clamps are specified as well as the discharge points for application of the HMM pulse. Using a current source, variations in the ESD gun waveform and pulse variation can be removed [1–3].

8.7.3 HMM Fixture Board

For the HMM test, test fixture boards can be integrated into the test method. Figure 8.6 is an example of a schematic layout of a 50 Ω coaxial source test fixture board. The system consists of a 50 Ω IEC pulse source whose output is connected to the component under test. The component under test has a bypass capacitor, an isolation inductor, and a DC power supply connected to a 54 dB attenuator. A 50 Ω coaxial cable is also connected from the pulse source to the 54 dB attenuator. The output of the attenuator is connected to a 50 Ω oscilloscope [1–3].

For the HMM test, the layout of the circuit board is shown in Figure 8.7 [1–3]. In the test, the ground locations are specified as well as the discharge points. The definitions include the following [1–3]:

- Board ground plane
- Ground trace on the board
- Ground trace in the component
- Ground pin on socket
- Discharge point – direct to ground plane
- Discharge point – through the device
- Discharge point – to the pin to be stressed.

For the HMM test, the test can be performed with a powered board. Figure 8.8 is an example of an HMM test configuration with a powered board. Figure 8.8 illustrates a pulse source, a component under test, and a power source. A bypass capacitor is also placed in parallel with the component under test.

Figure 8.8(b) shows an example of a test fixture board for the HMM test from the Grund Technical Solutions Tarvos HMM Device Test System [40]. In the case of an ESD HMM

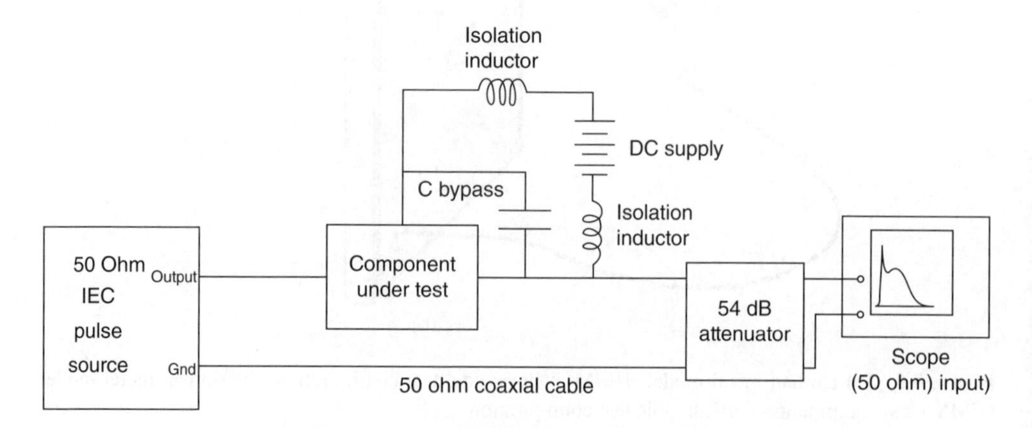

Figure 8.6 Schematic layout of IEC 50 Ω coaxial source test fixture board

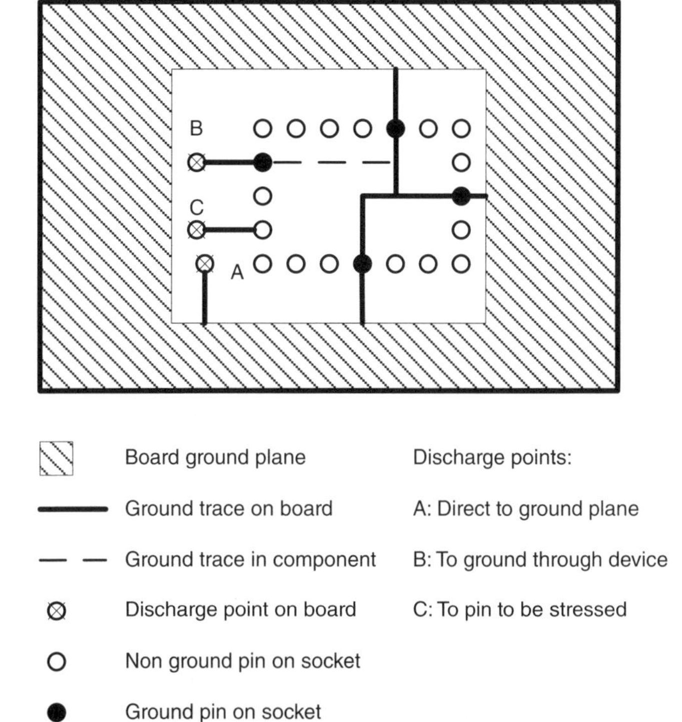

Figure 8.7 Layout of IEC 50 Ω coaxial source test fixture board showing ground connections

source event, the test fixture board is used. The GTS Tarvos HMM test system applies the IEC 61000-4-2 testing pulse to packaged devices, boards, and subassemblies with the use of a 50 Ω cable. This allows for a reproducible and repeatable pulse waveform.

8.8 Test Sequence and Procedure

Figure 8.9 shows the HMM test sequence and procedure. The procedure is as follows [1–3]:

- Choose a test configuration.
- Wire only external ports in the test fixture.
- Mount the DUT.
- Choose the reference and the pin under test (PUT).
- Charge the HMM source to a specified test level.
- Discharge HMM source to the PUT.
- Evaluate the leakage.
- Evaluate the leakage failure criteria.
- If no failure occurs, retest to a higher current level; if failure occurs, discontinue testing of the PUT.

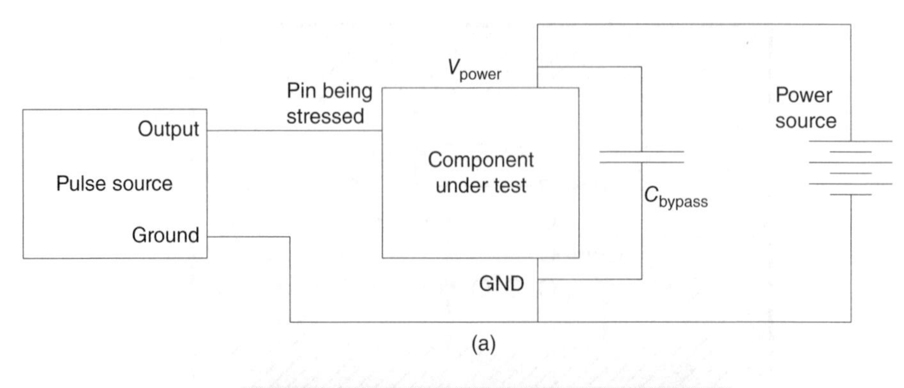

(b)

Figure 8.8 (a) HMM with powered board. (b) Human metal model (HMM) source test fixture board (Reproduced with permission from Grund Technical Solutions)

8.8.1 Current Waveform Verification

For the testing procedure, it is important to verify the current waveform. For this to be performed correctly, the physical arrangement for testing is critical. Figure 8.10 shows an example of the physical arrangement for HMM current waveform verification [1–3].

8.8.2 Current Probe Verification Methodology

For the current probe verification methodology, current probes must be evaluated for proper testing of the HMM event. Figure 8.11 shows a procedure of simultaneous measurement of the IEC probe, and the Fischer Custom Communication current probe F65-A [28]. The two probes are actually utilized in a simultaneous fashion [1–3].

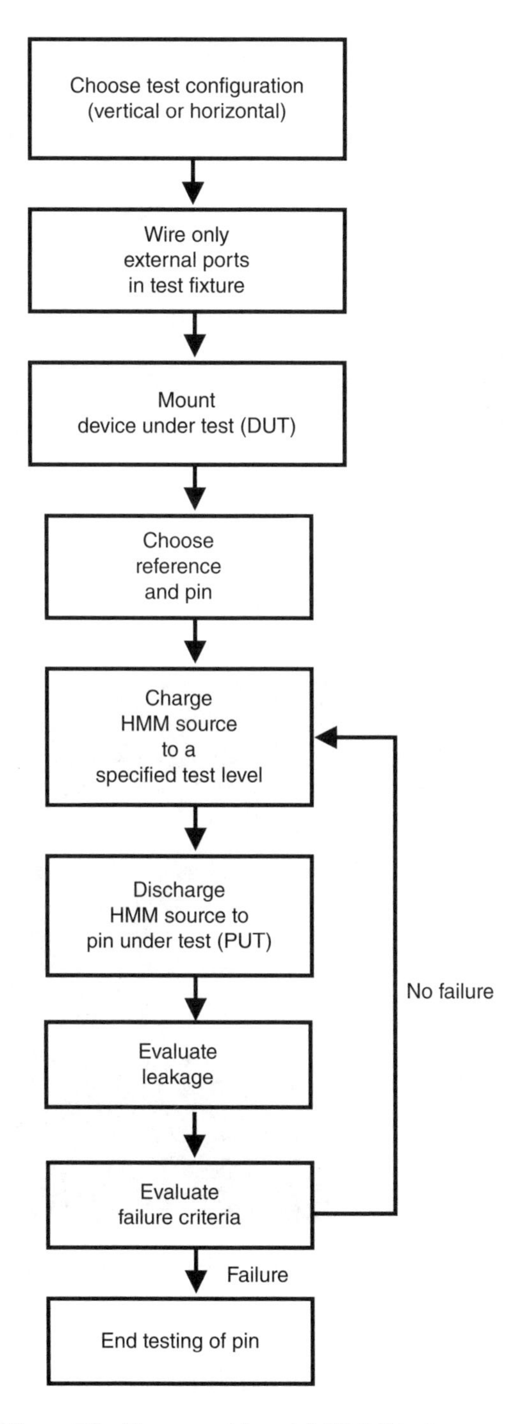

Figure 8.9 Human metal model (HMM) test sequence

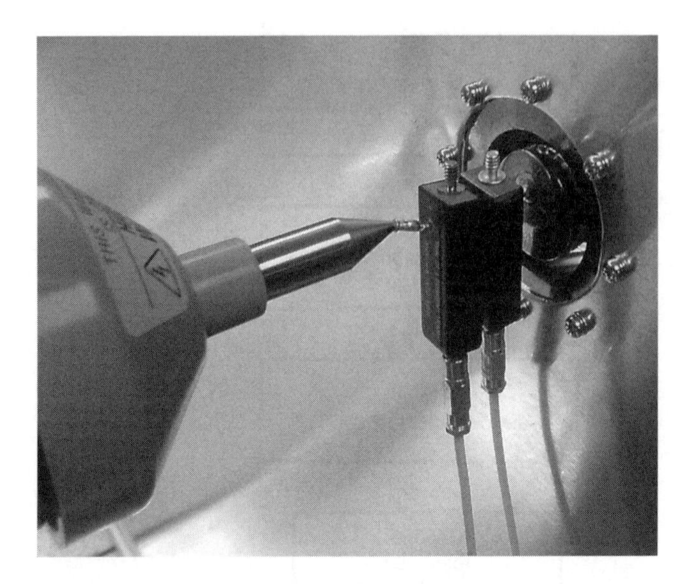

Figure 8.10 Human metal model (HMM) current waveform verification – physical arrangements

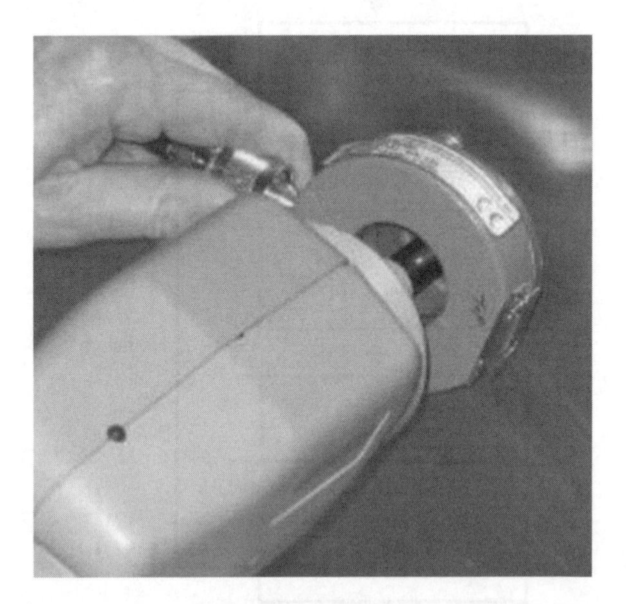

Figure 8.11 Simultaneous measurement of current probe IEC and the Fischer Custom Communication current probe F-65A

8.8.3 Current Probe Waveform Comparison

For the current probe waveform comparison, three tests were performed:

- IEC current probe only

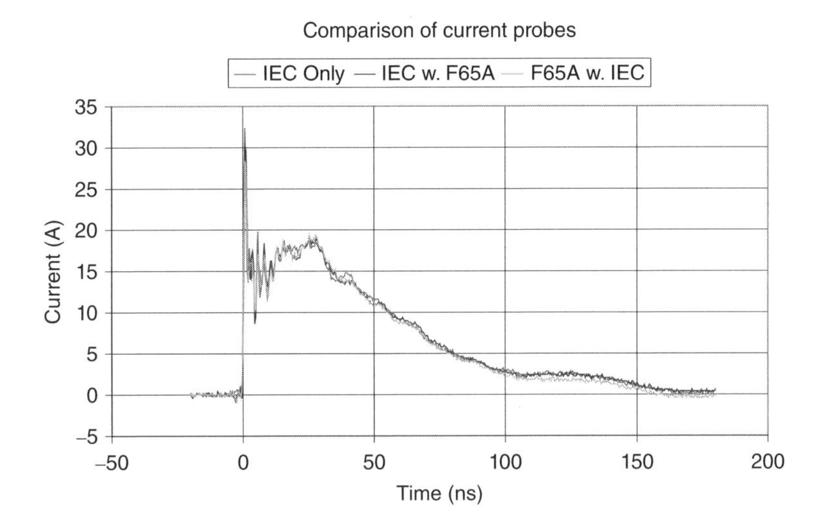

Figure 8.12 Human metal model (HMM) current probe waveform comparison

- IEC with F-65A current probes [28]
- F-65A with IEC current probes [28].

Figure 8.12 shows a procedure of simultaneous measurement of the IEC probe and the Fischer Custom Communication current probe F65-A. The two probes are actually utilized in a simultaneous fashion.

8.9 Failure Mechanisms

The failure mechanisms for the HMM event are equivalent to the failures that occur in the IEC 61000-4-2, with a few distinctions:

- The HMM pulse is applied only to specific pins that are connected to external ports.
- The HMM pulse failures can occur on pins that are not connected to external ports.

In the case of pins connected to external ports, HMM failures can occur in the following:

- Bond pad
- Interconnects between bond pad and ESD network
- ESD network
- I/O networks
- V_{DD} power rail
- V_{SS} power rail.

In the case of pins not connected to external ports, HMM failures can occur through the common power rails.

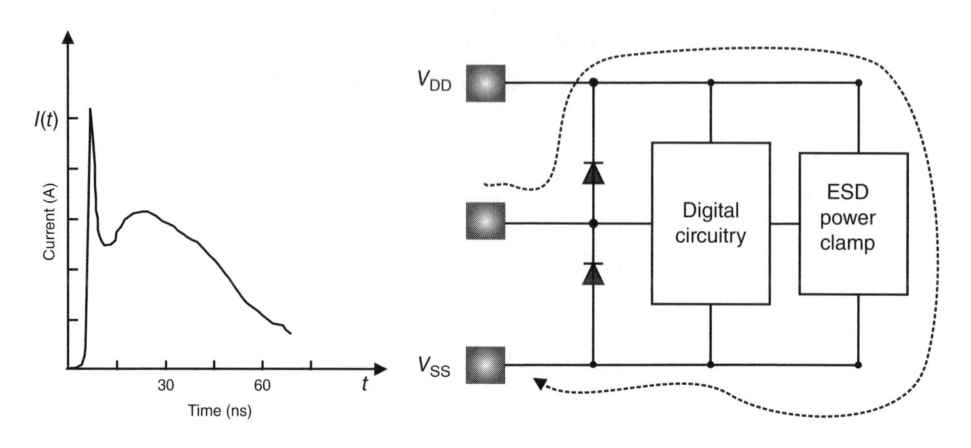

Figure 8.13 HMM ESD event current paths – V_{SS} positive polarity

8.10 ESD Current Paths

In the case of pins not connected to external ports, HMM failures can occur through the common power rails. Figure 8.13 shows an HMM event flowing into the designated signal pin (which is an external port). The distinction in this event from a typical HBM event is that the current can be significant due to the fast high current pulse event at the onset of the event. Second, this is only the testing of an external port signal pin. Another distinction is that the current into the V_{SS} ground rail can impact the other internal circuitry not shown in the picture [31, 36, 38].

8.11 ESD Protection Circuit Solutions

Figure 8.14 shows a solution to provide ESD protection to the semiconductor component [29–40]. The key issue is that a low resistance path between the external port pin and its local power connection is desired to discharge the current. The width of the interconnects must be able to survive the current magnitude of the HMM pulse event. In addition, it is desired that the current does not flow into the power grid adjacent to the other nonexternal port pins. This is achievable by providing the following design layout features:

- A dedicated HMM V_{DD} power pad adjacent to the external port signal pin
- A dedicated power HMM V_{DD} interconnect power bus
- A dedicated HMM V_{SS} power pad adjacent to the external port signal pin
- A dedicated HMM V_{SS} interconnect power bus
- Electrical connection between the HMM V_{DD} power bus and the V_{DD} power bus (e.g., a resistive connection)
- Electrical connection between the HMM V_{SS} power bus and the V_{SS} power bus (e.g., a resistive connection).

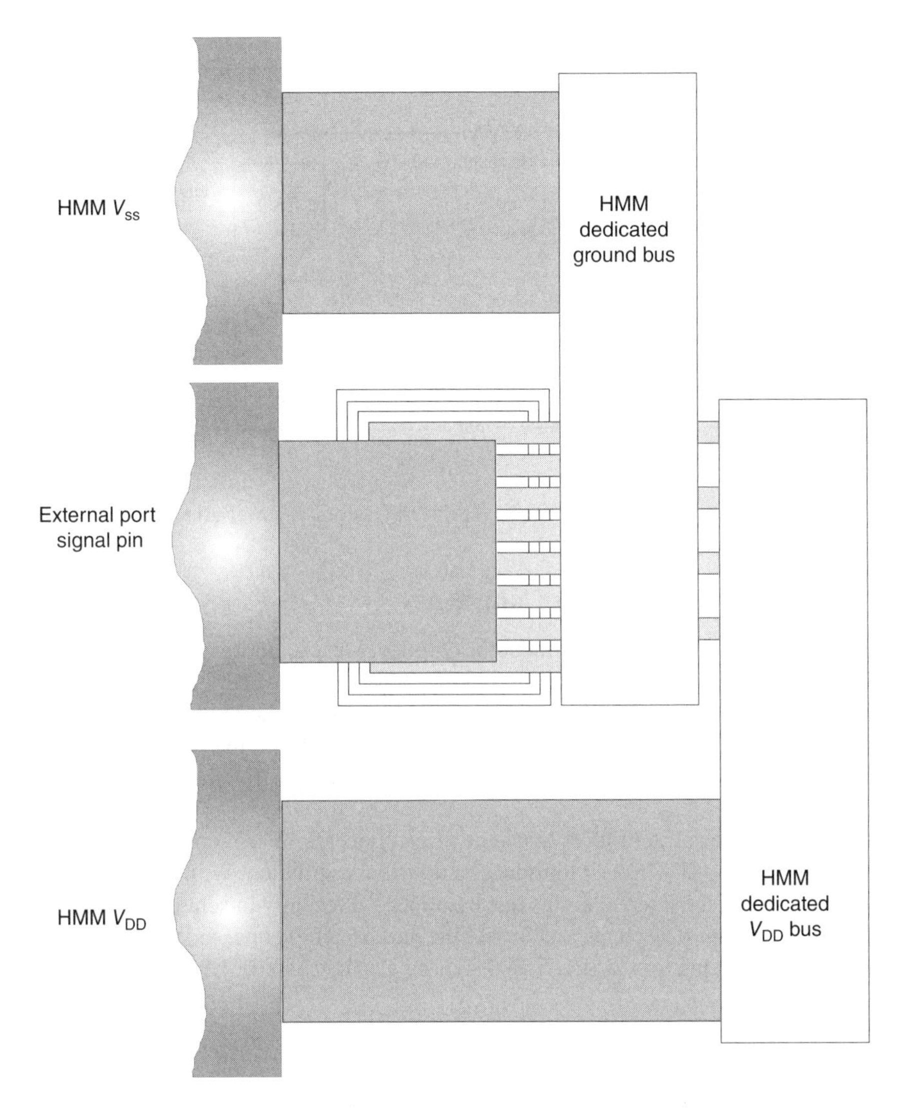

Figure 8.14 Human metal model (HMM) ESD protection solution – dedicated V_{DD} and V_{SS} power bus

The electrical connections are highlighted in Figure 8.15. The HMM bus resistance terms are designed significantly lower to allow current to flow to the respective power rails. In addition, the bus resistance to the power clamp and the resistance internal to the core are critical on how the HMM current distributes through the semiconductor component. The resistance to the power clamp local to the external port should have a low enough resistance to allow operability, but not to allow the high current peak to lead to electrical overstress (EOS) of the MOSFET gate structure.

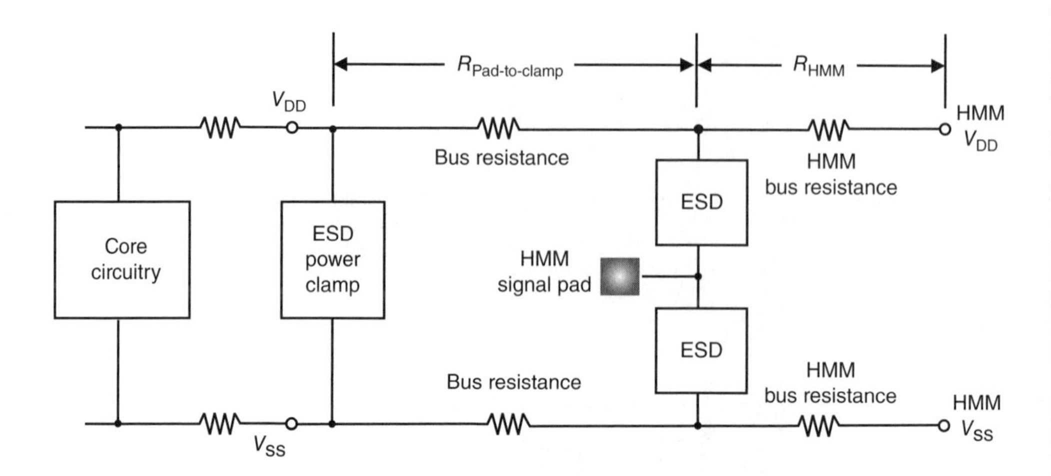

Figure 8.15 Human metal model (HMM) ESD event – power grid definition

8.12 Summary and Closing Comments

This chapter addressed a new semiconductor chip level test to address IEC 61000-4-2 pulse events into external ports of a semiconductor chip. This test, the HMM, introduces a fast transient followed by a slower HBM-like waveform that is only applied to specific ports exposed on a system level. With the growing interest in system level EOS, the interest in this test has emerged recently due to the request of system level developer for the semiconductor chip suppliers.

Chapter 9 addresses system level transient surge concerns in semiconductor development known as IEC 61000-4-5. This test introduces a transient oscillation. With the growing interest in system level EOS, the interest in this test has emerged recently in semiconductor chip suppliers to test equipment. The chapter addresses the purpose, the scope, and the pulse waveform. ESD protection concepts for the IEC 61000-4-5 are also discussed.

Problems

1. The IEC 61000-4-2 can be an air discharge event. What parameter variables influence the pulse waveform when it is an air discharge event? The HMM pulse event is a direct contact methodology. What parameter variables influence the pulse waveform of the HMM event?
2. The IEC 61000-4-2 discharge for an air discharge initiated an E- and H-field outside of the electronic system. Estimate the energy in the E-field and the H-field assuming a current pulse of current I. The HMM event is a direct contact event. Calculate the ratio of energy in the E and H field over the total energy of the system.
3. The IEC 61000-4-2 discharge for an air discharge initiated an E- and H-field outside of the electronic system. Estimate the energy in the E-field and the H-field assuming a current pulse of current I. The HMM event is a direct contact event. Calculate the ratio of energy of the HMM event over the total energy.
4. Why should we exclude the energy of the E-field and H-field for evaluation of the ESD robustness of semiconductor chips that are contained within systems?

5. An ESD gun can be used to initiate a direct HMM discharge to a cell phone that contains external port for charging of the cell phone. What factors influence the discharge event?
6. An ESD gun can be used to initiate a direct HMM discharge to a laptop. What are potential sources of system failures? Assume the discharge is to the laptop USB ports. What factors influence the discharge event?
7. An ESD gun is used to zap an antennae of a cell phone. The cell phone is electrically connected to a gallium arsenide (GaAs) bipolar transistor power amplifier. Show the ESD event. What ESD protection circuitry can be used to avoid failure of the GaAs cell phone power amplifier?

References

1. ESD Association. ESD-SP 5.6 – 2008. *ESD Association Standard Practice for the Protection of Electrostatic Discharge Sensitive Items – Electrostatic Discharge Sensitivity Testing – Human Metal Model (HMM) Testing Component Level*, Standard Practice (SP) document, 2008.
2. ESD Association. *Electrostatic Discharge Sensitivity Testing – Human Metal Model (HMM)*, ESD Standard Practice Document, 2009.
3. ESD Association. HMM – system level ESD pulses to components: application and Interpretation. Workshop C.4, *Proceedings of the Electrical Overstress/Electrostatic Discharge (EOS/ESD) Symposium*, 2009; 436.
4. IEC. *IEC 61000-4-2: Electromagnetic Compatibility (EMC) – Part 4–2: Testing and Measurement Techniques – Electrostatic Discharge Immunity Test*, IEC International Standard, 2007.
5. P. Richman. Classification of ESD hand/metal current waves versus approach speed, voltage, electrode geometry, and humidity. *Proceedings of the IEEE International Symposium on Electromagnetic Compatibility*, San Diego, 1986; 451–460.
6. R. Saini, and K.G. Balmain. Human/metal ESD and its physical limitations. *Proceedings of the Electrical Overstress/Electrostatic Discharge (EOS/ESD) Symposium*, 1995; 91–94.
7. D. Pommerenke, and M. Aidam. To what extent do contact mode and indirect ESD test methods reproduce reality? *Proceedings of the Electrical Overstress/Electrostatic Discharge (EOS/ESD) Symposium*, 1995; 101–109.
8. L. Loeb. Statistical factors in spark discharge mechanisms. *Review of Modern Physics*, Vol. 20, (1), 1948; 151–160.
9. J. Meek. A theory of spark discharge. *Physical Review*, Vol. 57, 1940; 722–728.
10. E. Grund, K. Muhonen, and N. Peachey. Delivering IEC 61000-4-2 current pulses through transmission lines at 100 and 330 ohm system impedances. *Proceedings of the Electrical Overstress/Electrostatic Discharge (EOS/ESD) Symposium*, 2008; 132–141.
11. R. Chundru, D. Pommerenke, K. Wang, T. Van Doren, F.P. Centola, J.S. Huang. Characterization of human metal ESD reference discharge event and correlation of generator parameters to failure levels – Part I: reference event. *IEEE Transactions on Electromagnetic Compatibility*, Vol. 46, (4), Nov 2004; 498–504.
12. K. Wang, D. Pommerenke, R. Chundru, T. Van Doren, F.P. Centola, and J.S. Huang. Characterization of human metal ESD reference discharge event and correlation of generator parameters to failure levels – Part II: correlation of generator parameters to failure levels. *IEEE Transactions on Electromagnetic Compatibility*, 46, (4), Nov 2004; 505–511.
13. J. Barth, D. Dale, K. Hall, H. Hyatt, D. McCarthy, J. Nuebel, and D. Smith. Measurement of ESD HBM events, simulator radiation, and other characteristics toward creating a more repeatable simulation or simulator should simulate. *Proceedings of the Electrical Overstress/Electrostatic Discharge (EOS/ESD) Symposium*, 1996; 211–222.
14. W. Stadler, S. Bargstadt-Franke, T. Brodbeck, R. Gaertner, M. Goroll, H. Gossner, N. Jensen, and C. Muller. From the ESD robustness of products to system ESD robustness. *Proceedings of the Electrical Overstress/Electrostatic Discharge (EOS/ESD) Symposium*, 2004; 67–74.
15. R. Ashton. System level ESD testing – the waveforms. *Conformity Magazine*, Vol. 12, (10), 2007; 28–34.
16. N. Peachey. ESD Open Forum – ESD stressing of devices and components. *Conformity Magazine*, Vol. 13, (6), 2008; 22–23.
17. D. Robinson-Hahn. Evaluating IC components utilizing IEC 61000-4-2. *Proceedings of the International Electrostatic Discharge (ESD) Workshop*, 2008; 274–283.

18. R. Ashton, and L. Lescouzeres. Characterization of off chip protection devices. *Proceedings of the Electrical Overstress/Electrostatic Discharge (EOS/ESD) Symposium*, 2008; 21–29.
19. J. Koo, Q. Cai, K. Wang, J. Maas, T. Takahasi, A. Martwick, and D. Pommerenke. Correlation between EUT failure levels and ESD generator parameters. *IEEE Transactions on Electromagnetic Compatibility*, Vol. 50, (4), Nov 2008; 794–801.
20. K. Muhonen, J. Dunihoo, E. Grund, N. Peachey and A. Brankov. Failure detection with HMM waveforms. *Proceedings of the Electrical Overstress/Electrostatic Discharge (EOS/ESD) Symposium*, 2009; 396–404.
21. K. Muhonen, N. Peachey, and A. Testin. Human metal model (HMM) testing challenges to using ESD guns. *Proceedings of the Electrical Overstress/Electrostatic Discharge (EOS/ESD) Symposium*, 2009; 387–395.
22. S. Marum, C. Duvvury, J. Park, A. Chadwick, and A. Jahanzeb. Protecting circuits from transient voltage suppressor's residual pulse during IEC 61000-4-2 stress. *Proceedings of the Electrical Overstress/Electrostatic Discharge (EOS/ESD) Symposium*, 2009; 377–386.
23. M. Scholz, G. Vandersteen, D. Linten, S. Thijs, G. Groeseneken, M. Sawada, T. Nakaei, T. Hasebe, D. LaFonteese, V. Vashchenko, and P. Hopper. On-wafer human metal model measurements for system-level ESD analysis. *Proceedings of the Electrical Overstress/Electrostatic Discharge (EOS/ESD) Symposium*, 2009; 405–413.
24. T. Vheriakoski, J. Hilberg, and L. Sillanpaa. ESD event receiver for system level testing. *Proceedings of the Electrical Overstress/Electrostatic Discharge (EOS/ESD) Symposium*, 2009; 414–426.
25. H. Heidler. Analytische blitzstromfunktion zur LEMP-Berechnung. International Conference on Lightning Protection (ICLP), Munich Germany, 1995.
26. F.E. Asimakopoulous, G.P. Fotis, I.F. Gonos, and I.A. Stathopulos. Parameter determination of Heidler's equation for the ESD current. *15th International Symposium on High Voltage Engineering*. August 27–31, 2007, T2-208, 1–4.
27. Barth Electronics. http://www.barthelectronics.com. Barth Model 4702 HMM+TM test system, 2016.
28. Fischer Custom Communication. http:/www.fischercc.com. Fischer Custom Communication current probe F65-A, 2016.
29. S. Voldman. *ESD: Physics and Devices*. Chichester, England: John Wiley and Sons, Ltd., 2004.
30. S. Voldman. *ESD: Circuits and Devices*. Chichester, England: John Wiley and Sons, Ltd., 2005.
31. S. Voldman. *ESD: Circuits and Devices 2nd Edition*. Chichester, England: John Wiley and Sons, Ltd., 2015.
32. S. Voldman. *ESD: RF Circuits and Technology*. Chichester, England: John Wiley and Sons, Ltd., 2006.
33. S. Voldman. *Latchup*. Chichester, England: John Wiley and Sons, Ltd., 2007.
34. S. Voldman. *ESD: Circuits and Devices*. Beijing, China: Publishing House of Electronic Industry (PHEI), 2008.
35. S. Voldman. *ESD: Failure Mechanisms and Models*. Chichester, England: John Wiley and Sons, Ltd., 2009.
36. S. Voldman. *ESD: Design and Synthesis*. Chichester, England: John Wiley and Sons, Ltd., 2011.
37. S. Voldman. *ESD Basics: From Semiconductor Manufacturing to Product Use*. Chichester, England: John Wiley and Sons, Ltd., 2012.
38. S. Voldman. *Electrical Overstress (EOS): Devices, Circuits, and Systems*. Chichester, England: John Wiley and Sons, Ltd., 2013.
39. S. Voldman. *ESD: Analog Design and Circuits*. Chichester, England: John Wiley and Sons, Ltd., 2014.
40. Grund Technical Solutions. http://www.grundtech.com. Tarvos HMM Device Test System, 2016.

9

IEC 61000-4-5

9.1 History

In this chapter, the focus is on the transient surge phenomena, associated with the standard IEC 61000-4-5 Electromagnetic Compatibility (EMC) – Part 4–5: Testing and Measurement Techniques – Surge Immunity Test standard [1].

Transient surges, switching, and lightning-induced transients have been known to introduce failure of components and systems due to electrical overvoltage (EOV). A test method was needed to find the reaction of the equipment under test (EUT) from transients and lightning surges [2–23]. Transient surges occur on AC power mains as a result of switching operations. Lightning strikes also induce transients to the power distribution systems or electrical ground. Electrical surges also lead to inductive and capacitive coupling into signal lines (e.g., I/O lines) that can lead to EOV. As a result, surge test standards are required to evaluate transient surges (Figure 9.1).

Electrical power lines are subjected to transient events from switching and lightning. Power system transient events can include major power system switching from capacitor banks. Load changes in the power distribution system can also introduce power transients. Power system disruption can also occur from system faults. System faults can include short circuits, and arcing during installation or operation.

Lightning can lead to transient surge. These include direct lightning and indirect events. Direct lightning events can strike outdoor electronics or systems that introduce high currents (e.g., electrical overcurrent (EOC)) and EOV. Direct lightning events also introduce lightning ground currents. A direct lightning discharge to earth can couple into the ground path of a grounding system. In addition, electromagnetic (EM) emissions from lightning local to systems can induce electrical voltage and currents on conductors inside or outside of buildings. Even the lightning protection devices can lead to electromagnetic interference (EMI) in adjacent systems.

For evaluation of electrical overstress (EOS), standard has been developed for "surge testing" [1]. A standard that is being utilized presently for evaluation of components or populated

ESD Testing: From Components to Systems, First Edition. Steven H. Voldman.
© 2017 John Wiley & Sons, Ltd. Published 2017 by John Wiley & Sons, Ltd.

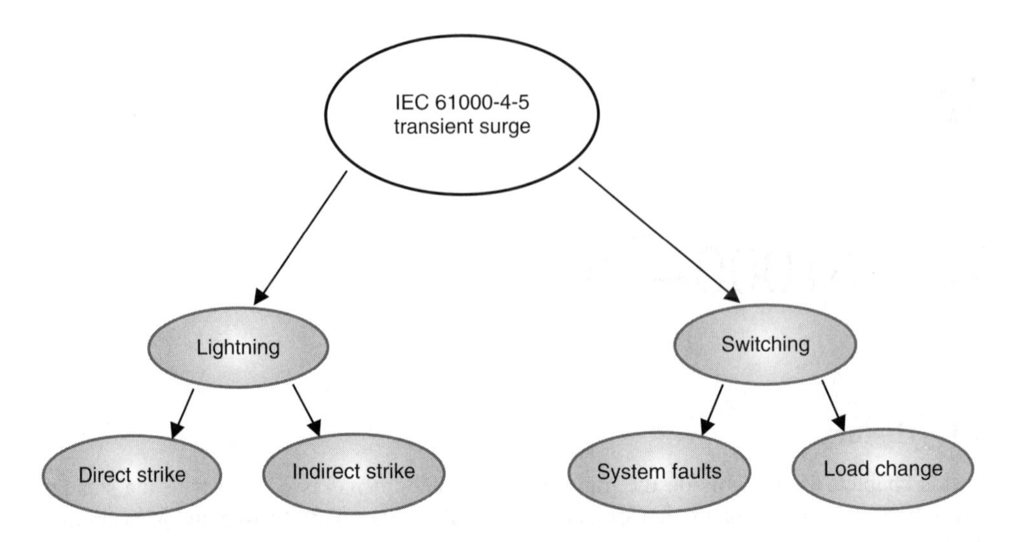

Figure 9.1 IEC 61000-4-5

printed circuit boards is the IEC 61000-4-5 EMC – Part 4–5: Testing and Measurement Techniques – Surge Immunity Test standard [1].

9.2 Scope

The scope of the IEC 61000-4-5 electrostatic discharge test is for the testing, evaluation, and classification of systems to transient surge and lightning. The test is to quantify the sensitivity or susceptibility of these components to damage or degradation to the defined IEC 61000-4-5 test [1].

9.3 Purpose

The purpose of the IEC 61000-4-5 transient surge test is for establishment of a test methodology to evaluate the repeatability and reproducibility of systems to a defined transient pulse surge event in order to classify or compare ESD sensitivity levels of systems. The purpose of the IEC 61000-4-5 test is to establish a common and reproducible basis for evaluation of electronic and electrical systems to transient surge events. The test method describes a consistent method to assess the immunity of equipment or systems. The IEC 61000-4-5 test defines the following (Figure 9.1) [1]:

- Pulse waveform
- Range of test levels
- Test equipment
- Test setup and configuration
- Test procedure
- Test calibration.

9.4 Pulse Waveform

Figure 9.2 shows the IEC 61000-4-5 surge test pulse open-circuit waveform. This pulse waveform is nonoscillatory for an open circuit. A surge generator is used that delivers a 1.2 μs rise time and a 50 μs full width half maximum voltage waveform [1].

Figure 9.3 shows the IEC 61000-4-5 surge test pulse short circuit waveform. This pulse waveform undergoes a negative transition after the first peak. A surge generator is used that delivers an 8 μs rise time and a 20 μs full width half maximum voltage waveform [1].

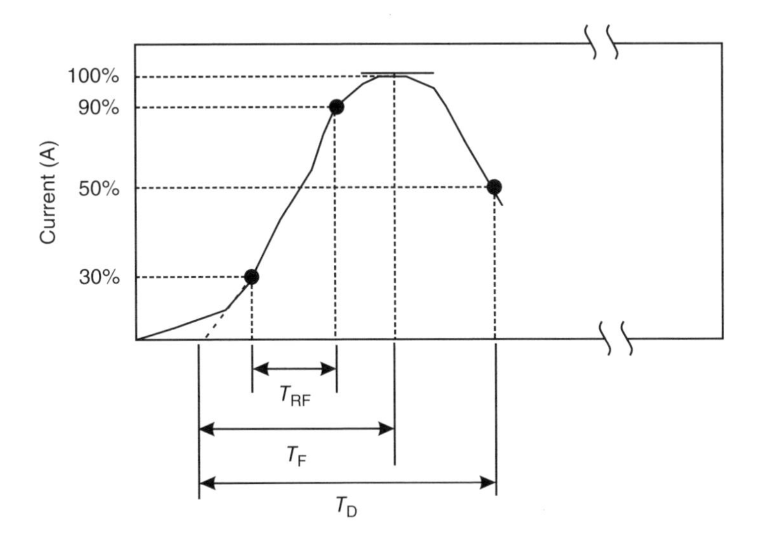

Figure 9.2 IEC 61000-4-5 surge test pulse open-circuit waveform

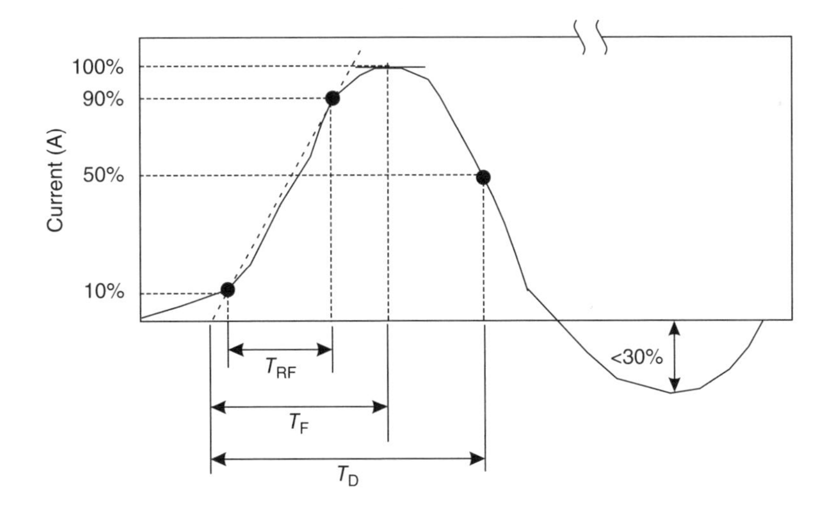

Figure 9.3 IEC 61000-4-5 surge test pulse short-circuit waveform

Table 9.1 IEC 61000-4-5 class and voltage levels

Class	Electrical environment	Voltage level
0	Well-protected environment	25 V
1	Partially protected environment	500 V
2	Environment with cables separated and short runs	1 kV
3	Environment with parallel signal and power cables	2 kV
4	Environment including outdoor cables, power cable, and electronics	4 kV
5	Environment for electronic equipment connected to telecommunication cables and power lines	Test level 4

Table 9.1 illustrates the classes and associated voltage levels. In the table, the electrical environment is also highlighted for each class and voltage level.

9.5 Equivalent Circuit

Figure 9.4 shows the IEC 61000-4-5 surge generator equivalent circuit. The surge generator is known as a combination wave generator [1].

9.6 Test Equipment

Commercial test equipment exists to support the IEC 61000-4-5 specification. Test equipment includes the Haefely AXOS 8 EMC immunity test system, Haefely PSURGE 8000 8 kV surge system, Teseq NSG3060 6 kV surge generator, and Thermo KeyTek ECAT 10 kV surge generator [23–26]. Figure 9.5(a) and (b) is an example of a present-day test system for IEC 61000-4-5 transient surge testing.

An example of a test system, the Thermo KeyTek ECAT 10 kV surge generator (i.e., compliant with the IEC 61000-4-5 specification) can generate a voltage waveform of 1.2 μs rise time and a 50 μs full width half maximum voltage waveform. The peak voltage can achieve

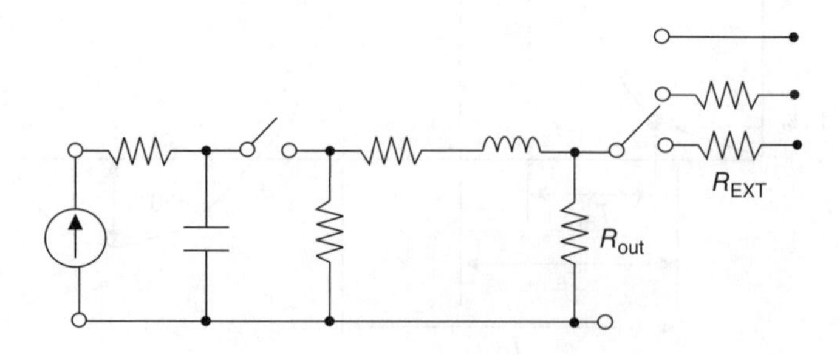

Figure 9.4 IEC 61000-4-5 equivalent circuit of a transient surge generator

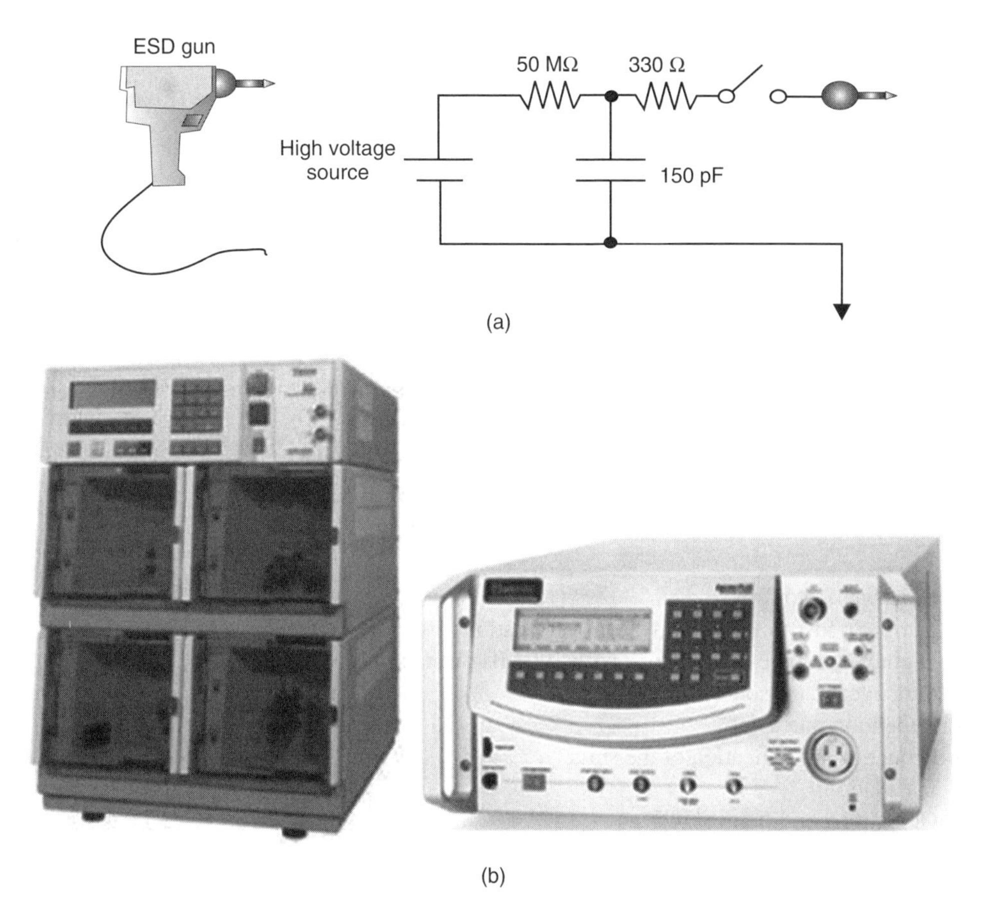

Figure 9.5 (a) IEC 61000-4-5 high voltage transient surge commercial test equipment. (b) IEC 61000-4-5 high voltage transient surge commercial test equipment – EMCPro Plus™ EMC Test System (Reproduced with permission of Thermo Fisher Scientific)

250 V to 6.6 kV in a 12 Ω mode of operation. The peak current of 125 A to 3.3 kA can be sourced. The repetition rate of these test systems can provide multiple pulses in a minute (e.g., maximum of four pulses in 1 min). For an open-circuit voltage, a voltage waveform of 1.2 μs rise time and a 50 μs duration, and a 30% undershoot.

A second test system used for precompliance, quality assurance, and certification is the EMCPro Plus™ EMC test system that satisfies the IEC 61000-4-5 standard [27]. This test system is the only test system capable of monitoring the IEC 61000-4-5 surge voltage and current at the output terminals; measurement at the generator can lead to large measurement errors. The system allows for surge testing to the IEC 61000-4-5 6 kV limit with combination, telecom, and ring waves. The combination wave surge requirement for 1.2/5 μs, the telecom wave surge requirement (10/700 ms), and the ring oscillatory requirement can be fulfilled with the test system [27].

9.7 Test Sequence and Procedure

The recording of the test results can be classified based on the response (Figure 9.6). The response of the test can be defined as follows [1]:

- Normal operation within the operational specification of the system
- Loss of operation, loss of function, or temporary degradation that is recoverable by system self-recovery or "self-healing" function
- Loss of operation, loss of function, or temporary degradation that is nonrecoverable by system self-recovery or "self-healing" function requiring operator intervention (e.g., power down, power up, or system reset)
- Loss of operation, loss of function, or nonrecoverable degradation that is nonrecoverable and damage to the system, components, software, or data.

The IEC 61000-4-5 standard is required to be performed according to the manufacturer test plans. In the test, the following items are specified [1]:

- Surge generator
- Current transformer for waveform verification
- Digital or storage oscilloscope with 100 MHz bandwidth
- Surge generator source impedance
- Repetition rate
- Sequence of application of the surge
- Installation
- Operation condition of EUT
- Number of tests
- Test levels
- Polarity
- Input and outputs tested
- Phase angle of coupling to alternating current (AC) mains
- Internal or external trigger.

Waveform verification is also required periodically. High voltage differential surge probes are needed for verification of the open-circuit voltage.

9.8 Failure Mechanisms

Failure mechanisms on a system level can be destructive and nondestructive events. The following responses of the IEC 61000-4-2 test can be defined as failure mechanisms [1]:

- Loss of operation, loss of function, or temporary degradation that is recoverable by system self-recovery or "self-healing" function
- Loss of operation, loss of function, or temporary degradation that is nonrecoverable by system self-recovery or "self-healing" function requiring operator intervention (e.g., power down, power up, or system reset)

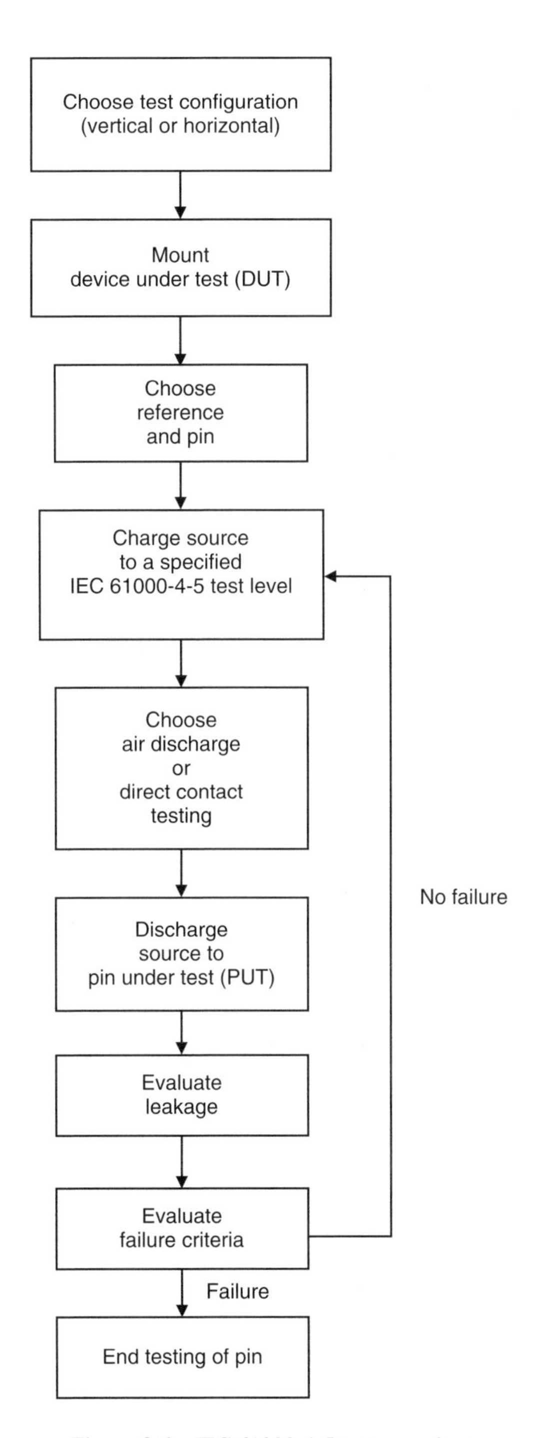

Figure 9.6 IEC 61000-4-5 test procedure

- Loss of operation, loss of function, or nonrecoverable degradation that is nonrecoverable and damage to the system, components, software, or data
- Permanent damage to the components and system.

9.9 IEC 61000-4-5 ESD Current Paths

Current paths from the IEC 61000-4-5 involve both signal pins and power rails. In this test, a significant focus exists on interaction of the circuitry and the ground connections. The ground connection can be a source of the transient phenomena. In the following section, solutions used for protection in the ground connections are discussed.

9.10 ESD Protection Circuit Solutions

ESD protection circuit solutions include solutions between pins and ground connections. Diodes are used to provide protection for positive and negative transient surges [28–40]. Figure 9.7 shows an example of a diode protection element. The circuit element must survive the peak pulse current (I_{PP}). In addition, the protection device is to limit the surge voltage (V_{CL}). Figure 9.7 shows the positive region and the negative region with the defined electrical parameters of the diode protection element.

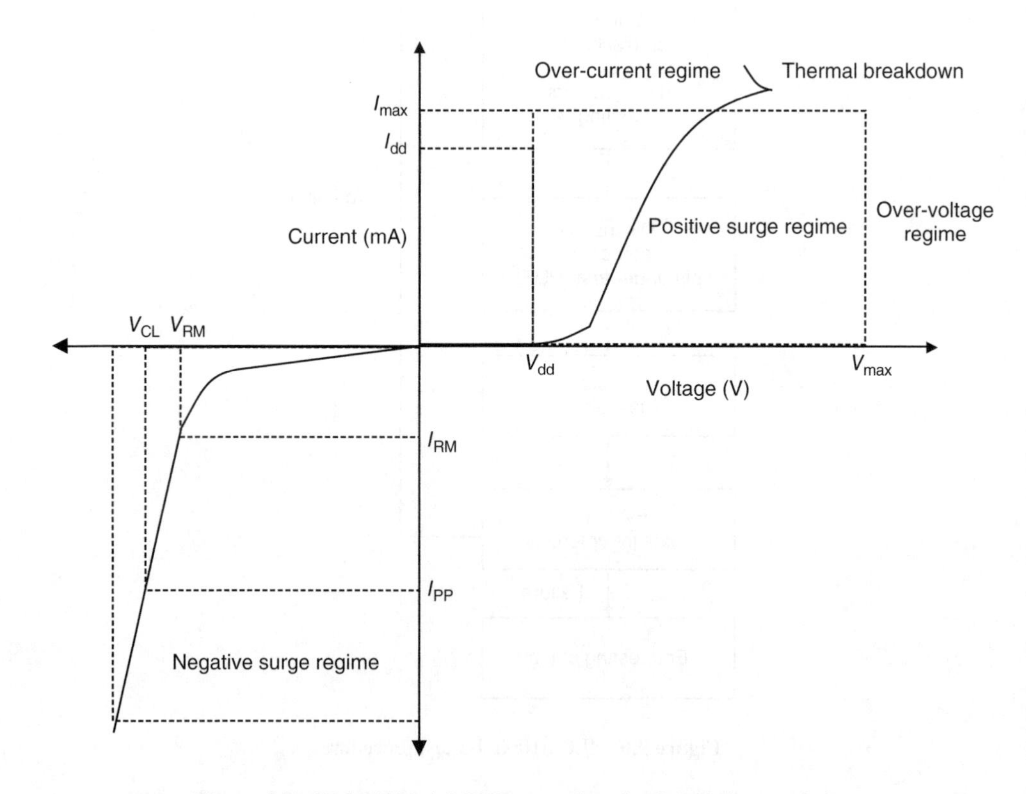

Figure 9.7 IEC 61000-4-5 ESD protection circuit *I–V* characteristics

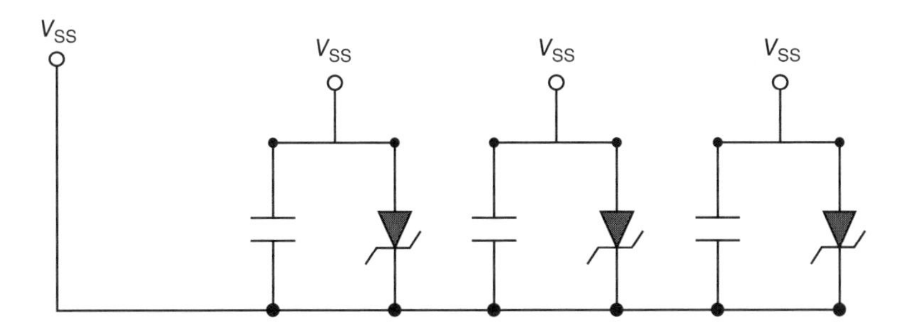

Figure 9.8 IEC 61000-4-5 ESD protection circuit for ground connections

The peak pulse current, I_{PP}, should be above the current in the IEC 61000-4-5 standard that requires an 8 μs/20 μs response.

Figure 9.8 is an example of a protection strategy between the ground connections using a capacitor and a breakdown diode. The ground connections are all connected to a common ground point.

9.11 Alternate Test Methods

For telecommunication lines, the test equipment is modified to address the distinction compared to other events. Telecommunication lines are vulnerable to lightning through both direct and indirect. Telecommunication lines can also be disturbed by power induction, or power transfer with the main AC lines.

An example of a telecom compliant test system that is compliant with the IEC 61000-4-5 specification can generate a voltage waveform of 10 μs rise time and a 700 μs duration. The peak voltage can achieve 250 V to 6.6 kV. The peak current of 6.25–165 A can be sourced in a 40 Ω mode. The repetition rate of these test systems can provide multiple pulses in a minute (e.g., maximum of four pulses in 1 min). For an open-circuit voltage, a voltage waveform of 7–11.7 μs rise time and a 576–840 μs duration, and a 30% undershoot. For a short circuit voltage, a voltage waveform of 3.5–6.5 μs rise time and a 256–384 μs duration. External couplers are also used for telecommunication lines. For a Telcordia GR-1089 core, specific surges are defined as 2 μs rise time and a 10 μs duration, and a second 10 μs rise time and a 1000 μs duration. Alternate standards exist for the rise time and pulse duration for telecommunication systems.

9.12 Summary and Closing Comments

This chapter addressed system level transient surge concerns in semiconductor development, known as IEC 61000-4-5. This test introduces a transient oscillation. With the growing interest in system level EOS, the interest in this test has emerged recently in semiconductor chip suppliers, to test equipment. The chapter addresses the purpose, the scope, and the pulse waveform. ESD protection concepts for the IEC 61000-4-5 were also discussed.

Chapter 10 addresses a new semiconductor chip level test to addresses a cable discharge event (CDE). This test, the CDE, introduces a fast transient followed by a slower square pulse waveform. With the growing interest of ports and smaller systems, cable discharge events are of increase for both chip developers and system designers.

Problems

1. The IEC 61000-4-5 is a direct contact transient test. How does the IEC 61000-4-5 test compare to the human metal model (HMM) test? What are the differences?
2. What are the differences of the IEC 61000-4-2 test and the IEC 61000-4-5 test? Are the purpose and scope of the tests different?
3. The IEC 61000-4-5 test is an oscillatory waveform. What is the difference between the IEC 61000-4-5 test and the machine model (MM) test?
4. The IEC 61000-4-5 test was developed for transient surge phenomena. How does it differ from a transient latchup (TLU) waveform?
5. Can the IEC 61000-4-5 test cause system level latchup? Why or why not?
6. The human metal model (HMM) was a direct contact version of the IEC 61000-4-2 standard. Is there an equivalent for the IEC 61000-4-5 test?
7. The IEC 61000-4-5 was defined to be used on a system. What are the potential failure mechanisms when injected into a printed circuit board (PCB)?
8. The IEC 61000-4-5 was defined to be used on a system. What are the potential failure mechanisms when injected into a semiconductor chip?
9. Can the IEC 61000-4-5 test induce CMOS latchup inside a chip mounted on a PCB? What are the potential issues of board-chip capacitance coupling and how will that influence the response?

References

1. International Electro-technical Commission (IEC). *IEC 61000-4-5 Electromagnetic Compatibility (EMC): Part 4–5: Testing and Measurement Techniques – Surge Immunity Test*, 2000.
2. D. Pommerenke, and M. Aidam. To what extent do contact mode and indirect ESD test methods reproduce reality? *Proceedings of the Electrical Overstress / Electrostatic Discharge (EOS/ESD) Symposium*, 1995; 101–109.
3. D. Lin, D. Pommerenke, J. Barth, L.G. Henry, H. Hyatt, M. Hopkins, G. Senko, and D. Smith. Metrology and methodology of system level ESD testing. *Proceedings of the Electrical Overstress/Electrostatic Discharge (EOS/ESD) Symposium,* 1998; 29–39.
4. E. Grund, K. Muhonen, and N. Peachey. Delivering IEC 61000-4-2 current pulses through transmission lines at 100 and 330 ohm system impedances. *Proceedings of the Electrical Overstress/Electrostatic Discharge (EOS/ESD) Symposium,* 2008; 132–141.
5. R. Chundru, D. Pommerenke, K. Wang, T. Van Doren, F.P. Centola, J.S. Huang. Characterization of human metal ESD reference discharge event and correlation of generator parameters to failure levels – Part I: reference event. *IEEE Transactions on Electromagnetic Compatibility*, Vol. 46, (4), Nov 2004; 498–504.
6. K. Wang, D. Pommerenke, R. Chundru, T. Van Doren, F.P. Centola, and J.S. Huang. Characterization of human metal model ESD reference discharge event and correlation of generator parameters to failure levels – Part II: correlation of generator parameters to failure levels. *IEEE Transactions on Electromagnetic Compatibility (EMC)*, Vol. 46, (4), Nov 2004; 505–511.
7. ESD Association. ESD-SP 5.6 – 2008. *ESD Association Standard Practice for the Protection of Electrostatic Discharge Sensitive Items – Electrostatic Discharge Sensitivity Testing – Human Metal Model (HMM) Testing Component Level*, Standard Practice (SP) document, 2008.
8. J. Barth, D. Dale, K. Hall, H. Hyatt, D. McCarthy, J. Nuebel, and D. Smith. Measurement of ESD HBM events, simulator radiation, and other characteristics toward creating a more repeatable simulation or simulator should simulate. *Proceedings of the Electrical Overstress/Electrostatic Discharge (EOS/ESD) Symposium*, 1996; 211–222.

9. W. Stadler, S. Bargstadt-Franke, T. Brodbeck, R. Gaertner, M. Goroll, H. Gossner, N. Jensen, and C. Muller. From the ESD robustness of products to system ESD robustness. *Proceedings of the Electrical Overstress/Electrostatic Discharge (EOS/ESD) Symposium*, 2004; 67–74.
10. ESD Association. *Electrostatic Discharge Sensitivity Testing – Human Metal Model (HMM)*, ESD Standard Practice Document, 2009.
11. International Electro-technical Commission (IEC). *IEC 61000-4-2: Electromagnetic Compatibility (EMC) – Part 4–2: Testing and Measurement Techniques – Electrostatic Discharge Immunity Test*, IEC International Standard, 2007.
12. R. Ashton. System level ESD testing – the waveforms. *Conformity Magazine*, Vol. 12, (10), Aug 2007; 28–34.
13. N. Peachey. ESD Open Forum – ESD stressing of devices and components. *Conformity Magazine*, Vol. 13, (6), 2008; 22–23.
14. D. Robinson-Hahn, et al. Evaluating IC components utilizing IEC 61000-4-2. *Proceedings of the International Electrostatic Discharge (ESD) Workshop*, Sept 2008; 274–283.
15. R. Ashton, and L. Lescouzeres. Characterization of off chip protection devices. *Proceedings of the Electrical Overstress/Electrostatic Discharge (EOS/ESD) Symposium*, 2008; 21–29.
16. J. Koo, Q. Cai, K. Wang, J. Maas, T. Takahasi, A. Martwick and D. Pommerenke. Correlation between EUT failure levels and ESD generator parameters. *IEEE Transactions on Electromagnetic Compatibility*, Vol. 50, (4), Nov 2008; 794–801.
17. K. Muhonen, J. Dunihoo, E. Grund, N. Peachey and A. Brankov. Failure detection with HMM waveforms. *Proceedings of the Electrical Overstress/Electrostatic Discharge (EOS/ESD) Symposium*, 2009; 396–404.
18. K. Muhonen, N. Peachey, and A. Testin. Human metal model (HMM) testing challenges to using ESD guns. *Proceedings of the Electrical Overstress/Electrostatic Discharge (EOS/ESD) Symposium*, 2009; 387–395.
19. S. Marum, C. Duvvury, J. Park, A. Chadwick, and A. Jahanzeb. Protecting circuits from transient voltage suppressor's residual pulse during IEC 61000-4-2 stress. *Proceedings of the Electrical Overstress/Electrostatic Discharge (EOS/ESD) Symposium*, 2009; 377–386.
20. M. Scholz, G. Vandersteen, D. Linten, S. Thijs, G. Groeseneken, M. Sawada, T. Nakaei, T. Hasebe, D. LaFonteese, V. Vashchenko, and P. Hopper. On-wafer human metal model measurements for system-level ESD analysis. *Proceedings of the Electrical Overstress/Electrostatic Discharge (EOS/ESD) Symposium*, 2009; 405–413.
21. T. Vheriakoski, J. Hilberg, and L. Sillanpaa. ESD event receiver for system level testing. *Proceedings of the Electrical Overstress/Electrostatic Discharge (EOS/ESD) Symposium*, 2009; 414–426.
22. Haefely Corporation. http://www.haefely-hiptronics.com. Haefely AXOS 8 EMC immunity test system, 2016.
23. Haefely Corporation. http://www.haefely-hiptronics.com. Haefely PSURGE 8000 8 KV surge system, 2016.
24. Teseq Corporation. http://www.teseq.com. Teseq NSG3060 6 kV surge generator, 2016.
25. Thermo Fisher Scientific. http://www.thermofisher.com. Thermo KeyTek ECAT 10 kV surge generator, 2016.
26. Thermo Fisher Scientific. http://www.thermofisher.com. EMCPro Plus™ EMC test system, 2016.
27. ESD Association. HMM – system level ESD pulses to components: application and interpretation. Workshop C.4, *Proceedings of the Electrical Overstress/Electrostatic Discharge (EOS/ESD) Symposium*, 2009; 436.
28. M. Hernandez. Electrical overstress, and transient latchup pulse generator system, circuit, and method. U.S. Patent No. 7,928,737, April 19, 2011.
29. S. Voldman. *ESD: Physics and Devices*. Chichester, England: John Wiley and Sons, Ltd., 2004.
30. S. Voldman. *ESD: Circuits and Devices*. Chichester, England: John Wiley and Sons, Ltd., 2005.
31. S. Voldman. *ESD: Circuits and Devices 2nd Edition*. Chichester, England: John Wiley and Sons, Ltd., 2015.
32. S. Voldman. *ESD: RF Circuits and Technology*. Chichester, England: John Wiley and Sons, Ltd., 2006.
33. S. Voldman. *Latchup*. Chichester, England: John Wiley and Sons, Ltd., 2007.
34. S. Voldman. *ESD: Circuits and Devices*. Beijing, China: Publishing House of Electronic Industry (PHEI), 2008.
35. S. Voldman. *ESD: Failure Mechanisms and Models*. Chichester, England: John Wiley and Sons, Ltd., 2009.
36. S. Voldman. *ESD: Design and Synthesis*. Chichester, England: John Wiley and Sons, Ltd., 2011.
37. S. Voldman. *ESD Basics: From Semiconductor Manufacturing to Product Use*. Chichester, England: John Wiley and Sons, Ltd., 2012.
38. S. Voldman. *Electrical Overstress (EOS): Devices, Circuits, and Systems*. Chichester, England: John Wiley and Sons, Ltd., 2013.
39. S. Voldman. *ESD: Analog Design and Circuits*. Chichester, England: John Wiley and Sons, Ltd., 2014.
40. C. Duvvury, and H. Gossner. *System Level ESD Co-Design*. Chichester, England: John Wiley and Sons, Ltd., 2015.

10

Cable Discharge Event (CDE)

10.1 History

With the static charging of transmission line cables, electronic systems have been vulnerable to the current pulse that comes from cable events.

Historically, cable discharge event (CDE) due to large computer system was a problem. In large mainframes, cables extended to the length of a room. When dragged on the floor, cables were charged due to triboelectric charging. When inserted into the ports of a large system, the cable would discharge into the port, leading to system failure. This was solved in early times by handling procedures of system operators to provide system level operation control. In these systems, "touch pads" were placed on the chassis of large systems, where the operator would discharge the cable during installation.

In the world today, the assembly, installation, and handling of cables are not a controlled process. This leads to the requirement of ESD solutions embedded in the systems, from the connectors, printed circuit boards (PCB), semiconductor components, and systems. With the growth of Ethernet networks, Ethernet interfaces require protection.

Ethernet network connections contain unshielded twisted pair (UTP). The telecommunication industry introduced registered jack (RJ) for connectors. An RJ is a standardized telecommunication network interface for connecting voice and data equipment.

Since 2000, it was clear that there was a need to develop a standard for CDE, leading to the document ESD Association. DSP 14.1 – 2003. *ESD Association Standard Practice for the Protection of Electrostatic Discharge Sensitive Items – System Level Electrostatic Discharge Simulator Verification Standard Practice.* Standard Practice (SP) document, 2003 [1]. In addition, there was a requirement to develop a measurement process for a cable, ESD Association. DSP 14.3 – 2006 *ESD Association Standard Practice for the Protection of Electrostatic Discharge Sensitive Items – System Level Cable Discharge Measurements Standard Practice.* Standard Practice (SP) document, 2006 [2] as well as ESD Association, DSP 14.4 – 2007, *ESD Association Standard Practice for the Protection of Electrostatic Discharge Sensitive Items – System Level Cable Discharge Test Standard Practice.* Standard Practice (SP) document, 2007 (Figure 10.1) [3].

ESD Testing: From Components to Systems, First Edition. Steven H. Voldman.
© 2017 John Wiley & Sons, Ltd. Published 2017 by John Wiley & Sons, Ltd.

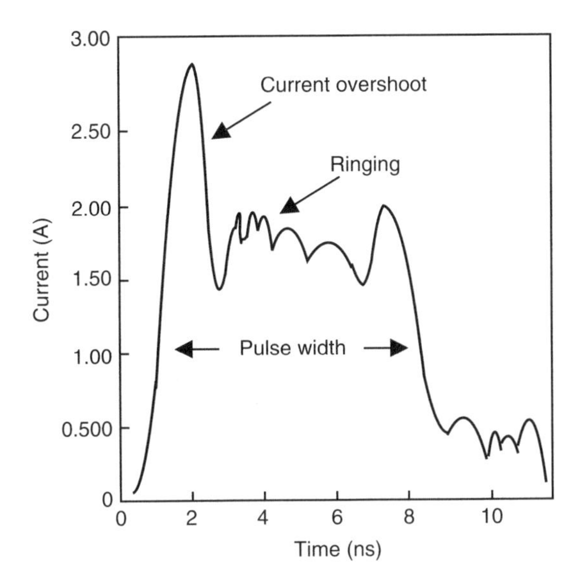

Figure 10.1 Cable discharge event (CDE)

10.2 Scope

The scope of the CDE electrostatic discharge (ESD) test is for the testing, evaluation, and classification of cables. The test is to quantify the sensitivity or susceptibility of these components or systems to damage or degradation to the defined CDE test [1–3].

10.3 Purpose

The purpose of the CDE ESD test is for the establishment of a test methodology to evaluate the repeatability and reproducibility of systems to a defined cable discharge pulse event in order to classify or compare ESD sensitivity levels of systems [1–3].

10.4 Cable Discharge Event – Charging, Discharging, and Pulse Waveform

CDEs can impact the cable, the connector termination, the PCB, and internal components [4–12]. For example, a UTP CDE can discharge a large current due to its low source impedance. Other ESD models, such as human body model (HBM), machine model (MM), and charged device model (CDM), have high impedance. As a result, the charge from a cable discharge can be significant. In many cases, a CDE can exceed the latchup robustness of the components that are mounted on the PCB [13, 14].

10.4.1 *Charging Process*

The charging process of a cable can occur from two sources. A common source of charging is triboelectric charging between the cable and a ground surface. Triboelectric charge transfer can occur due to physical contact between the cable and flooring, conduits, or adjacent cables.

As an example, a Category 5 (CAT5) UTP cable can be placed on a floor or ground surface (Figure 10.2(a)). Figure 10.2 shows an example of only one pair of connectors. Dragging the cable across the floor can lead to charge transfer leading to charge accumulation on the cable when the cable is unterminated. The amount of charge transfer is dependent on the two materials in the triboelectric charge transfer process, cable characteristics, and the length of the cable. The charge will remain in the cable until leakage occurs, or termination. CAT5 and CAT6 UTP cables have low dielectric leakage characteristics, leading to storage of the charge along time. Environmental conditions (e.g., relative humidity) can also influence the storage time of the accumulated charge [6].

A second source of charging is electromagnetic induction. Electromagnetic induction can charge the cable without physical contact or triboelectric effect.

10.4.2 *Discharging Process*

To demonstrate the CDE, a telecommunication system is used as an example where the circuit is a media-dependent interface (MDI) transceiver. Figure 10.2(b) shows the CAT5 cable and its associated ground plane. The UTP connections are attached to the registered jack 45 (RJ45) connectors. An isolation transformer is connected on the chassis side. The input of the isolation transformer is connected to a Bob Smith termination network [19]. The CDE can follow many paths in the network leading to a failure in the system [6].

10.4.3 *Pulse Waveform*

A CDE forms a square pulse waveform. Figure 10.2(c) provides an example of a typical CDE. The cable event typically has a fast rise time and an overshoot at the beginning of the event. The waveform settles to a plateau. The plateau of the waveform has some structure associated with ringing. The length of the pulse event is a function of the length of the cable. The current magnitude is associated with the accumulated charge in the cable [1].

10.4.4 *Comparison of CDE and IEC 61000-4-2 Pulse Waveform*

The CDE and the IEC 61000-4-2 test are used on systems in the automobile industry. Comparing the cable discharge model to the IEC 61000-4-2, the CDE waveform and the IEC 61000-4-2 both pulses contain a fast event at the onset of the pulse. But the IEC 61000-4-2 exhibits an HBM-like waveform with a higher series impedance [13].

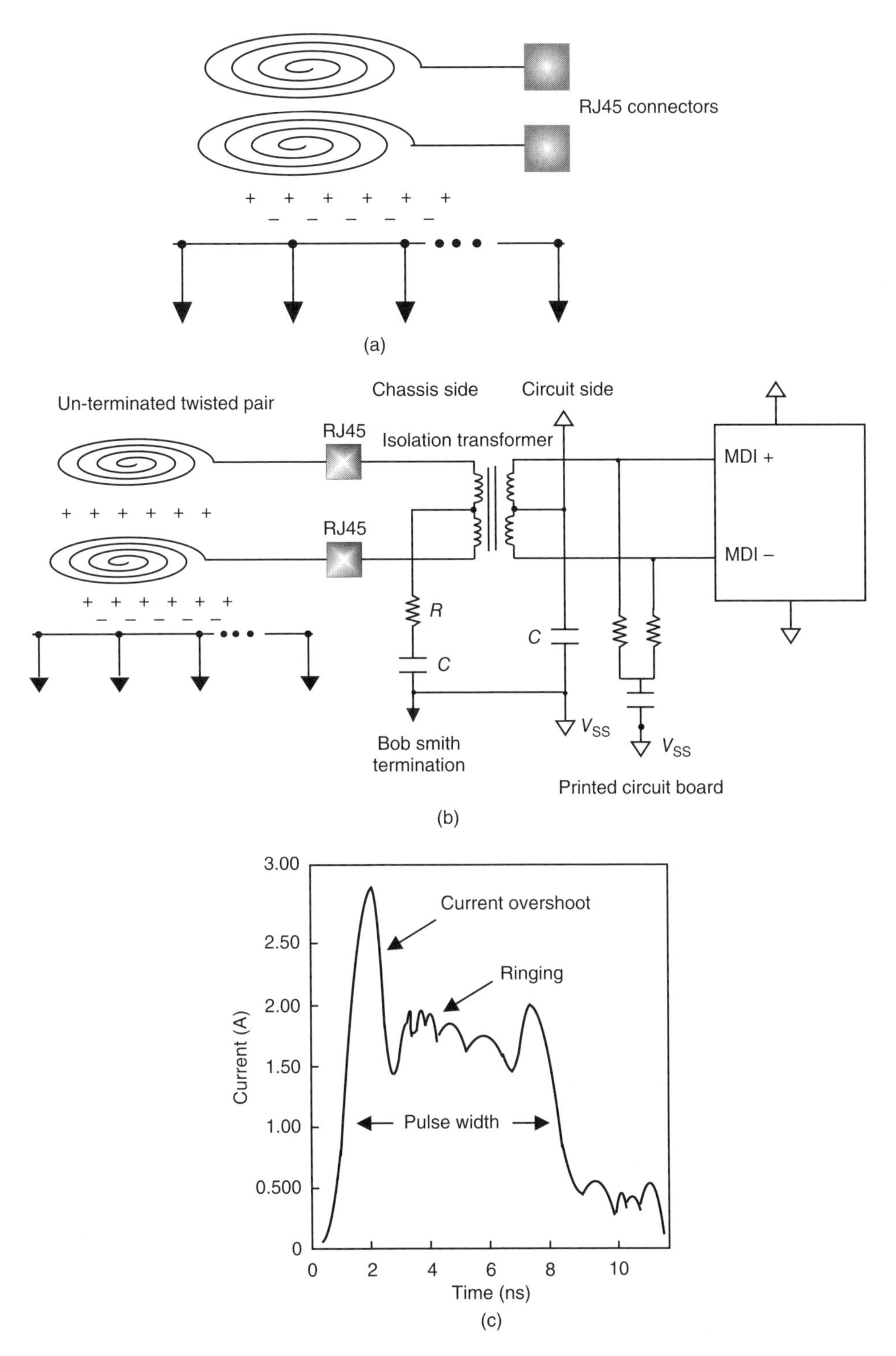

Figure 10.2 (a) Charging process on Category 5 (CAT 5) cable. (b) Cable discharge event (CDE) discharging process and example test setup. (c) Cable discharge event (CDE) model pulse waveform

10.5 Equivalent Circuit

An equivalent circuit can be defined that represents the CDE forms a square pulse waveform. The CDE can be represented as an inductor–capacitor transmission line (Figure 10.3(a)). The incremental model has an inductor per unit length between the input and the output of the cable. The incremental model of the cable is also represented by a capacitance per unit length. As an individual circuit element, this can be represented as a transmission line symbol.

The CDE can be represented as an inductor–resistor–capacitor transmission line (Figure 10.3(b)). The incremental model has an inductor and a series resistor per unit length between the input and the output of the cable. The incremental model of the cable also is represented by a capacitance per unit length. As an individual circuit element, this can be represented as a lossy transmission line symbol. In these test systems, the losses associated with the resistance per unit length is low.

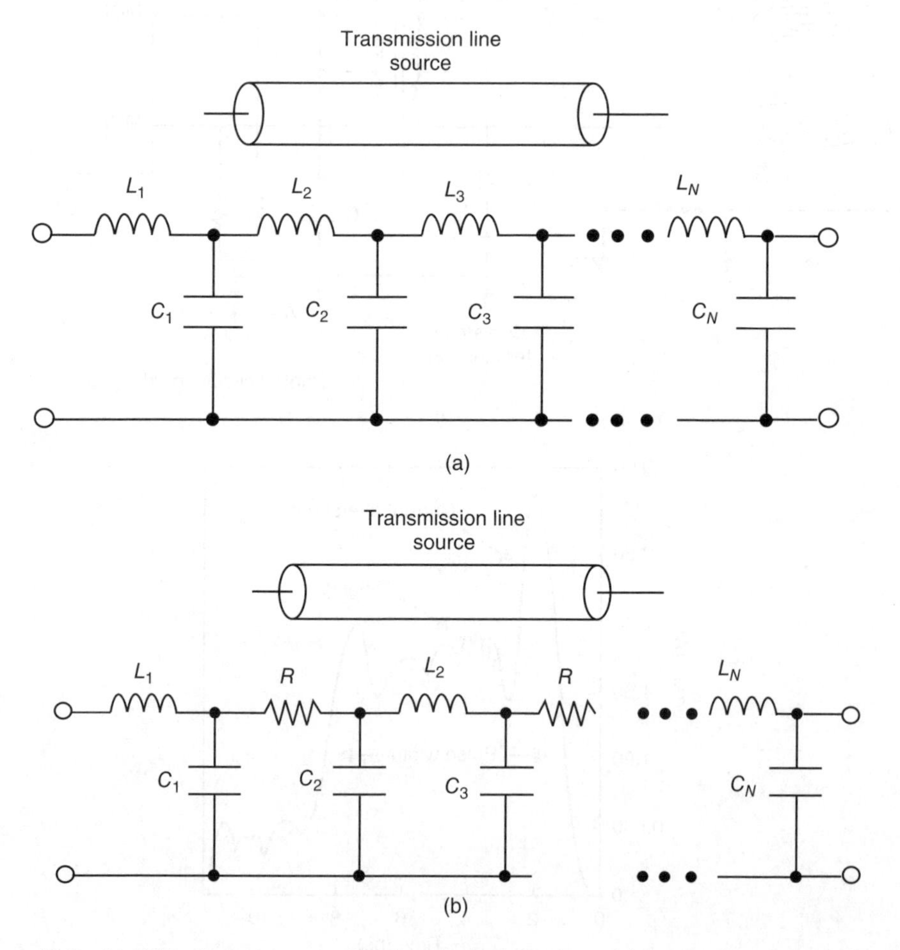

Figure 10.3 (a) Cable discharge event (CDE) equivalent circuit model as a lossless transmission line. (b) Cable discharge event (CDE) equivalent circuit with parasitics as a lossy transmission line

10.6 Test Equipment

Commercial test systems exist for CDE evaluation. Commercial test systems were developed for standardization of the cable discharge test. These are discussed in the following sections.

10.6.1 Commercial Test Systems

Commercial test systems exist for CDE evaluation such as the ESDEMC Corporation test system. The ESDEMC Model ES631-LAN CDE Evaluation System is shown in Figure 10.4 [21]. The ES631-LAN CDE Evaluation System is a comprehensive test system designed to simulate the effects of charged Ethernet cable discharge to any Ethernet-based network system or device. The system allows for charging of the Ethernet cable to 4 kV, or differentially 2 kV. The system includes embedded ESD current probes for discharging lines. The ES-631 system allows for four different test modes [21]:

- Single line discharge mode
- Differential pair discharge mode
- Contact sequence discharge mode
- "Through traffic" running mode.

This system allows the operator to be able to set up and analyze the CDE in different combinations. It also provides a "traffic" mode to evaluate through the CDE system itself. The test system allows for evaluation of different applications [21]:

- Ethernet CDE and analysis

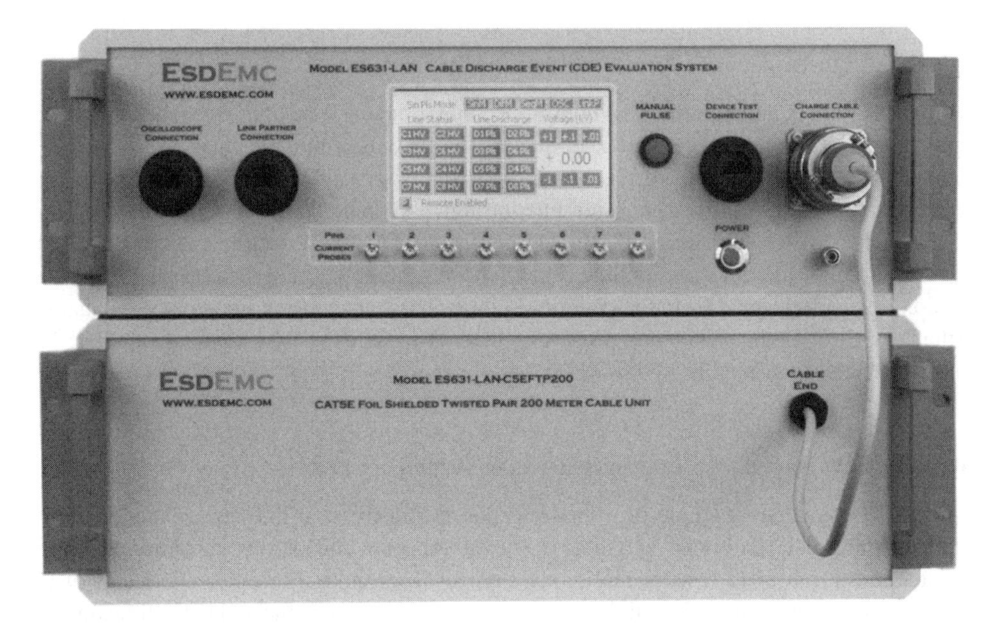

Figure 10.4 Cable discharge event (CDE) commercial test system – ESDEMC ES631 – LAN cable discharge event (CDE) evaluation system (Reproduced with permission of ESDEMC Technology LLC)

- PHY ESD protection test and analysis
- Network switch/router ESD robustness test and analysis
- Automatic test equipment (ATE) for Ethernet system ESD

10.7 Test Measurement

For the CDE, the test equipment, sequence, and procedures are important for the measurement process. These are discussed in the following sections.

10.7.1 Measurement

For the CDE, the test equipment is important for the measurement process. Figure 10.5(a) shows an example of a pulse generator serving as the cable source. The pulse generator has different pulse lengths for testing. In addition, the test system requires a fast pulse head module and an attenuator. Figure 10.5(b) shows the target and target adapter [2].

For the CDE, measurement of the cable characteristics is important to the evaluation of the impact to the device under test (DUT). For the CDE, measurement of the cable characteristics is important to the evaluation of the impact to the DUT.

10.7.2 Measurement –Transmission Line Test Generators

For the CDE, transmission line test generators exist for different impedance levels.

10.7.2.1 High-Impedance Transmission Line Test Generator

Figure 10.6 shows an example of a high-impedance transmission line test generator. In the figure, a transmission line of length L is shown. The far end of the transmission line is unterminated. The outer conductor of the transmission line is grounded. A high-voltage charging source is connected to the transmission line through a switch and a high-impedance resistor element. The resistor element is 10 MΩ. During the charging process, the switch is closed to allow charging of the transmission line cable. After the completion of the charging, the transmission line is decoupled from the charging source. The transmission line is then discharged to the DUT through the switch and resistor network. The series resistor element is 300 Ω. The resistor element to ground is 58–60 Ω [2].

10.7.2.2 Low-Impedance Transmission Line Test Generator

Figure 10.7 shows an example of a low-impedance transmission line test generator. In the figure, a transmission line of length L is shown. The far end of the transmission line is terminated using a 50 Ω resistor and diode element to ground. The outer conductor of the transmission line is grounded. A high-voltage charging source is connected to the transmission line through a switch and a high-impedance resistor element. The resistor element is 10 MΩ. During the charging process, the switch is closed to allow charging of the transmission line

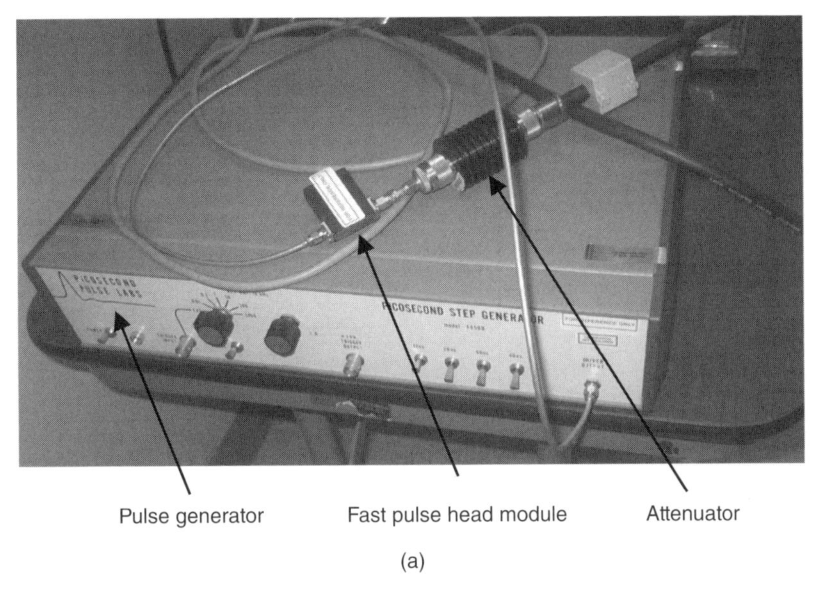

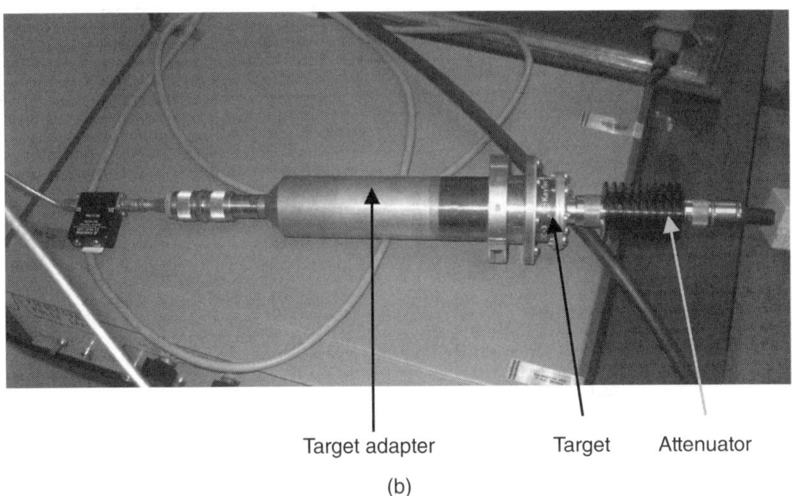

Figure 10.5 (a) Test equipment. (b) Test equipment target and adapter

cable. After the completion of the charging, the transmission line is decoupled from the charging source. The transmission line is then discharged to the DUT through the switch [1–3].

10.7.3 Measurement – Low-Impedance Transmission Line Waveform

An example of a waveform shows the characteristics of the CDE. Figure 10.8(a) is a waveform from the low-impedance transmission line tester. The characteristics of the waveform involve

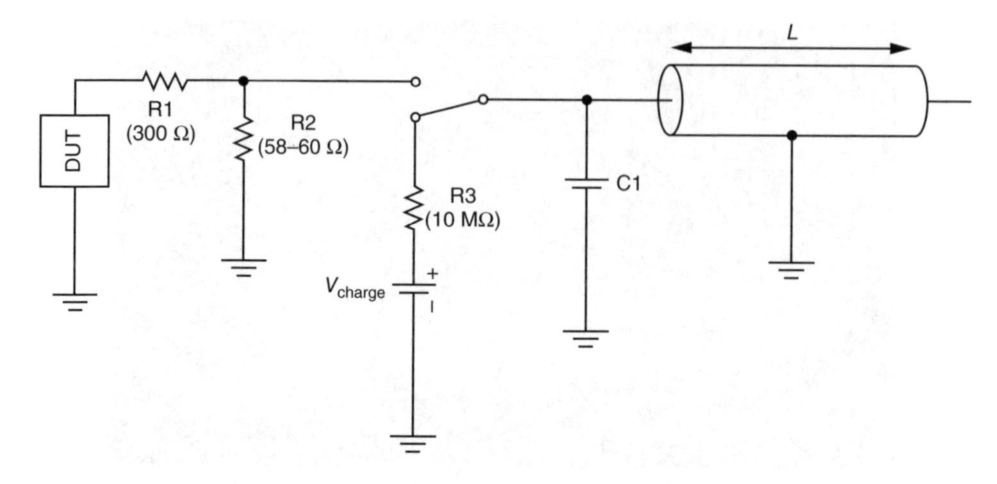

Figure 10.6 Cable discharge event (CDE) measurement – high-impedance test generator

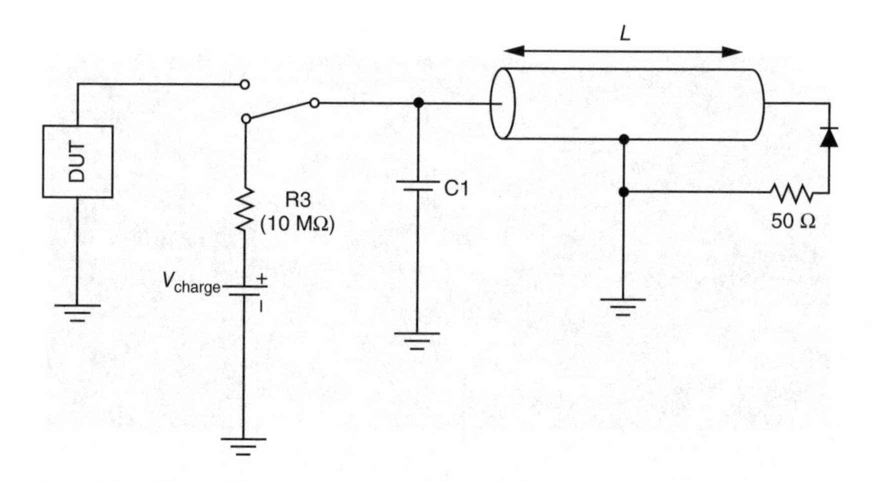

Figure 10.7 Cable discharge event (CDE) measurement – low-impedance test generator

a fast rise time, overshoot, and oscillation at the beginning of the pulse event. The waveform is followed by a voltage plateau until the end of the pulse event [1–3].

10.7.4 Schematic Capturing System Response to Reference Waveform

In the test system, to obtain highly accurate understanding of the system response, an ESD target and target adapter is key to the measurement system. Figure 10.8(b) shows a high-level

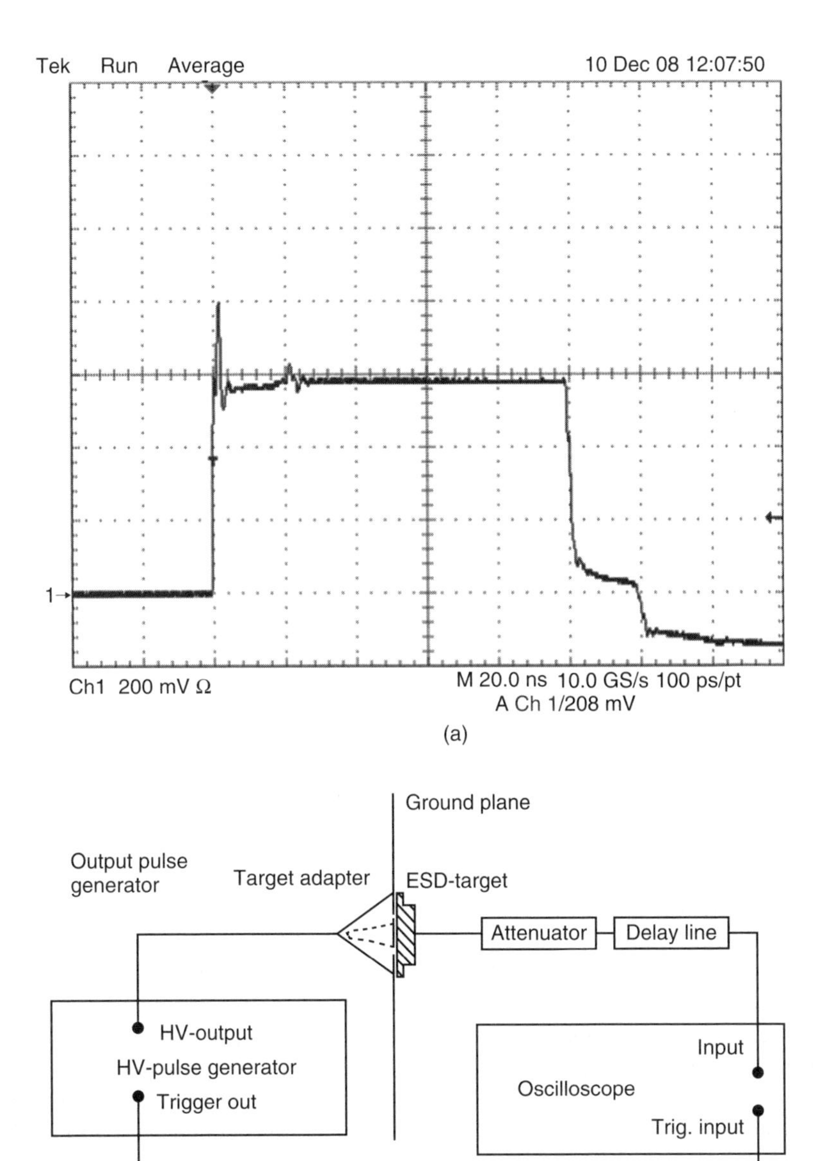

Figure 10.8 (a) Low-impedance transmission line waveform. (b) Capturing system response to reference waveform. (c) Capturing system response – alternate current transducer. (d) Capturing system response – alternate current transducer

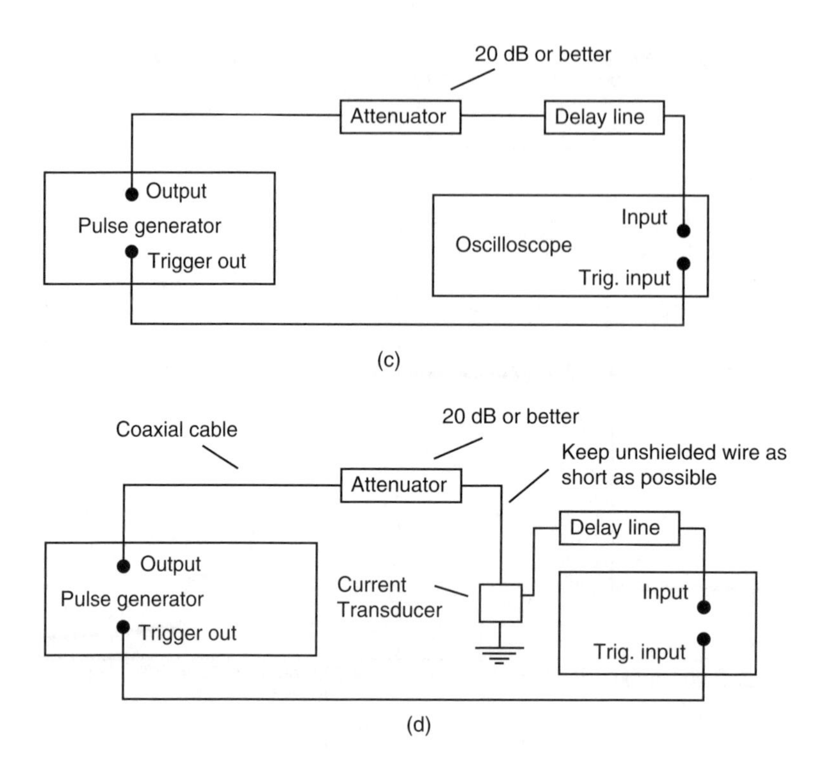

Figure 10.8 (*continued*)

schematic for capturing the system response to the reference waveform. In this case, a high voltage pulse generator is used to establish an output pulse. A high-quality target adapter is connected to the high voltage pulse generator output. The ESD target is connected to a ground plane. The input of the target adapter is connected to an attenuator and a delay line. The delay line is connected to the input of a high-speed oscilloscope. The output trigger of the high voltage pulse generator is coupled to the oscilloscope trigger input [1–3].

In the test system, an alternate current transducer method is used to capturing the reference waveform. Figure 10.8(c) shows a high-level schematic for capturing the system response to the reference waveform. In this case, a high voltage pulse generator is used to establish an output pulse. The pulse generator is connected to an attenuator and a delay line. The delay line is connected to the input of a high-speed oscilloscope. The output trigger of the high voltage pulse generator is coupled to the oscilloscope trigger input.

Figure 10.8(d) shows a high-level schematic for capturing the system response to the reference waveform where a current transducer is added to the network. A coaxial cable is introduced between the pulse generator and attenuator. The pulse generator is connected to an attenuator and a delay line. The delay line is connected to the input of a high-speed oscilloscope. The output trigger of the high voltage pulse generator is coupled to the oscilloscope trigger input. It is recommended that the unshielded wire between the attenuator and current transducer is made as short as possible.

10.7.5 *Tapered Transmission Lines*

For standards development, it is important to provide the highest quality of understanding possible. To obtain high-quality measurements of the CDE, sophisticated measurement equipment is used to define a 50 Ω transmission line. High-quality transmission lines and targets are specifically designed in fine details. Figure 10.9(a) shows an example of a constant impedance tapered transmission line adapter. Figure 10.9(b) shows a cross-sectional view of the tapered transmission line. In the design, the angles of the transmission line are specified. The location of the mechanical tapered are also specified at the two ends of the transmission line. Figure 10.9(c) shows a commercial 50 Ω transmission line target adapter (developed by the Barth Electronics Corporation) [30, 31].

10.7.6 *ESD Current Sensor*

For standards development, it is important to provide the highest quality of understanding possible in the design of the ESD current sensor. To obtain high-quality measurements of the CDE, the ESD current sensor construction has highly specified dimensions. Figure 10.10(a) shows the ESD current sensor physical dimensions of top view, bottom view, and cross section [1–3].

Figure 10.10(b) shows the construction elements of the ESD current sensor. This includes the SMA connector, the central brass part, the PTFE part I, SMD-resistor element, cover, top part of center conductor, and the PTFE part II. The diagram shows the placement and screw connections for assembly of the ESD current sensor [1–3].

Figure 10.10(c) shows cross-sectional views of the construction elements of the ESD current sensor in finer details. This includes the PTFE-part I, PTFE-part II, SMA connector, the center conductor, cover, top part of center conductor, and the SMA connector [1–3].

10.8 Test Procedure

Figure 10.11(a) shows the CDE test procedure. In the test procedure, the switch to the DUT must be opened prior to charging the cable. The switch between the high-voltage charging source and the cable is closed. The cable is then charged to a specified voltage level. The charged cable switch is then changed to apply the CDE to the DUT. The channels of the DUT are then evaluated. Depending on the failure criteria, the test is repeated and testing is continued to a higher voltage level.

In test systems for differential inputs, the test procedure can include introduction of different delays to the differential inputs using a delay control network.

10.9 Measurement of a Cable in Different Conditions

Measurement of a cable discharge is also a function of the test configuration and conditions. In the following section, the test configuration, the conditions, procedure, and sequence are discussed [30, 34].

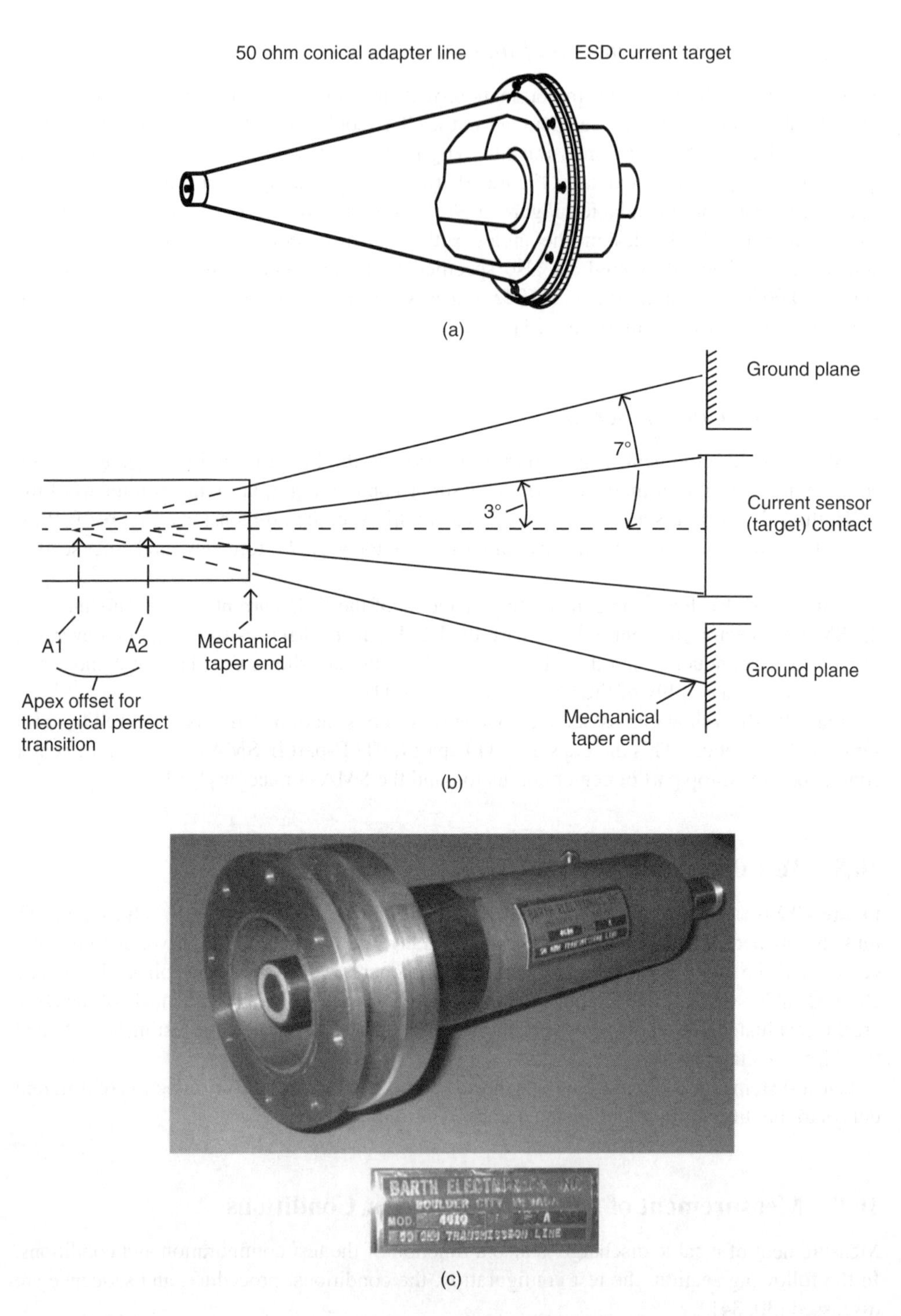

Figure 10.9 (a) Constant impedance tapered transmission line. (b) Constant impedance tapered transmission line. (c) Tapered transmission line (Reproduced with permission from Barth Electronics Corp.)

10.9.1 Test System Configuration and Diagram

Figure 10.11(b) shows a test configuration and diagram. The diagram addresses a key issue in cable measurement [30, 34]:

- Floor configuration

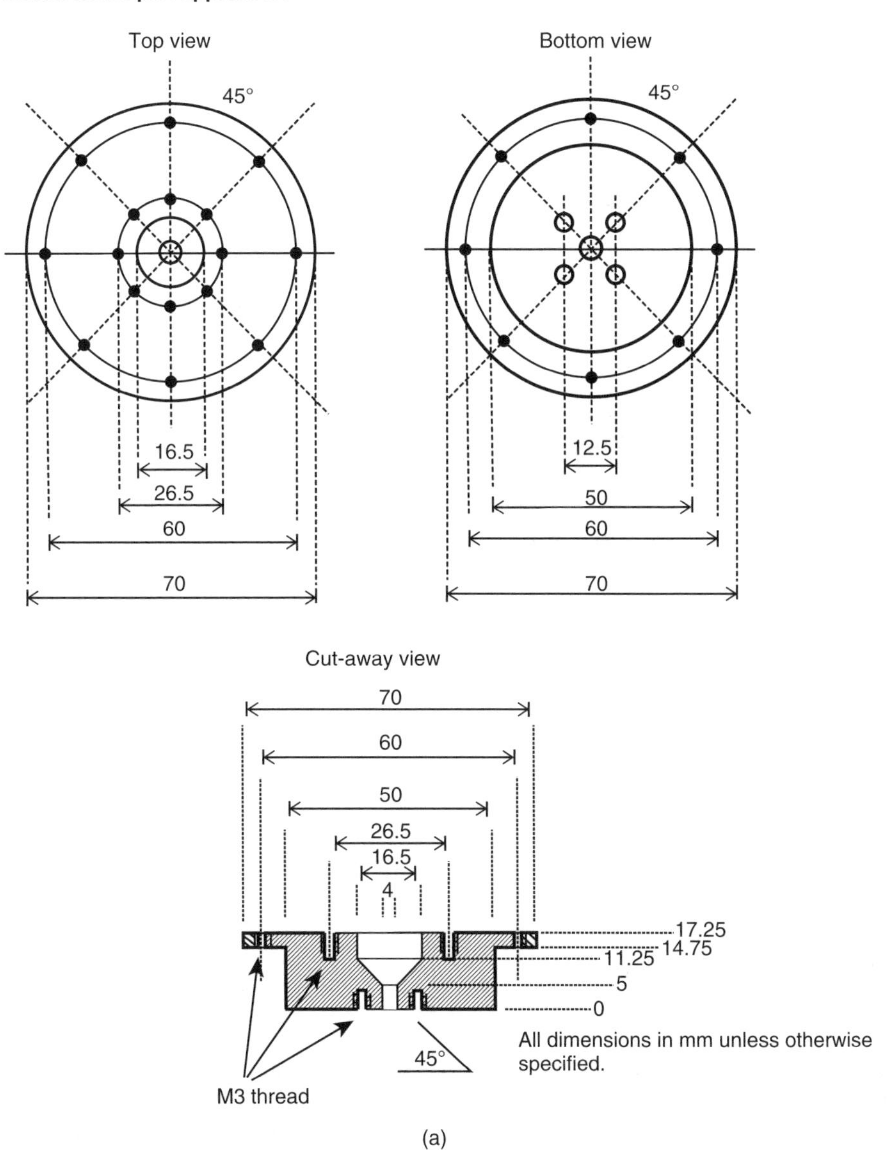

Figure 10.10 (a) ESD current sensor. (b) ESD current sensor. (c) ESD current sensor (Reproduced with permission from ESD Association)

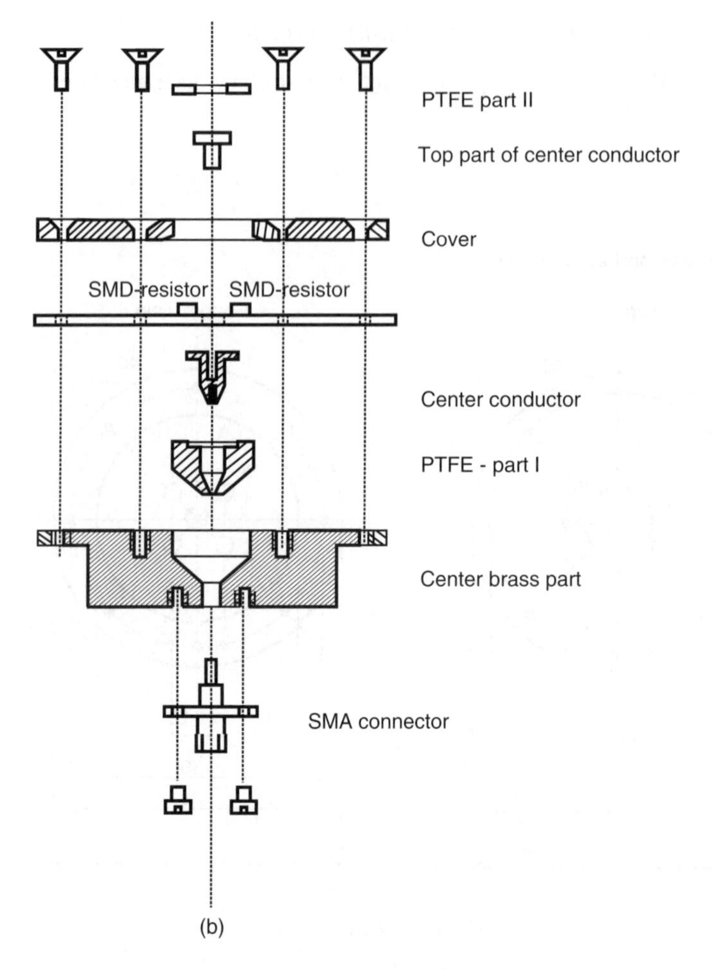

Figure 10.10 (*continued*)

- Floor material
- Ground plane configuration
- Cable type (e.g., UTP, CAT4, CAT5)
- Length of the cable
- Configuration of the cable (held, coiled, etc.)
- Cable-to-ground plane conditions
- Target.

The figure shows an example of a cable test configuration. A vertical ground plane and horizontal floor is shown. A 100 m cable is placed on the linoleum tile flooring. The first meter of the cable is raised near the vertical wall. An insulator is used to hold the cable and place the end of the cable near the target. An oscilloscope is connected to the target.

PTFE-part I approx. 1:3

16.5
10.5
5
12.25
11.25
6.25
5.25
45°
0
Part has symmetry of rotation
90°
4

Center conductor, brass approx. 1:3

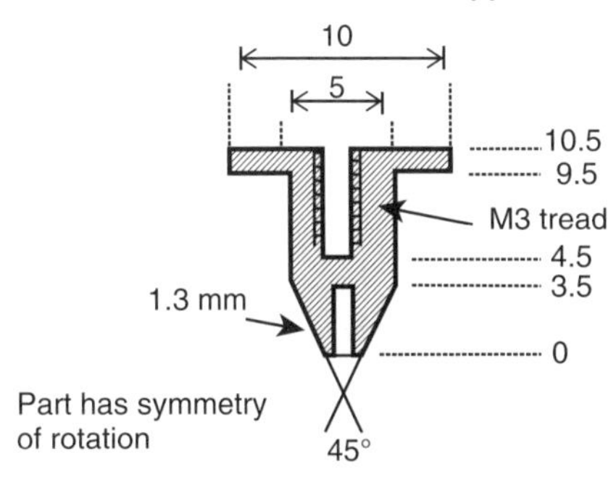

Top part of center conductor. stainless steel, approx. 1:3

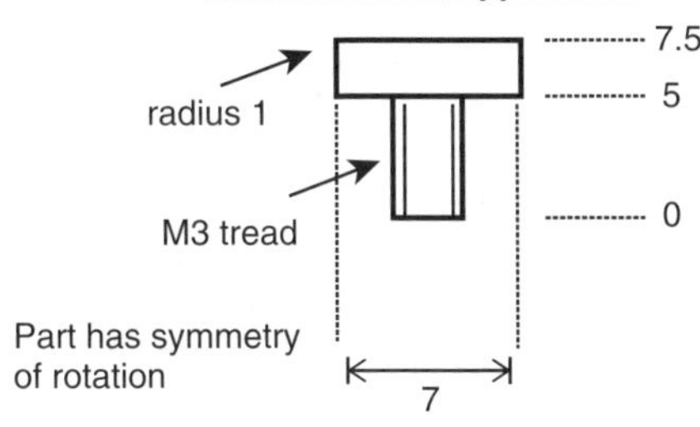

PTFE-part II, approx. 1:3

Top view

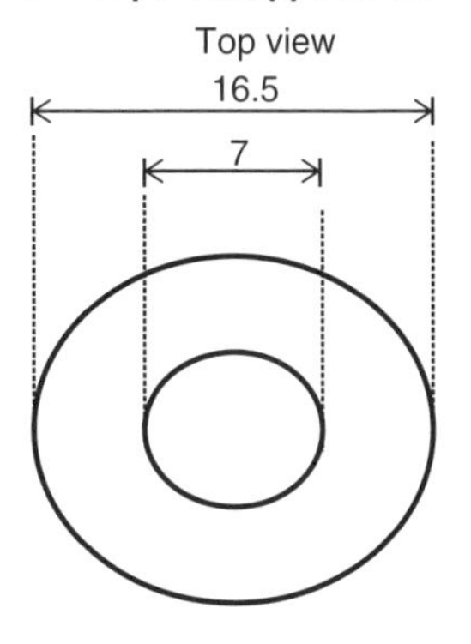

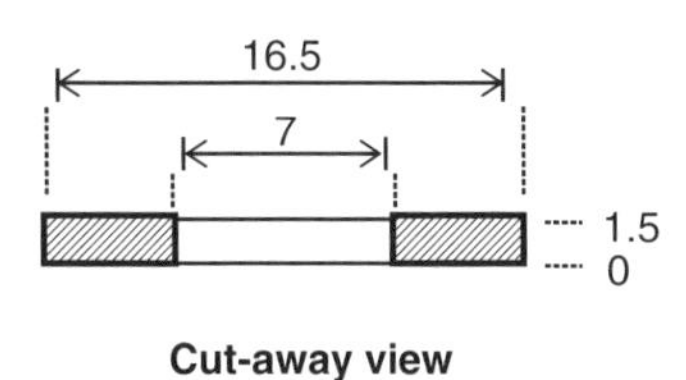

SMA-connector

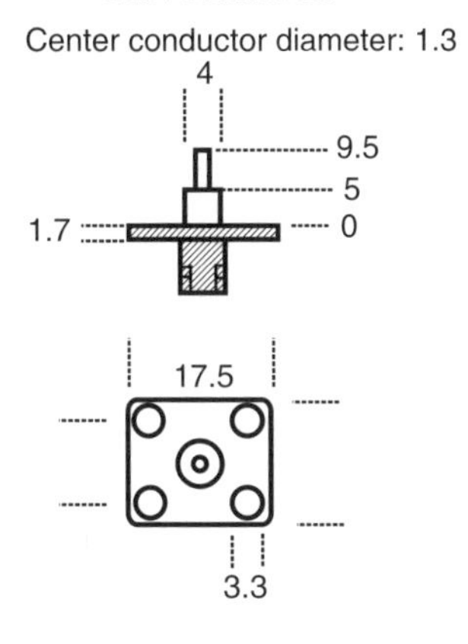

(c)

Figure 10.10 *(continued)*

10.9.2 Cable Configurations – Handheld Cable

The cable signal response is altered by the configuration. In this section, the pulse event from a handheld cable is shown. The first section of the cable is free floating and uncoiled

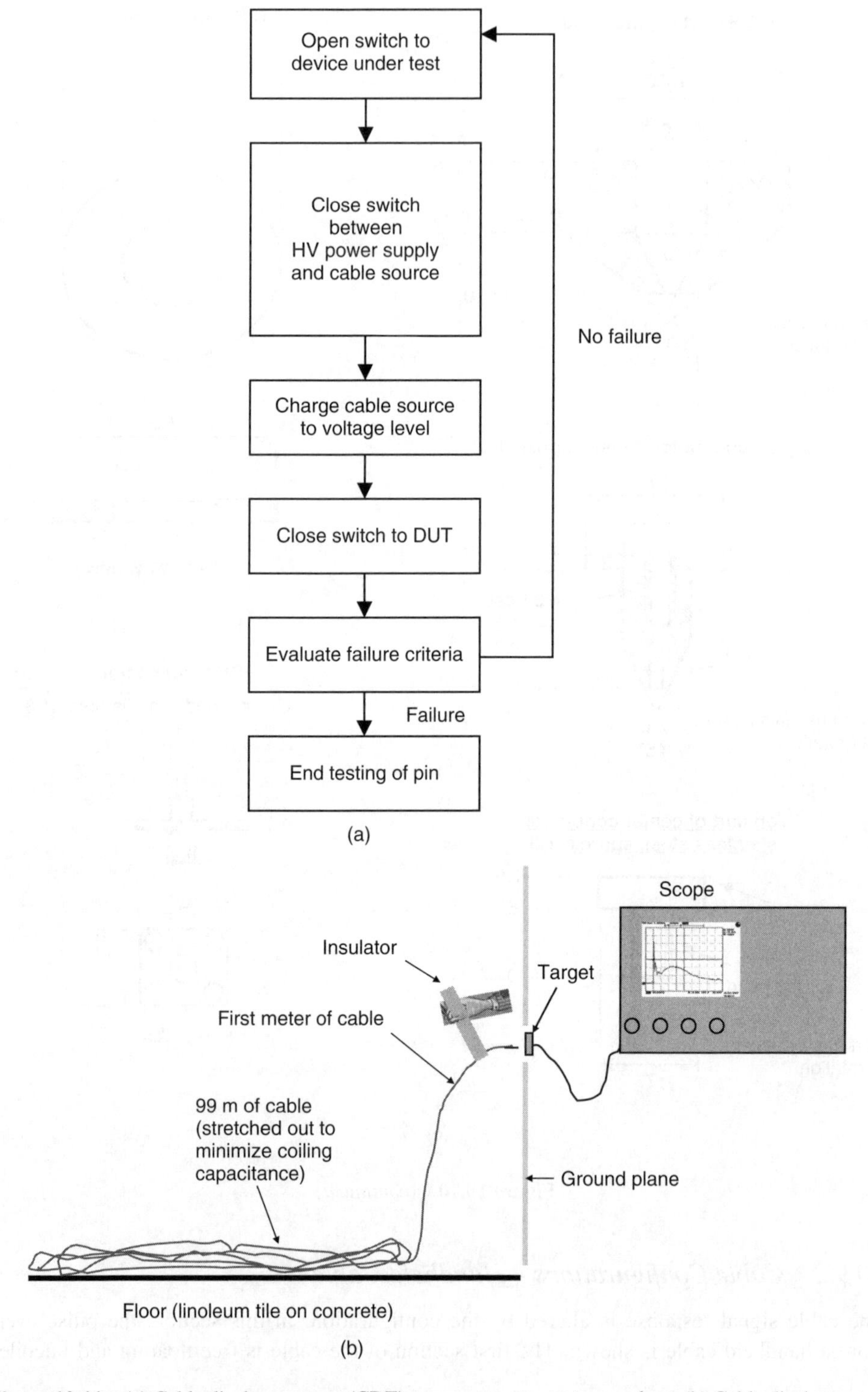

Figure 10.11 (a) Cable discharge event (CDE) measurement – test procedure. (b) Cable discharge test configuration (Reproduced with permission from ESD Association)

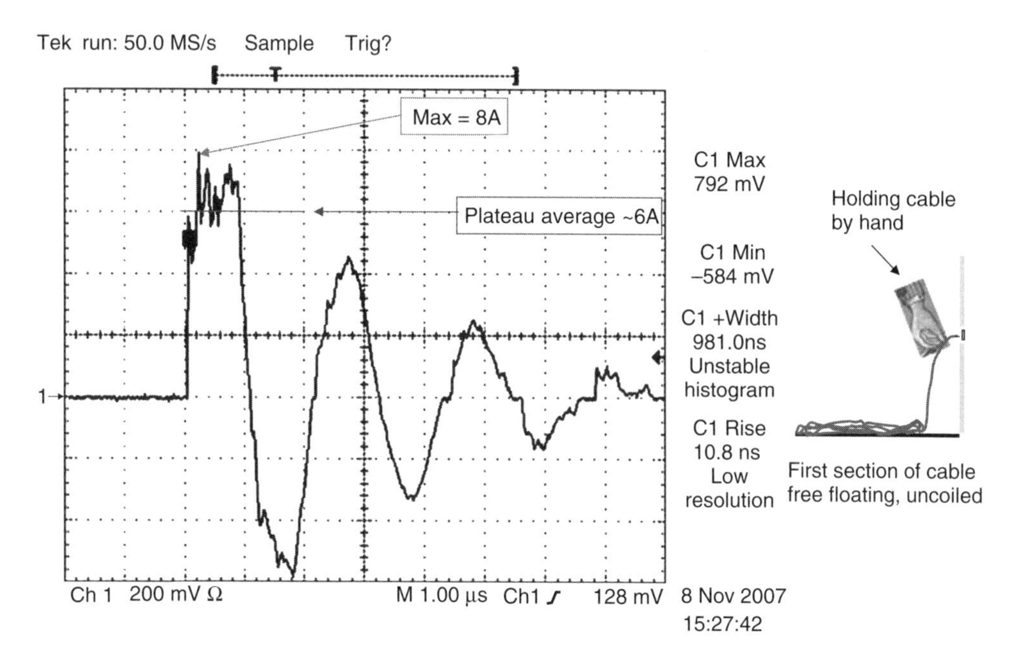

Figure 10.12 Cable discharge event (CDE) measurement – holding cable (Reproduced with permission from ESD Association)

(Figure 10.12). The pulse waveform is shown as well. A vertical ground plane and horizontal floor is shown. A 100 m cable is placed on the linoleum tile flooring. The first meter of the cable is raised near the vertical wall. No insulator is used to hold the cable and place the end of the cable near the target. An oscilloscope is connected to the target [34].

10.9.3 Cable Configuration – Taped to Ground Plane

In this section, the cable is taped to the vertical ground plane. The first section of the cable is taped to the ground plane, and the rest of the cable is coiled (Figure 10.13(a)). An insulator is used to hold the cable and place the end of the cable near the target. An oscilloscope is connected to the target. A peak current of 11.8 A is shown [30, 34].

Figure 10.13(b) shows a comparison of the cable discharge of a taped versus floating cable. The cable discharge of the initial peak current for the case of the cable taped to the wall is larger than the floating cable.

10.9.4 Cable Configurations – Pulse Analysis Summary

The initial peak current can be shown related to the initial charging voltage. Figure 10.14(a) shows the correlation between the peak current and the charging voltage.

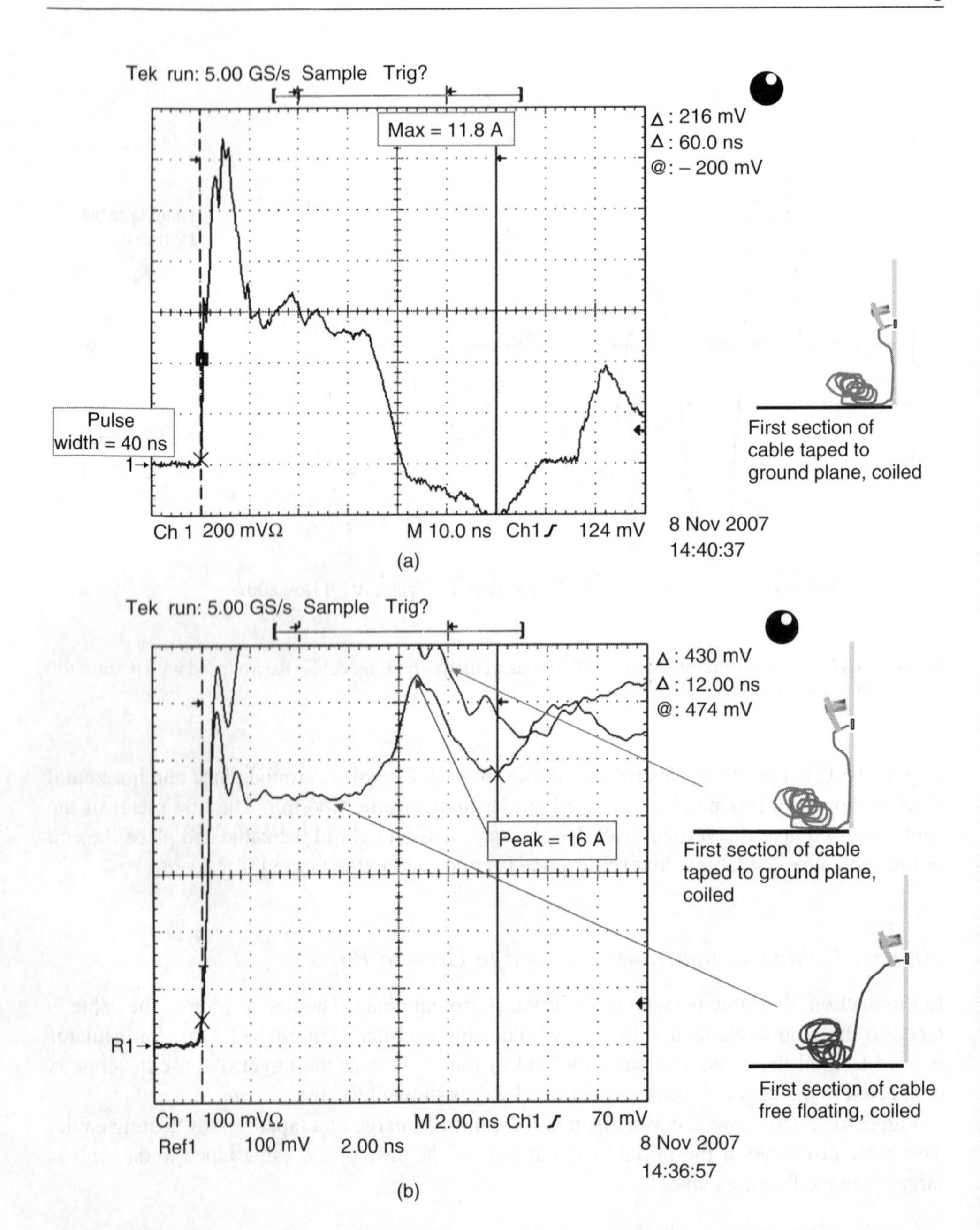

Figure 10.13 (a) Cable discharge event (CDE) measurement – taped to vertical wall. (b) Cable discharge event (CDE) measurement – taped to vertical wall versus floating cable (Reproduced with permission from ESD Association)

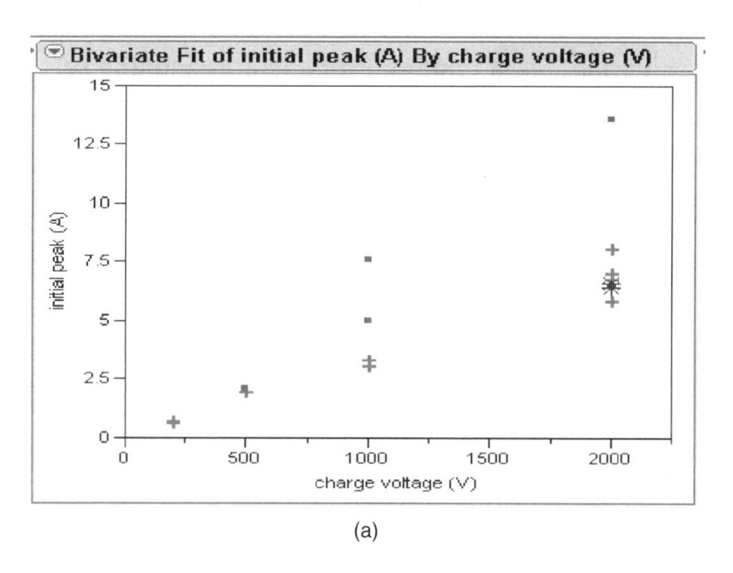

(a)

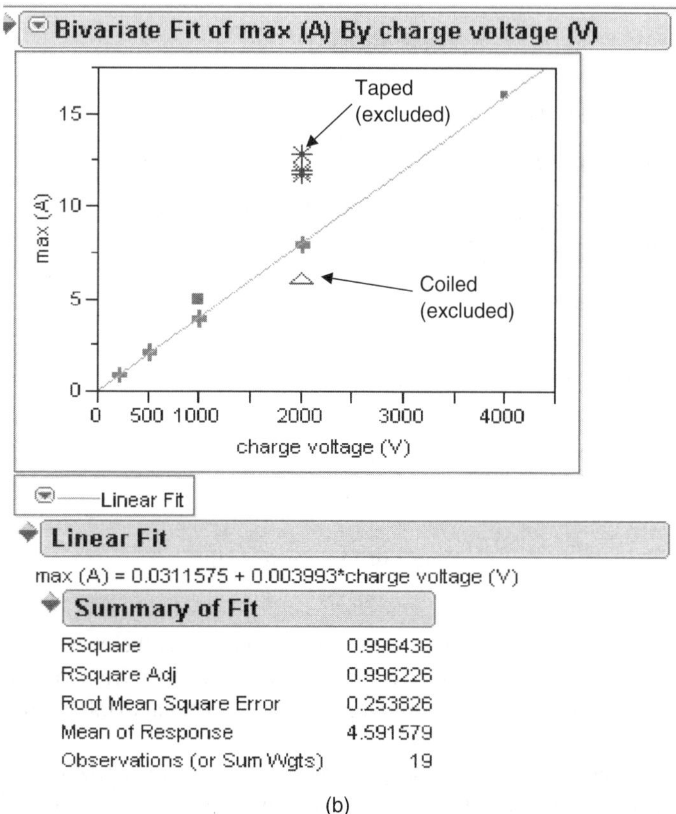

(b)

Figure 10.14 (a) Cable discharge event (CDE) measurement – pulse analysis summary. (b) Cable discharge event (CDE) measurement – pulse analysis summary – initial peak current. (c) Cable discharge event (CDE) measurement – pulse analysis summary – plateau current (Reproduced with permission from ESD Association)

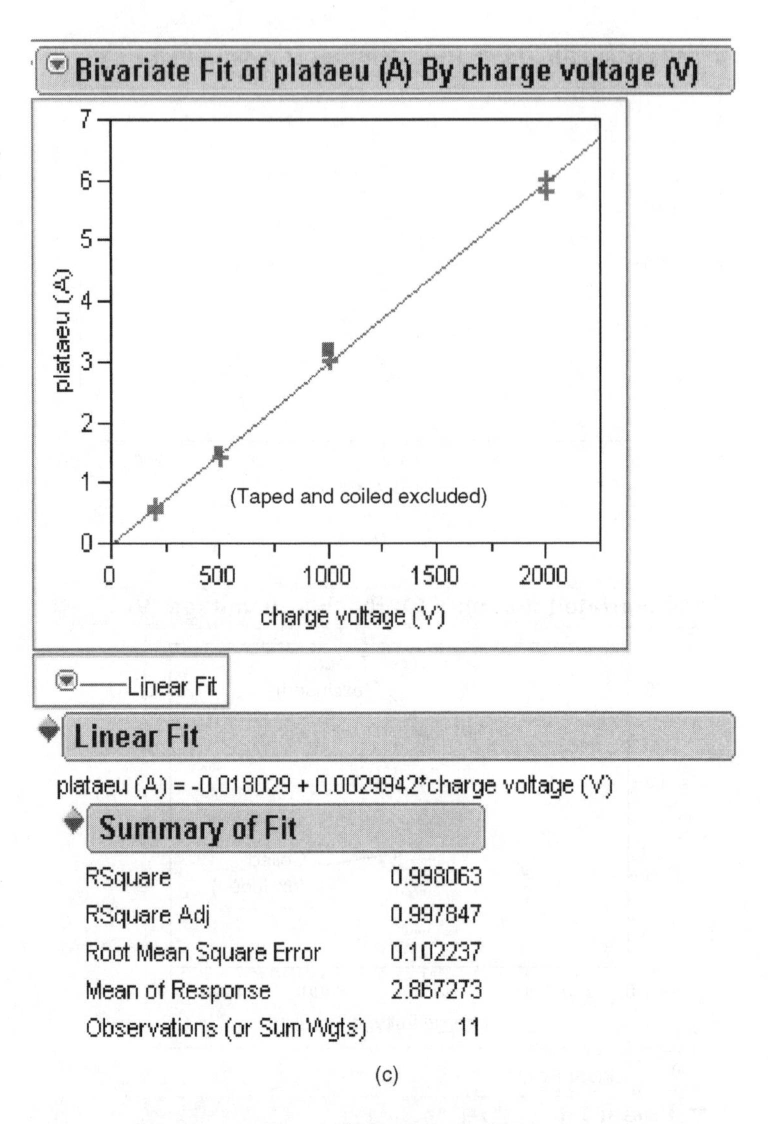

(c)

Figure 10.14 *(continued)*

Figure 10.14(b) shows the bivariate fit of the current versus the charging voltage. In the analysis, the case of the taped and coiled configurations is excluded from the correlation study. The simplification of the fit shows the initial peak current is equal to 0.004 times the charging voltage [34].

Figure 10.14(c) shows the bivariate fit of the plateau current versus the charging voltage. In the analysis, the case of the taped and coiled configurations is excluded from the correlation study. The simplification of the fit shows the plateau current is equal to 0.003 times the charging voltage.

10.10 Transient Field Measurements

During a CDE, there is both the current from the discharge and associated electromagnetic transient fields. For short-length CDEs, test configurations can be established to evaluate the transient fields [34].

10.10.1 Transient Field Measurement of Short-Length Cable Discharge Events

Figure 10.15(a) shows the test configuration and experimental setup for the discharge current. A 1.0 m-long cable is placed in a vertical configuration. The center conductor discharges through a switch to a 100-mm 50 Ω microstrip line that is terminated by a 50 Ω resistor. A Tektronix CT-6 current probe measures the current discharge and records it on an oscilloscope. A 20 dB attenuator is used at the input to the oscilloscope [34].

Figure 10.15(b) shows the test configuration and experimental setup for measurement of the magnetic field induced by the discharge current. A 1.0 m-long cable is placed in a vertical configuration. The center conductor discharges through a switch to ground. The cable is placed 100 mm from the H-field sensor. A 50 Ω resistor is placed on a 10-mm loop H-field sensor (Figure 10.15(c)). The current loop is placed 100 mm from the ground plane. A Tektronix CT-6 current probe measures the current in the loop sensor and records it on an oscilloscope. A 20 dB attenuator is used at the input to the oscilloscope [34].

Figure 10.15(d) shows the test configuration and experimental setup for measurement of the electric field induced by the discharge current. A 1.0 m-long cable is placed in a vertical configuration. The center conductor discharges through a switch to ground. The cable is placed 100 mm from the E-field sensor. The E-field sensor is a short monopole antenna. The short monopole antenna loop is placed 100 mm from the ground plane. The short monopole antenna loop is directly connected to the input of the oscilloscope. Examples of the monopole antenna loops are shown in Figure 10.15(e) [33].

10.10.2 Antenna-Induced Voltages

Antenna-induced voltages are generated from the CDE [34]. Figure 10.16 is an example of an antenna-induced voltage on the short monopole antenna detector. An oscillatory waveform is evident on the antenna detector. A peak-to-peak voltage of 2.3 V occurs on the 10 mm antenna at a 100 mm spacing between the detector and the discharge cable in a nanosecond time regime. Note that this voltage is significant enough to lead to failures in magnetic recording technology, as well as advanced CMOS technology.

10.11 Telecommunication Cable Discharge Test System

To demonstrate the CDE, a test system is needed to evaluate CDEs in a telecommunication system. As discussed, a telecommunication system includes CAT5 cables, a registered jack 45 (RJ45) connectors, an isolation transformer, a Bob Smith termination, and an MDI transceiver circuit (Figure 10.17). In the test system, a high-voltage charging source, relays, and switches

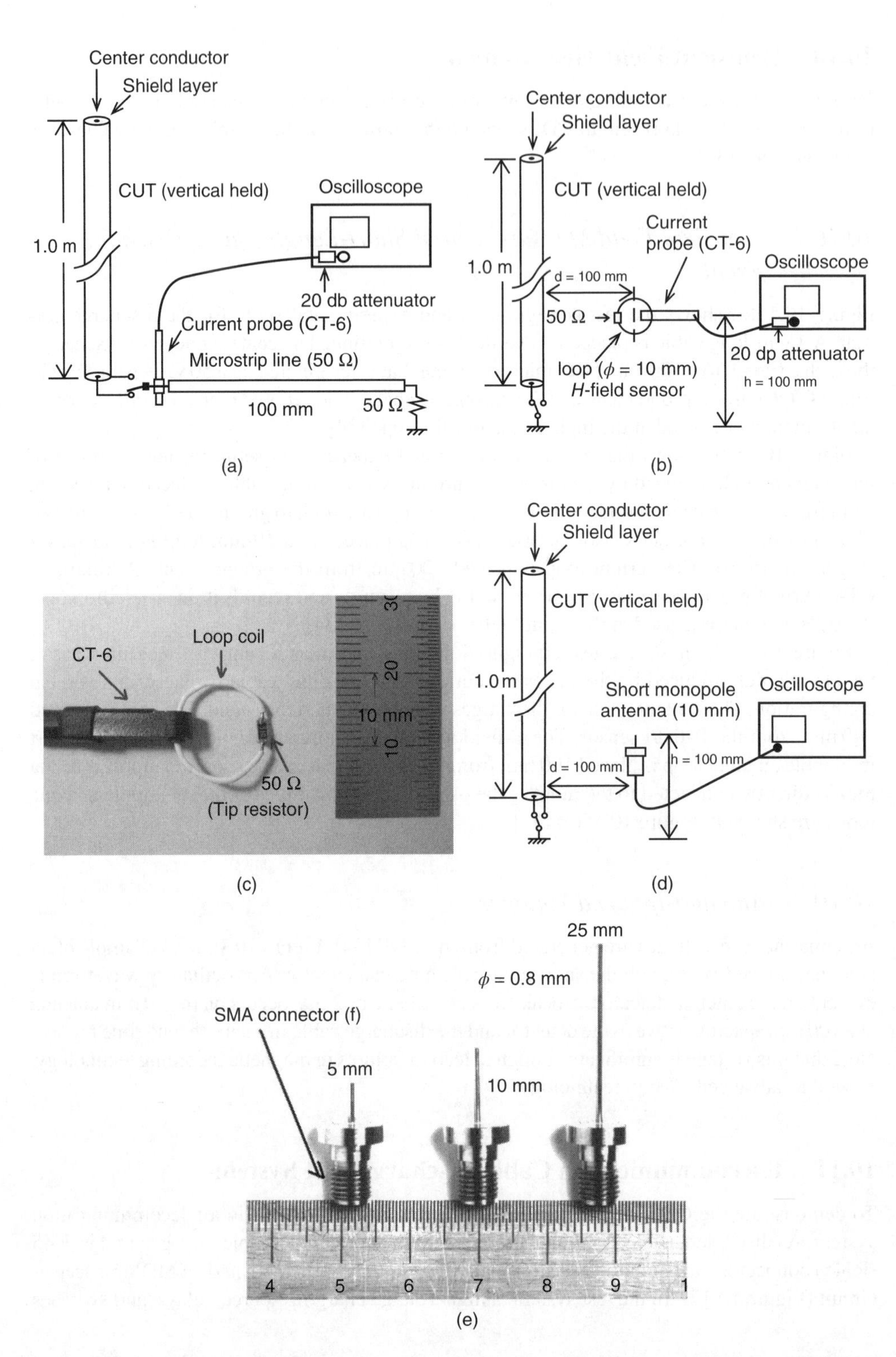

Figure 10.15 (a) Transient discharge – discharge current. (b) Transient discharge – magnetic field. (c) Magnetic current loop. (d) Transient discharge – electric field. (e) Monopole antenna loops (Reproduced with permission from ESD Association)

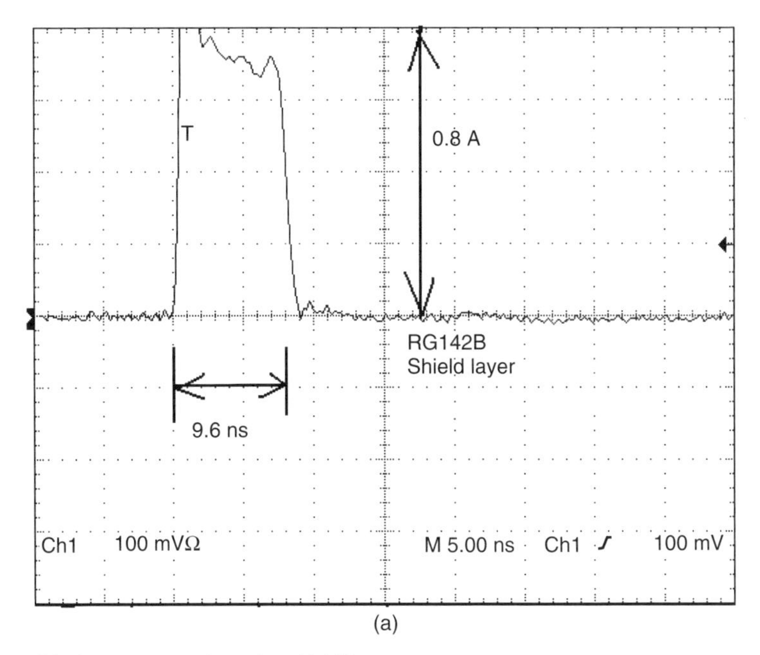

(a)

Discharge current from the shield layer
9.6 ns ≈ 2 × (1 m / *c* × WCR) *c* : light speed

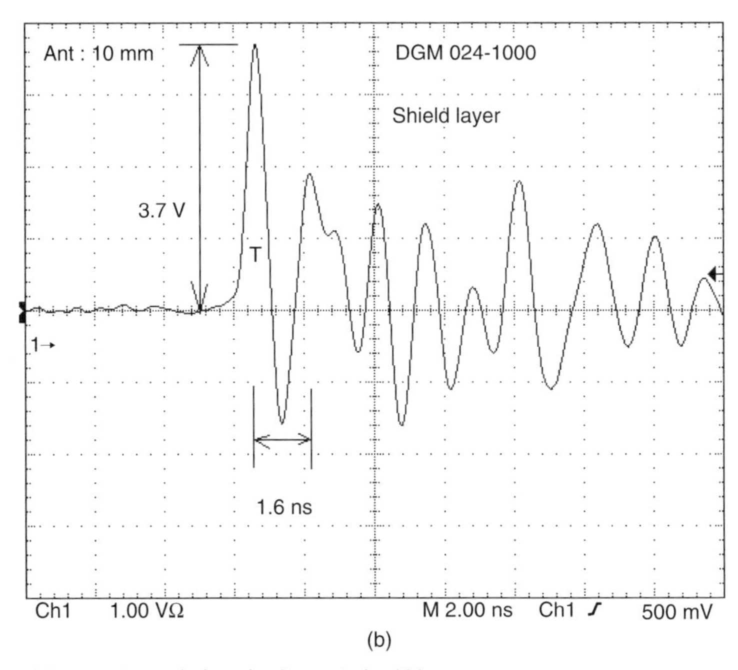

(b)

10 mm antenna-induced voltage at *d* = 100 mm

Figure 10.16 Antenna-induced voltage (Reproduced with permission from ESD Association)

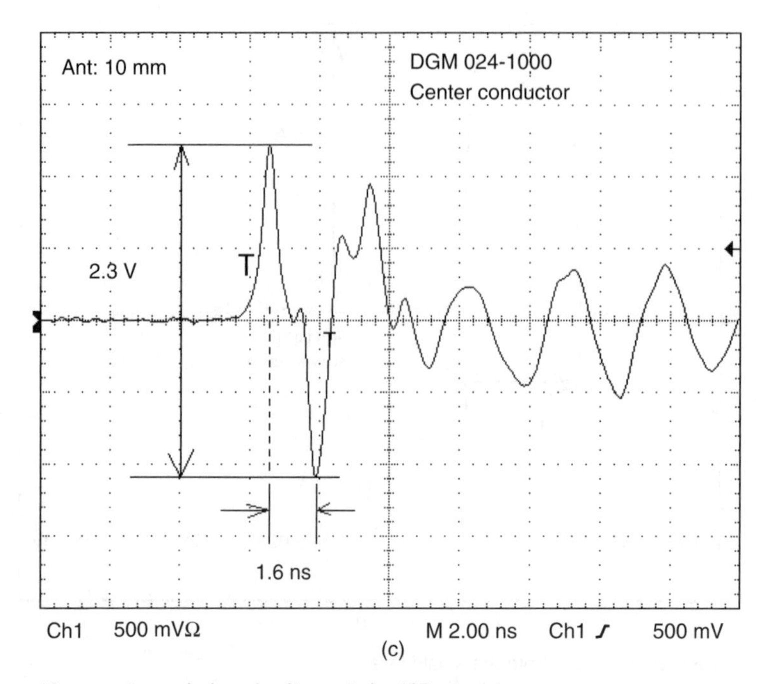

(c)

10 mm antenna-induced voltage at d = 100 mm

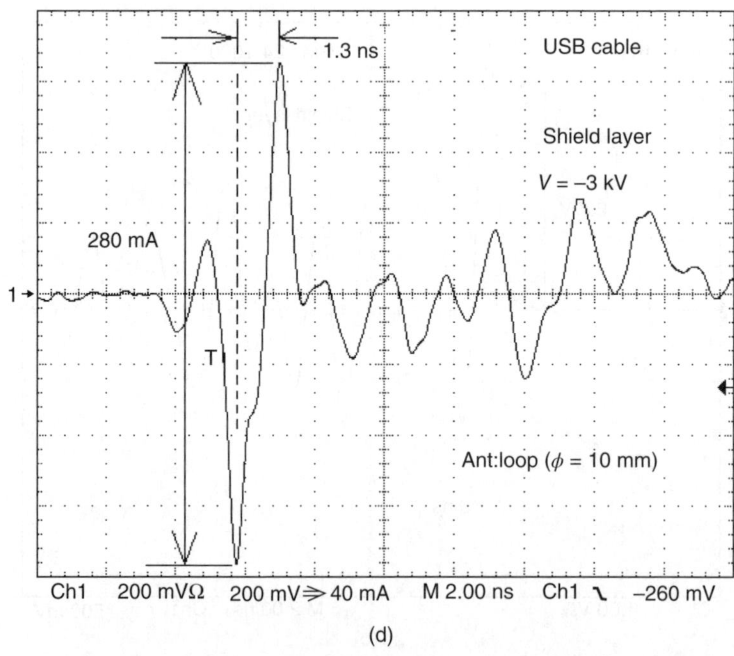

(d)

10 mm loop induced current measured by CT-6 at d = 100 mm

Figure 10.16 (*continued*)

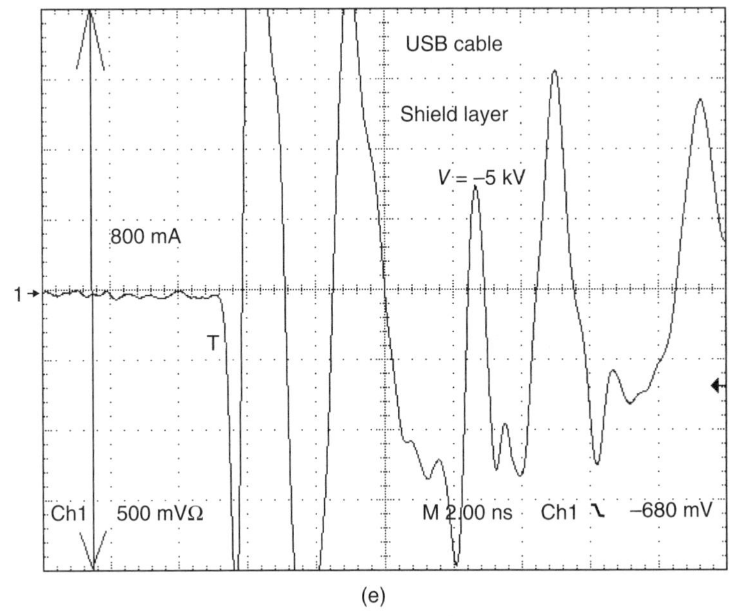

(e)

10 mm loop induced current measured by CT-6 at d = 100 mm

Figure 10.16 *(continued)*

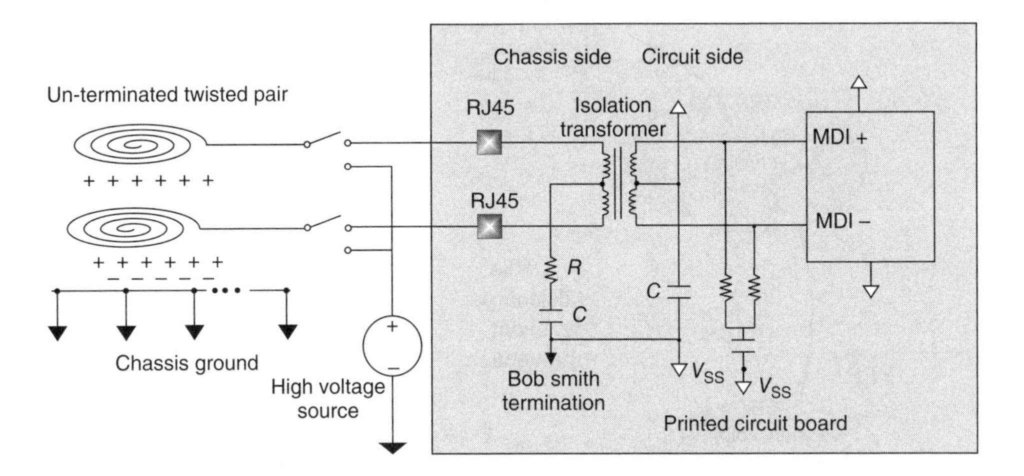

Figure 10.17 Telecommunication CDE test system

are needed to simulate the charging process. The switches are enabled to charge and discharge the cable source for evaluating the UTP cable installation.

The high-voltage source charges the cable to a given voltage level through the charging switches. For the discharge process, the switches to the RJ45 connectors are closed, allowing discharge to the system. After the discharge event, the transceiver network is tested for a full

DC and AC functional test. In the case of latent damage or permanent damage, the signal levels of the MDI transceiver circuit can be impacted.

10.12 Cable Discharge Current Paths

CDEs can follow different current paths depending on the system. The current path that will be taken by the cable discharge will be a path of low resistance and low impedance. This can involve the connector, the isolation transformer, termination, and circuitry.

10.13 Failure Mechanisms

Failure mechanisms from CDEs can be due to dielectric breakdown, thermal overstress, and melting (Figure 10.18).

10.13.1 Cable Discharge Event Failure – Connector Failure

CDE can fail in cable connectors. Cable connectors are designed to survive certain ESD levels. For example, a registered jack 45 (RJ-45) connector is rated to a given ESD robustness. Given that the current level or voltage level is exceeded, failure can occur in the RJ45 termination.

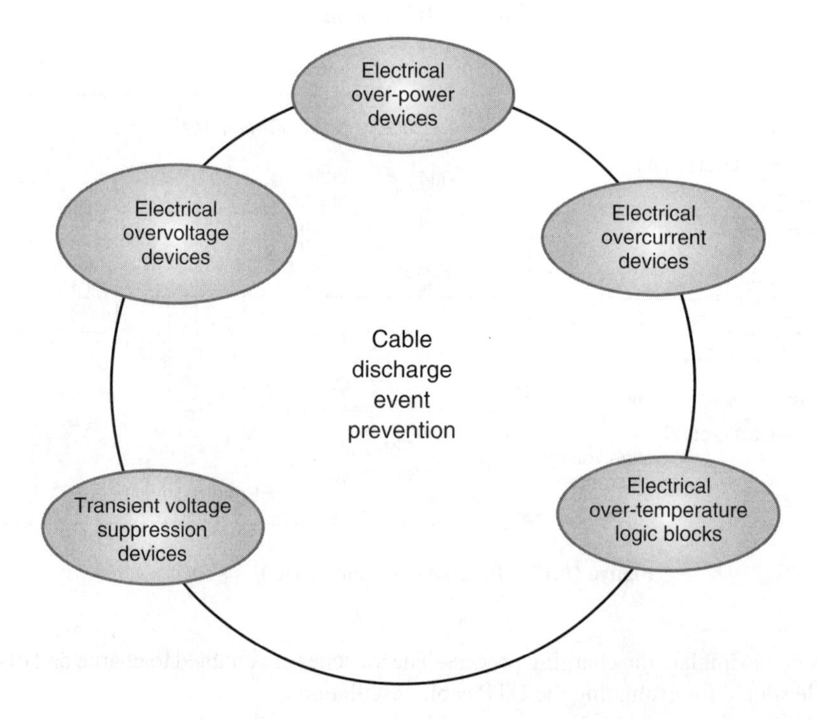

Figure 10.18 Telecommunication CDE test system

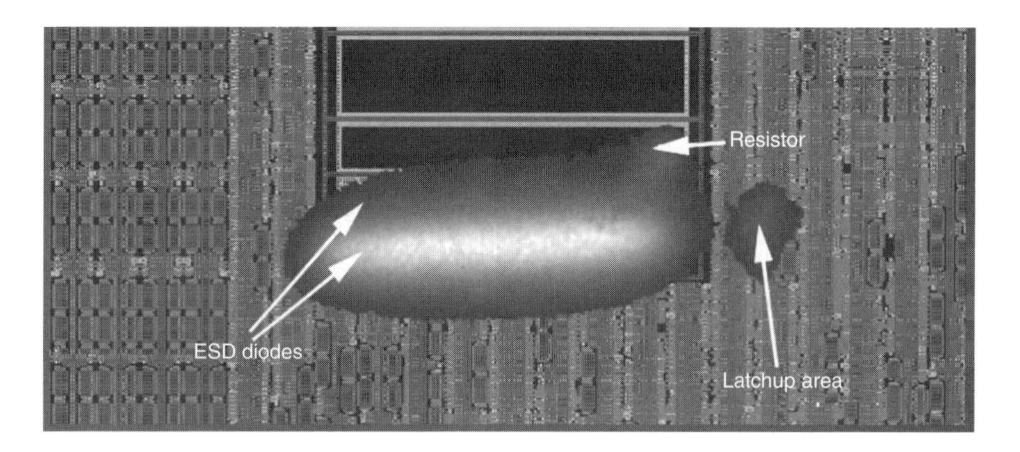

Figure 10.19 Cable discharge-induced latchup

10.13.2 Cable Discharge Event Failure – Printed Circuit Board

CDE can fail in the PCB. Failure can occur between the traces of a PCB leading to impact to the dielectric material or metallization patterns. The failure between adjacent traces will be a function of the physical spacing between the traces.

10.13.3 Cable Discharge Event Failure – Semiconductor On-Chip

CDE can lead to failure in semiconductor chips that are connected to the cable inputs. Semiconductor failure can occur in the ESD protection circuits or the input/output circuits. In the case of receiver elements, the receiver failure can include dielectric breakdown, and MOSFET second breakdown. Passive and active elements can fail from the CDE event.

10.13.4 Cable Discharge Event (CDE)-Induced Latchup

CDE can lead to failure in semiconductor chips due to CMOS latchup [26]. Given that the CDE exceeds the latchup tolerance of the semiconductor chip, latchup can occur. Given that the tetrode condition for latchup is satisfied, latchup can occur in I/O or internal circuitry. Latchup will lead to permanent damage to the semiconductor chip. This is discussed in Chapter 11 (Figure 10.19).

10.14 Cable Discharge Event (CDE) Protection

CDE system level solutions exist to avoid system level failures (Figure 10.20). Solutions can be provided with system level PCB design, touch pads, crowbars, and handling procedures to electrical connectors.

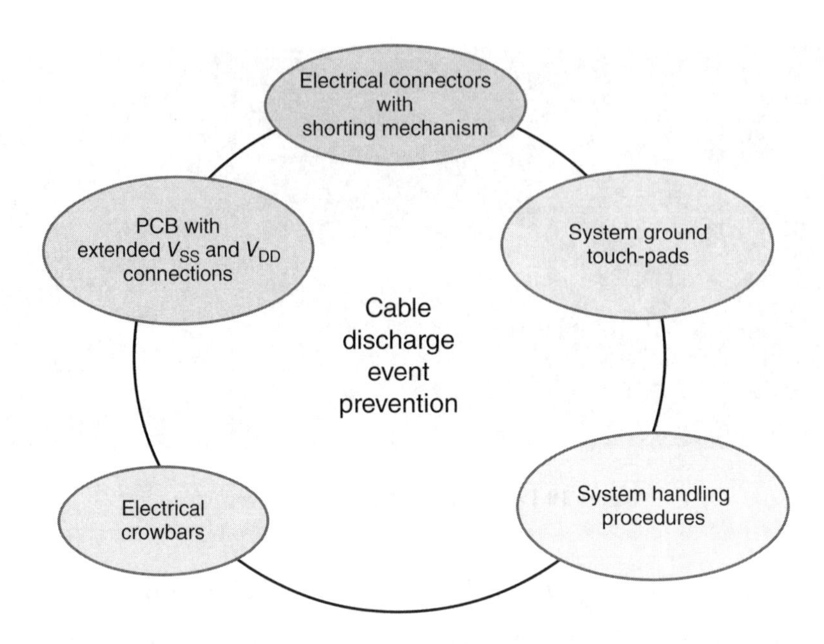

Figure 10.20 Cable discharge event (CDE) system level solutions

10.14.1 RJ-45 Connectors

CDE can lead to arcing in the RJ-45 connectors. RJ-45 connectors are used for UTP. RJ-45 connectors should be used that can withstand large transient voltages. The IEEE 802.3 standard requires for an isolation voltage of 2250 V direct current (DC) and a 1500 V alternating current (AC).

10.14.2 Printed Circuit Board Design Considerations

CDE can lead to PCB failures of dielectric breakdown and spark discharge. To avoid PCB failures, the traces from incoming signal of the UTP should be separated from the ground traces. The spacing of at least 250 mils is recommended to withstand 2000 V transient events.

10.14.3 ESD Circuitry

CDE can lead to high voltages in semiconductor components. The CDE that enters the semiconductor chip can lead to input and output circuit failure. ESD protection circuitry can be placed off-chip and on-chip to eliminate failure of the components [15–32]. CDE protection devices can include Schottky, Zener, or silicon diodes, as well as silicon controlled rectifiers networks. Due to the oscillatory nature of the CDE, it is best to provide an ESD circuit solution that is bidirectional (e.g., dual-diode or bidirectional SCR circuitry).

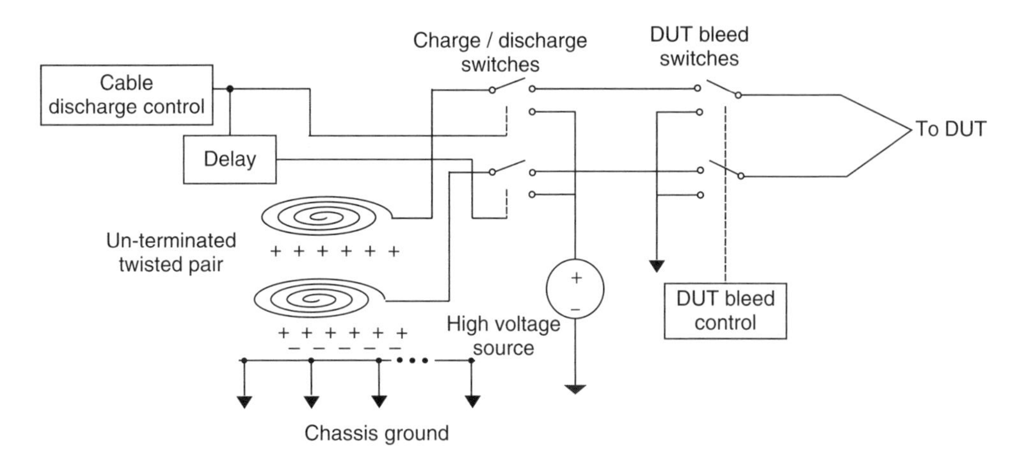

Figure 10.21 Cable discharge event (CDE) validation test system

10.14.4 Cable Discharge Event (CDE) ESD Protection Validation

To verify that the ESD protection is adequate, a validation process can be initiated for the CDE. A test procedure can be performed using a test system that allows for the charging of the cable, and discharging to the DUT through relays and switching system. In the test system, a delay can be introduced to evaluate the discharge with differential inputs.

Figure 10.21 shows a CDE validation test system. A test system is needed to evaluate both common mode and differential mode events. The test system comprises of a cable, a high-voltage generator source, and switches. Switches are needed for the charging process. In addition, a cable discharge control switch is placed in the system, as well as a delay network. At the DUT, switches are needed to discharge the device using a bleed circuit.

The high-voltage generator source charges the cable relative to the chassis ground. For a CAT5 cable, the relay switches are RJ-45 connectors. The switches to the DUT are then closed, applying the CDE. After testing, the residual charge is removed from the device through a bleed network.

The delay network simulates the insertion process of the cable allowing for evaluation of differential pair voltage differences. A contact delay is introduced that can be varied to evaluate different insertion events. The delay process introduces a differential voltage between the two connections of the differential pair, leading to a differential voltage at the inputs of the DUT.

10.15 Alternative Test Methods

Alternative test methods can exist for specific applications associated with automobile industry, military, and space applications. Alternative test methods can also exist for different cable lengths outside of the scope of this chapter. This discussion will provide a base of understanding for measurement, testing, and the issues of CDEs.

10.16 Summary and Closing Comments

This chapter addressed a new semiconductor chip level test to address a CDE. This test, the CDE, introduces a fast transient followed by a slower square pulse waveform. With the growing increase of ports and smaller systems, CDEs are of interest for both chip developers and system designers.

Chapter 11 discusses CMOS latchup. CMOS latchup physics and theory are briefly discussed, followed by test methods and issues. CMOS DC and transient latchup will be highlighted as well as "internal latchup" versus "external latchup."

Problems

1. Why are cables an issue in electronic systems?
2. How does one define a cable? What are the electrical characteristics that make a component a cable?
3. Draw a cable as a distributed LC electrical model highlighting the incremental nature of the cable.
4. Draw a cable as a distributed LRC electrical model highlighting the incremental nature of the cable.
5. A package can be represented as a Pi-network of a capacitor–inductor–capacitor. What is the difference between a package and a cable?
6. Explain the charging and discharging process of a cable into a system.
7. A cable has a BNC connector on the end of the cable, and the system also has sockets to socket the BNC connector. Explain the discharge process.
8. A system engineer drags a cable on the floor and sockets the cable into a system. Explain the charging process of the cable, the system engineer, and the charge transfer to the system.
9. In a corporation with large servers, procedures are established to avoid discharging cables into a large server. What procedure can be developed to avoid discharging the cable into the system?
10. In a corporation with large servers, procedures are established to avoid a system invention on the chassis of the large server. What was the solution?
11. Cable discharge current magnitude can exceed the JEDEC latchup test standard current magnitude. Explain how CMOS latchup can occur from a cable discharge event.

References

1. ESD Association. DSP 14.1 – 2003. *ESD Association Standard Practice for the Protection of Electrostatic Discharge Sensitive Items – System Level Electrostatic Discharge Simulator Verification Standard Practice*, Standard Practice (SP) document, 2003.
2. ESD Association. DSP 14.3 – 2006. *ESD Association Standard Practice for the Protection of Electrostatic Discharge Sensitive Items – System Level Cable Discharge Measurements Standard Practice*, Standard Practice (SP) document, 2006.
3. ESD Association. DSP 14.4 – 2007. *ESD Association Standard Practice for the Protection of Electrostatic Discharge Sensitive Items – System Level Cable Discharge Test Standard Practice*, Standard Practice (SP) document, 2007.
4. Intel Corporation. Cable discharge event in local area network environment. White Paper, Order No: 249812–001, 2001.

5. R. Brooks. A simple model for the cable discharge event. *IEEE 802.3 Cable-Discharge Ad-hoc Committee*, 2001.
6. Telecommunications Industry Association (TIA). Category 6 cabling: static discharge between LAN cabling and data terminal equipment. *Category 6 Consortium*, 2002.
7. L. Loeb. Statistical factors in spark discharge mechanisms. *Review of Modern Physics*, Vol. 20, (1), Sept 1948; 151–160.
8. J. Meek. A theory of spark discharge. *Physical Review*, Vol. 57, Sept 1940; 722–728.
9. J. Deatherage, D. Jones. Multiple factors trigger discharge events in Ethernet LANs. *Electronic Design*, Vol. 48, (25), Sept 2000; 111–116.
10. H. Geski. DVI compliant ESD protection to IEC 61000-4-2 level 4 standard. *Conformity Magazine*, Sept 2004; 12–17.
11. W. Stadler, T. Brodbeck, R. Gartner, and H. Gossner. Cable discharges into communication interfaces. *Proceedings of the Electrical Overstress/Electrostatic Discharge (EOS/ESD) Symposium*, 2006; 144–151.
12. USB-ESD Problem bei manchen Pentium-4 Mainboards. 12/2005.
13. S. Voldman. Cable discharge event and CMOS latchup. *Proceedings of the Taiwan ESD Conference (T-ESDC)*, 2005; 63–68.
14. D. Wang, S. Marum, W. Kemper, and D. McLain. System event triggered latchup in IC chips: test issues and chip level protection design. *Proceedings of the Electrical Overstress/Electrostatic Discharge (EOS/ESD) Symposium*, 2006; 1–7.
15. T. Brodbeck, W. Stadler, R. Gartner, H. Gossner, W. Hartung, and R. Losehand. ESD-schutznassnahmen fur schnittstellan in der kommunikationselektronik (ESD protection measures for communication IC interfaces). *Proceedings of the 9th ESD Forum*, 2005; 115.
16. B. Arndt, F. zur Nieden, R. Pohmerer, J. Edenhofer, and S. Frei. Comparing cable discharge events to IEC 61000-4-2 or ISO 10605 discharges, 2008.
17. W. Stadler, T. Brodbeck, J. Niemesheim, R. Gaertner, and K. Muhonen. Characterization and simulation of real-world cable discharge events. *Proceedings of the Electrical Overstress/Electrostatic Discharge (EOS/ESD) Symposium*, 2009; 419–426.
18. Texas Instruments. AN-1511 cable discharge event. Application Report SNLA087 – July 2006 – Revised April 2013.
19. R. Smith. Apparatus and method for terminating cables to minimize emissions and susceptibility. U.S. Patent No. 5,321,372, June 14, 1994.
20. S. Poon, and T. Maloney. Shielded cable discharge induced current on interior signal lines. *Proceedings of the Electrical Overstress/Electrostatic Discharge (EOS/ESD) Symposium*, 2007; 311–317.
21. ESDEMC Corporation. http://www.esdemc.com. ESDEMC model ES631-LAN cable discharge event (CDE) evaluation system.
22. S. Voldman. *ESD: Physics and Devices*. Chichester, England: John Wiley and Sons, Ltd., 2004.
23. S. Voldman. *ESD: Circuits and Devices*. Chichester, England: John Wiley and Sons, Ltd., 2005.
24. S. Voldman. *ESD: Circuits and Devices 2nd Edition*. Chichester, England: John Wiley and Sons, Ltd., 2015.
25. S. Voldman. *ESD: RF Circuits and Technology*. Chichester, England: John Wiley and Sons, Ltd., 2006.
26. S. Voldman. *Latchup*. Chichester, England: John Wiley and Sons, Ltd., 2007.
27. S. Voldman. *ESD: Circuits and Devices*. Beijing, China: Publishing House of Electronic Industry (PHEI), 2008.
28. S. Voldman. *ESD: Failure Mechanisms and Models*. Chichester, England: John Wiley and Sons, Ltd., 2009.
29. S. Voldman. *ESD: Design and Synthesis*. Chichester, England: John Wiley and Sons, Ltd., 2011.
30. S. Voldman. *ESD Basics: From Semiconductor Manufacturing to Product Use*. Chichester, England: John Wiley and Sons, Ltd., 2012.
31. S. Voldman. *Electrical Overstress (EOS): Devices, Circuits, and Systems*. Chichester, England: John Wiley and Sons, Ltd., 2013.
32. S. Voldman. *ESD: Analog Design and Circuits*. Chichester, England: John Wiley and Sons, Ltd., 2014.
33. J. Stephan, and M. Hopkins. SP14.3 verification, 07WW44, cable discharge verification, WG14.3. ESD Association Cable Discharge Standard Meeting, October 18, 2007.
34. M. Honda. WG 14.2 cable discharge standard meeting. Presentation, 2003.

11

Latchup

11.1 History

Complementary metal-oxide semiconductor (CMOS) latchup is a fundamental issue inherent in CMOS technology [1–37]. Latchup was discovered in CMOS technology in the early 1970s at the beginning of CMOS technology. In many corporations, latchup was first discovered with the first CMOS technology manufactured. CMOS was developed for low power, and many of the first efforts to focus on low power were space applications. A significant amount of understanding was first established at national laboratories and universities [9–17]. One of the first publications was in 1973 by B.L. Gregory, and B.D. Shafer of Sandia Laboratories, titled "Latchup in CMOS integrated circuits," in the IEEE Transaction on Nuclear Science [9]. The first textbook was introduced in 1985, by R. Troutman, titled "Latchup in CMOS Technology: The Problem and the Cure" [1]. The second book on latchup was published in 2007, capturing the advancements between 1984 and 2007 [2].

Figure 11.1 illustrates the evolution of latchup testing. In the 1960s, Fairchild Semiconductor invented CMOS technology. In the early 1970s, research continued on CMOS latchup to mainstream integration of CMOS. In the 1970s, laboratory test systems were constructed in laboratories to do latchup testing. There was no standard test structure to test or measure latchup phenomena. Latchup test structure standardization was established in the mid-1980s (within the IEEE) as a means of benchmarking technology for latchup robustness. In the 1980s, commercial test equipment was established for latchup testing. One of the first commercial latchup test systems was developed by Stag Microsystems Inc. called the Verifier. The Stag Microsystem's Verifier was developed for 256-pin testing.

Figure 11.2 illustrates a semiconductor chip cross section. Latchup occurs in semiconductors with the formation of a parasitic pnpn structure. The parasitic structure is present in a single-well or dual-well CMOS technology. From the p-channel MOSFET structure, there is a parasitic pnp bipolar junction transistor (BJT) formed from the p+ diffusion MOSFET implant, the n-well tub region, and the p-substrate. From the n-channel MOSFET structure, there is a parasitic npn BJT formed from the n+ diffusion MOSFET implant, the p-substrate, and the n-well region.

ESD Testing: From Components to Systems, First Edition. Steven H. Voldman.
© 2017 John Wiley & Sons, Ltd. Published 2017 by John Wiley & Sons, Ltd.

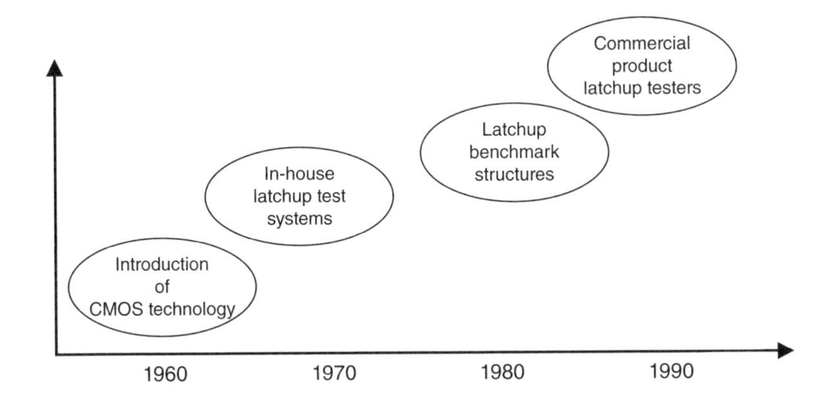

Figure 11.1 Latchup testing evolution

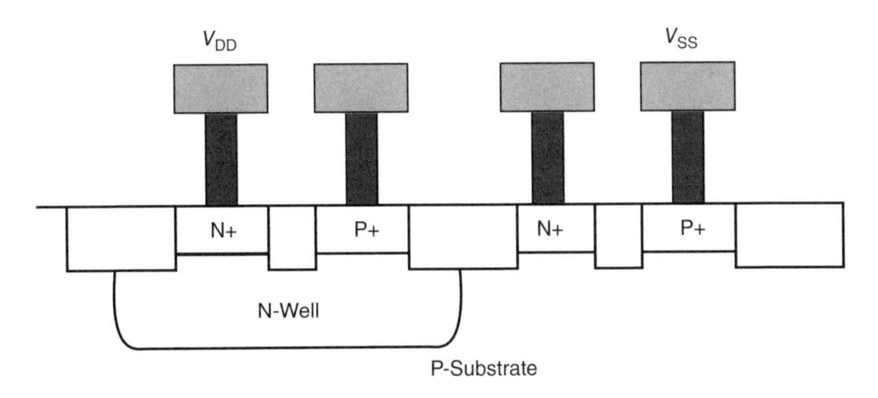

Figure 11.2 Latchup cross section

In this structure, the pnp and npn transistors form a four region device, forming a pnpn. Latchup occurs due to the regenerative feedback from the cross-coupled pnp and npn transistor.

Figure 11.3 illustrates the electrical schematic representation of a CMOS latchup structure [1, 2]. The schematic shows two transistor representations that are cross-coupled. Latchup occurs in semiconductors with the formation of a parasitic pnpn structure formed by the cross-coupled parasitic pnp BJT and the parasitic npn BJT. In the electrical circuit schematic, there is a first resistor element representing the n-well resistance, R_{NW}, and a second resistor element representing the p-well resistance, R_{PW}. The n-well resistor element is the resistance between the n-well contact and the PFET device. The p-well resistor element is the resistance between the p-well contact and the NFET device.

In order to characterize the latchup sensitivity of a technology, a simplified four-stripe structure can be formed representing the physical dimensions representing the four physical elements (e.g., two transistors and two resistors) [1, 2]. Figure 11.3(b) represents a simple test structure used to characterize CMOS latchup. The p+ diffusion stripe in the n-well represents the CMOS p-channel MOSFET diffusion. The n+ diffusion stripe in the p-well represents the

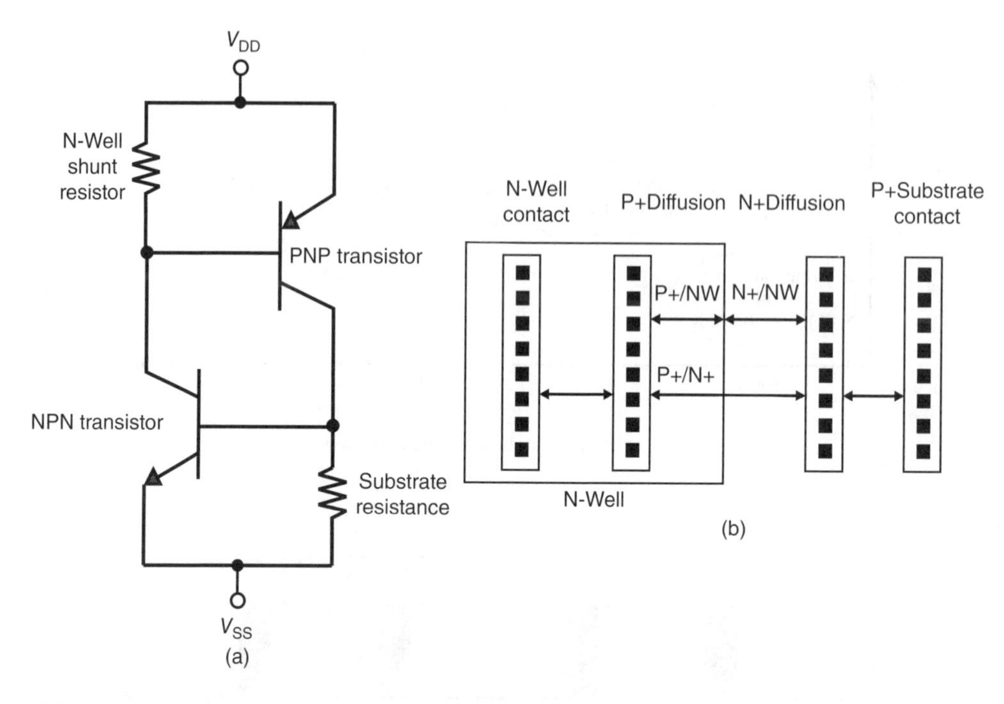

Figure 11.3 (a) Latchup two transistor representation. (b) Latchup four-stripe PNPN test structure

CMOS n-channel MOSFET diffusion. The n+ diffusion within the n-well region represents the n-well contact. The p+ diffusion within the p-well region represents the p-well (e.g., p-substrate) contact.

The electrical state of the circuit is a function of the electrical characteristics of the four physical elements (e.g., the two transistors and the two resistors). Latchup occurs when the state of the circuit switches from a low current/high voltage state to a high current/low voltage state. When the pnpn structure is in a high current state, Joule self-heating occurs leading to thermal failure of the circuit.

Figure 11.4 represents the current–voltage (*I–V*) characteristic of a CMOS latchup pnpn structure [1, 2]. The *I–V* characteristic has a first low current "off-state." This state continues with an increasing voltage, until the circuit is in an avalanche multiplication state. The latchup structure "switches" to a conducting region in the "negative resistance" region. The negative resistance state is an unstable state; hence, the circuit recovers into stable low voltage/high current state. When the circuit is in this state, the BJT's current gain increases from self-heating. The circuit can continue in this mode, leading to thermal breakdown.

11.2 Purpose

The purpose of the latchup test is for establishment of a test methodology to evaluate the repeatability and reproducibility of components or microelectronic circuitry to a defined pulse event in order to classify or compare latchup sensitivity levels of components [38].

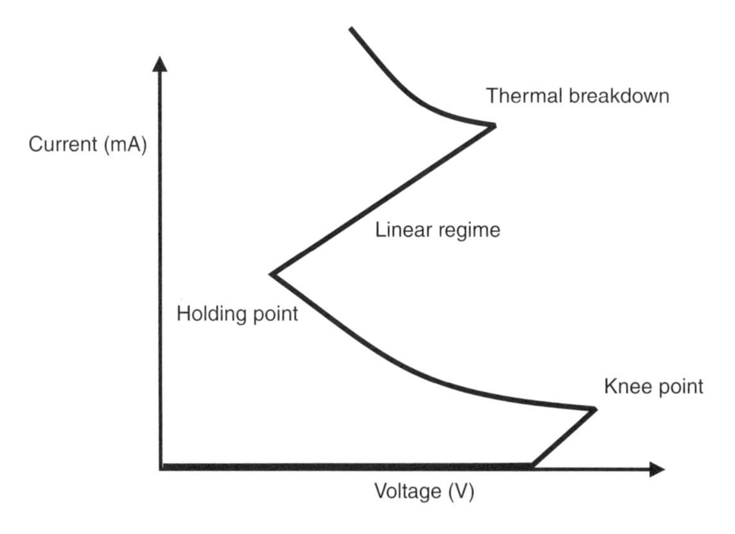

Figure 11.4 Latchup current–voltage (*I–V*) characteristic

11.3 Scope

The scope of latchup test is for the testing, evaluation, and classification of components and micro- to nanoelectronic circuitry. The test is to quantify the sensitivity or susceptibility of these components to damage or degradation to the defined latchup test [38].

11.4 Pulse Waveform

For latchup testing, specific waveforms and bias conditions are defined for testing. These bias conditions and states are specified in the JEDEC JESD78 standard specification [38]. Figure 11.5 represents the waveform specification for the CMOS latchup testing.

For transient latchup testing, a pulse is applied to the test structure or system. Figure 11.6 represents the waveform specification for the CMOS latchup testing.

11.5 Equivalent Circuit

An equivalent circuit representation for CMOS latchup can be defined from the test system. Figure 11.7 represents the equivalent circuit for the latchup test system and latchup structure.

11.6 Test Equipment

For CMOS latchup testing, there is testing that is completed on test structures for semiconductor process and device development, and second, latchup testing on products for reliability, quality, and product assurance qualification and release [2].

Latchup testing on test structures are typically performed on a wafer level [2]. Wafer level test systems are used to develop semiconductor process technology, design rules, and physical

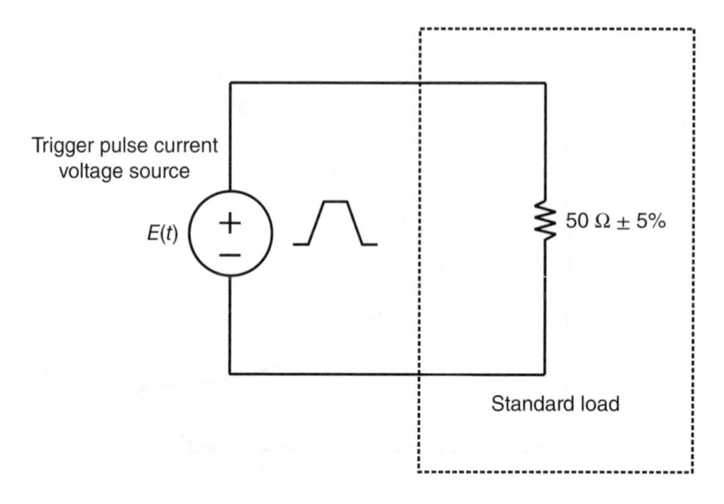

Figure 11.5 Latchup waveform for DC latchup

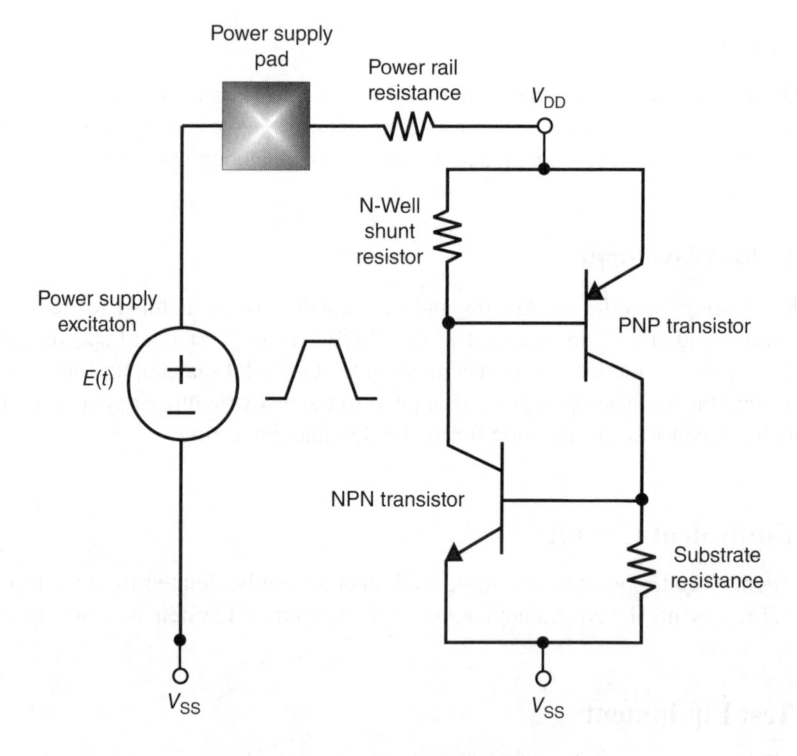

Figure 11.6 Transient latchup waveform

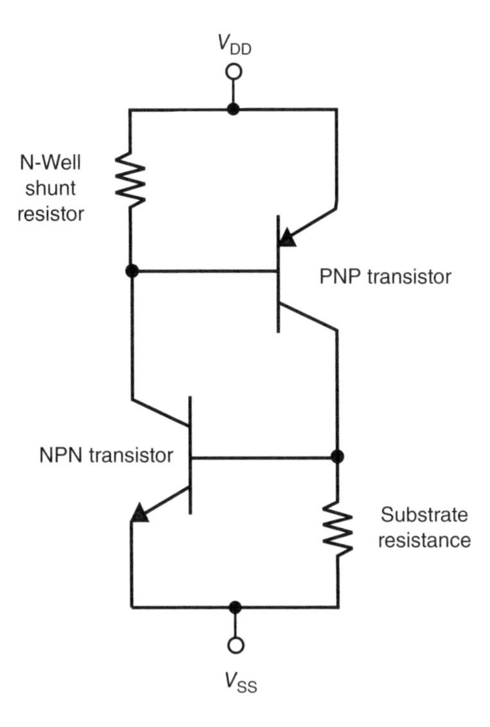

Figure 11.7 Latchup equivalent circuit model

dimensions. The wafer level systems are used to quantify the dimensional rules that influence the parasitic pnp transistor, the npn transistor, the n-well resistance, and the p-well resistance.

Wafer level testing can be performed on different types of test structures, from simple four-stripe structure to more complex structures [1, 2]. Figure 11.8 represents an example of a wafer level test system used for testing a four-stripe structure.

Figure 11.9 is an example of a commercial test system, the Thermo Scientific MK4™ for component level latchup testing [39]. This latchup test system satisfies the JEDEC JESD 78D and AEC Q100-004 testing standards. The tester includes preconditioning, state read-back, and full control. It has six independent vector voltage levels allowing for support testing of complex I/O circuitry and multicore products. The test system provides the latchup stimulus and the device biasing.

11.7 Test Sequence and Procedure

Figure 11.10(a) illustrates a latchup test procedure and test flow as shown in the JEDEC JESD78 specification [38]. Prior to the latchup test, the device needs to be at stable state with a reproducible nominal current. The supply current must be stable enough and low enough to detect the supply current increase from CMOS latchup. A minimum of three devices are to be tested using the "I-test" and the supply overvoltage tests. The devices used are required to pass the ATE testing to the device specification. The devices are subjected to the I-test first. If they pass, the test continues to the power supply overvoltage test.

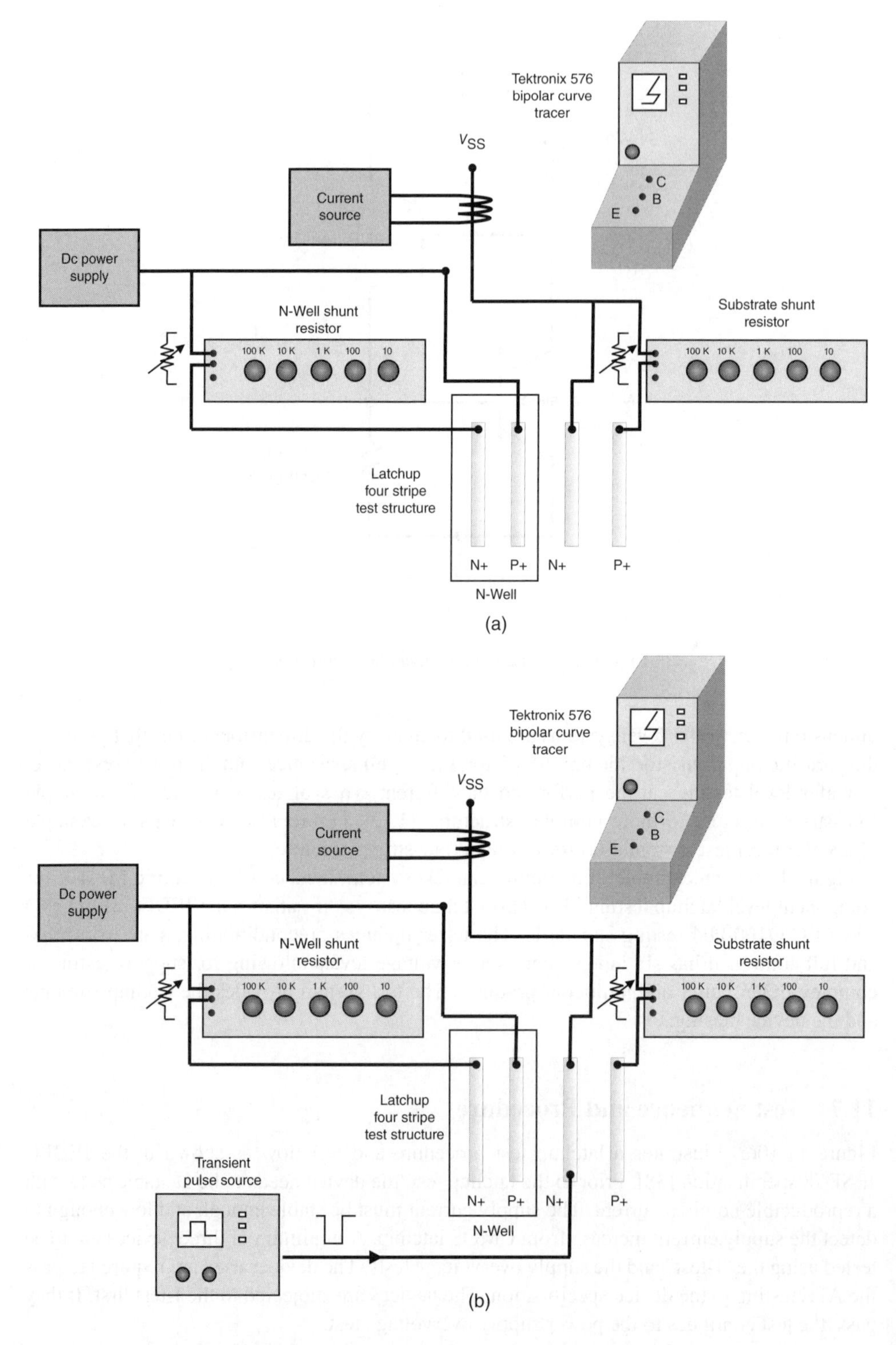

Figure 11.8 Example wafer level latchup test system

Figure 11.9 Thermo Scientific MK4™ commercial latchup test system (Reproduced with permission of Thermo Fisher Scientific)

Figure 11.10(b) illustrates a latchup test procedure for a positive I-test. The supply voltage is raised to a maximum supply voltage, and the pin under test is set to the maximum logic high state. A positive current trigger is then applied to determine if latchup occurs. Note that during the test, overshoots may lead to thermal overstress (TOS).

Figure 11.10(c) is the test procedure for a negative I-test.

Figure 11.10(b) illustrates a latchup test procedure for a positive I-test. The supply voltage is raised to a maximum supply voltage, and the pin under test is set to the minimum logic low state. A negative current trigger is then applied to determine if latchup occurs. Note that during the test, TOS may be evident in the pin under test response.

The second test, the power supply overvoltage test is initiated if the chip passes the I-test. The overvoltage test is performed on each supply pin or pin group. The device under test (DUT) is biased during this test. All input pins are set to the maximum logic high level. A voltage trigger source is then applied to the power supply pin. After the trigger source is removed, the supply pin is returned to the state it was before the application of the trigger pulse. The supply current is then measured to determine if latchup occurred in the DUT.

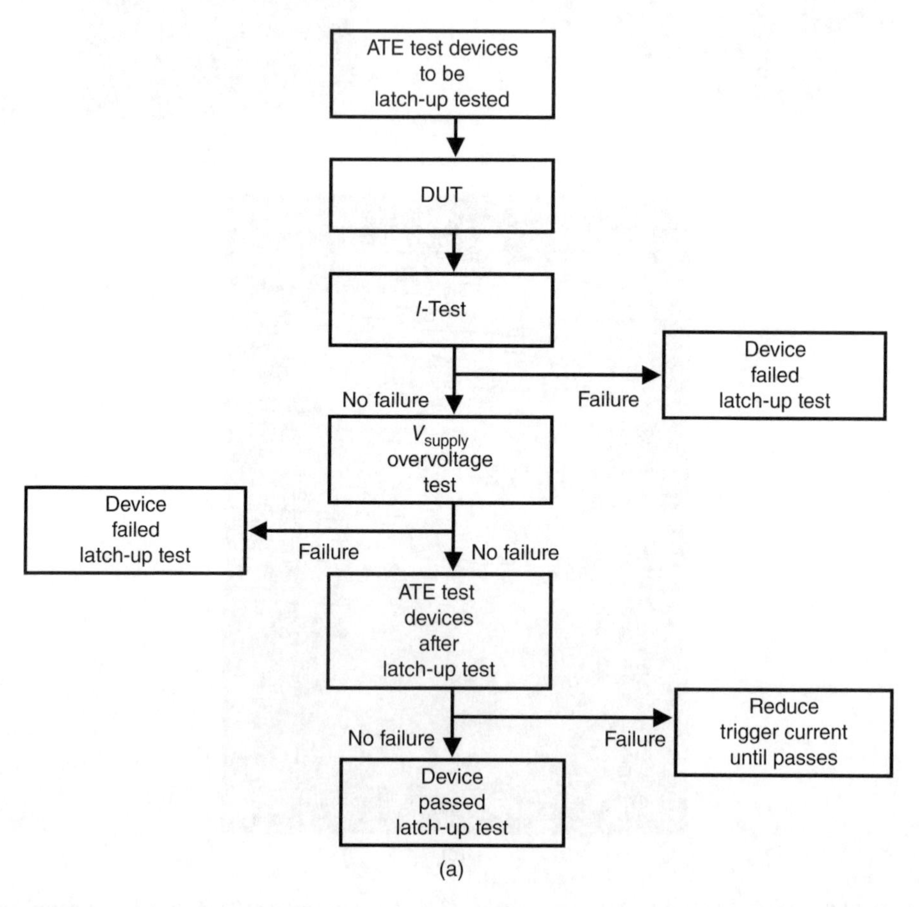

Figure 11.10 (a) Latchup test procedure and test flow. (b) Latchup test procedure – positive I-test. (c) Latchup test procedure – negative I-test

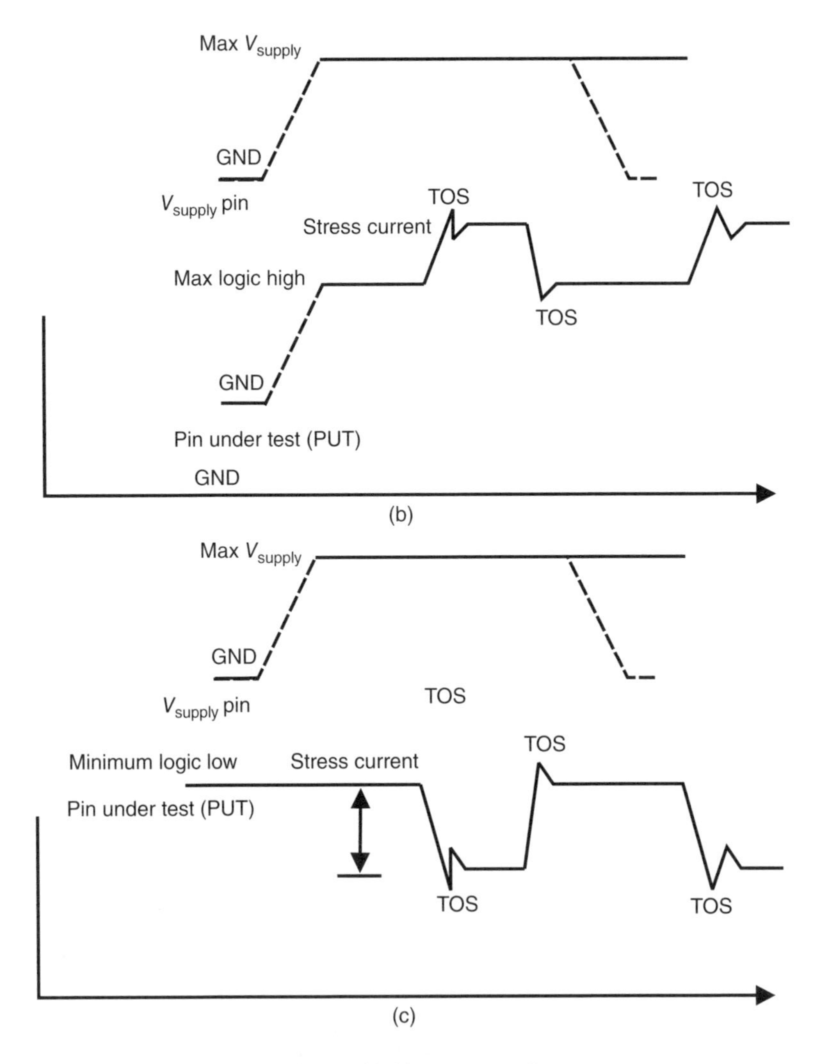

Figure 11.10 *(continued)*

11.8 Failure Mechanisms

Latchup can lead to destructive damage in components to systems. In a semiconductor chip, latchup failure can occur in the following [2, 3, 5]:

- Silicon devices
- Isolation
- Silicides
- Interlevel dielectric (ILD)
- Interconnects
- Bond pads.

Figure 11.11(a) is an example of silicon device failure due to a latchup event. In silicon devices, latchup can occur in the parasitic pnp or npn BJT. The silicon damage will occur from junction failure or when the temperature exceeds the melting temperature of silicon. Other failure mechanisms can occur from silicide penetration of the junctions or metallurgy failure. Interconnect failure can occur in the contacts, vias, or interconnect layers. Interlevel dielectric cracking can lead to failure from the thermal stress of the latchup event.

Latchup failure can lead to semiconductor component package failure [3, 5]. The failure in the package can be the wirebond, lead frame, or package resin materials. The current magnitude from a latchup event can lead to melting of wirebonds and package material. The Joule heating from the latchup event can lead to the physical package ablation, material separation, and physical package melting. A latchup event can lead to melting of the package, preventing separation from the printed circuit board. Figure 11.11(b) is an example of package failure from a latchup event.

11.9 Latchup Current Paths

In latchup, the current path for the latchup event is a function of whether the event occurs intracircuit, in adjacent circuits, or intercircuit. Within a given circuit or adjacent circuits, CMOS latchup currents are contained within the pnpn structure and the resistor elements associated with the pnpn structure. With two adjacent circuits, it is possible that the pnpn structure exists between the two circuits. But there is no current low external to the pnpn and the resistor elements. Figure 11.12(a) is an example of a latchup current path that occurs within a single circuit. When the structure undergoes a high current state, the majority of the current exists near and under the isolation where the pnp and the npn are cross-coupled.

In latchup, the current path for the latchup event can be between adjacent circuits. For example, the pnp can be in a first circuit and an adjacent n-well structure can be in an adjacent circuit. In this case, the npn transistor is formed between the first and second circuit. Hence, the pnpn is formed between two circuits. This can occur in the I/O of a semiconductor chip, where I/O are adjacent. In addition, the I/O can be next to a decoupling capacitor that has an n-well structure.

A second case of interest is intercircuit where the injection source of current comes from another region of the chip, outside of the pnpn structure and its corresponding well taps.

In this case, the current stimulus is external to the pnpn and the resistor elements. Figure 11.12(b) is an example of a latchup current path that occurs external to the latchup structure.

11.10 Latchup Protection Solutions

Protection solutions exist for avoiding CMOS latchup [2]. Figure 11.13 highlights different classes of protection from CMOS latchup. CMOS latchup can be addressed by semiconductor process solutions, design layout, circuitry, or system level.

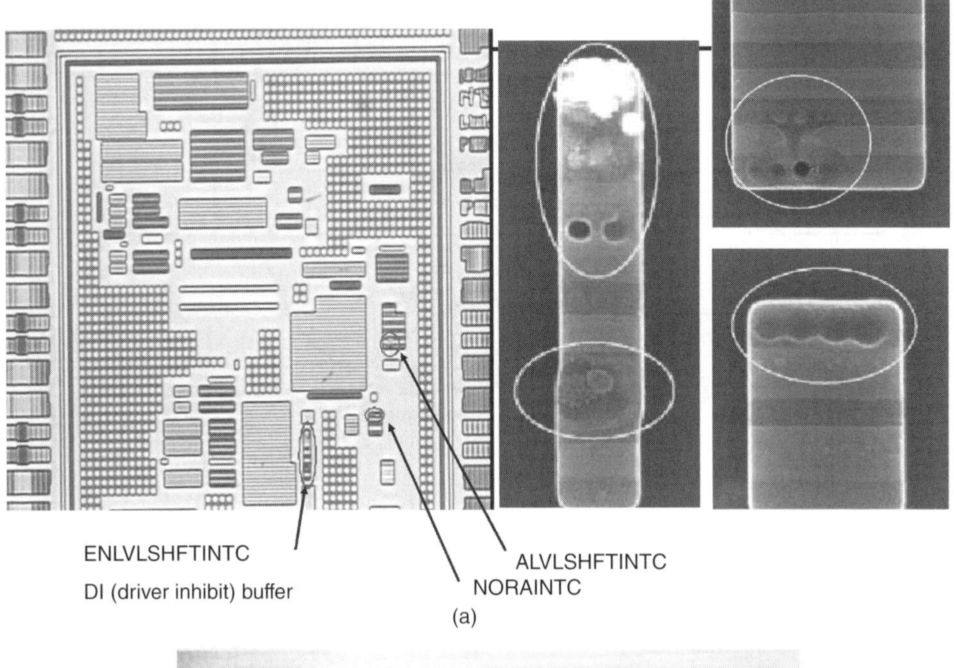

(a)

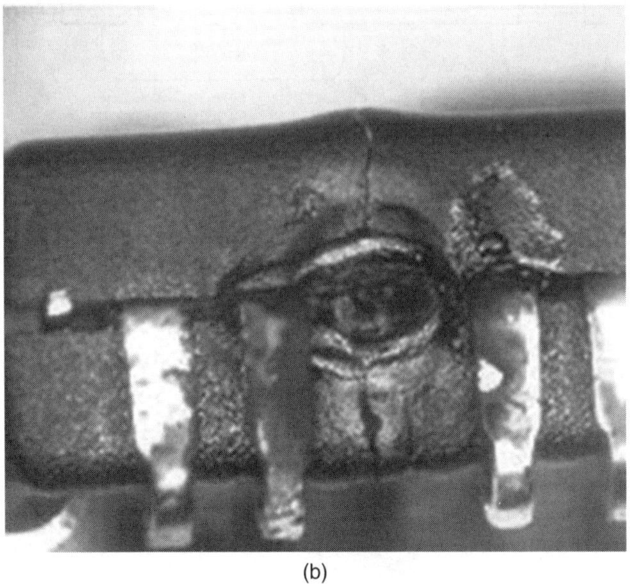

(b)

Figure 11.11 (a) Latchup failure mechanism – chip level damage. (b) Latchup failure mechanism – package level damage

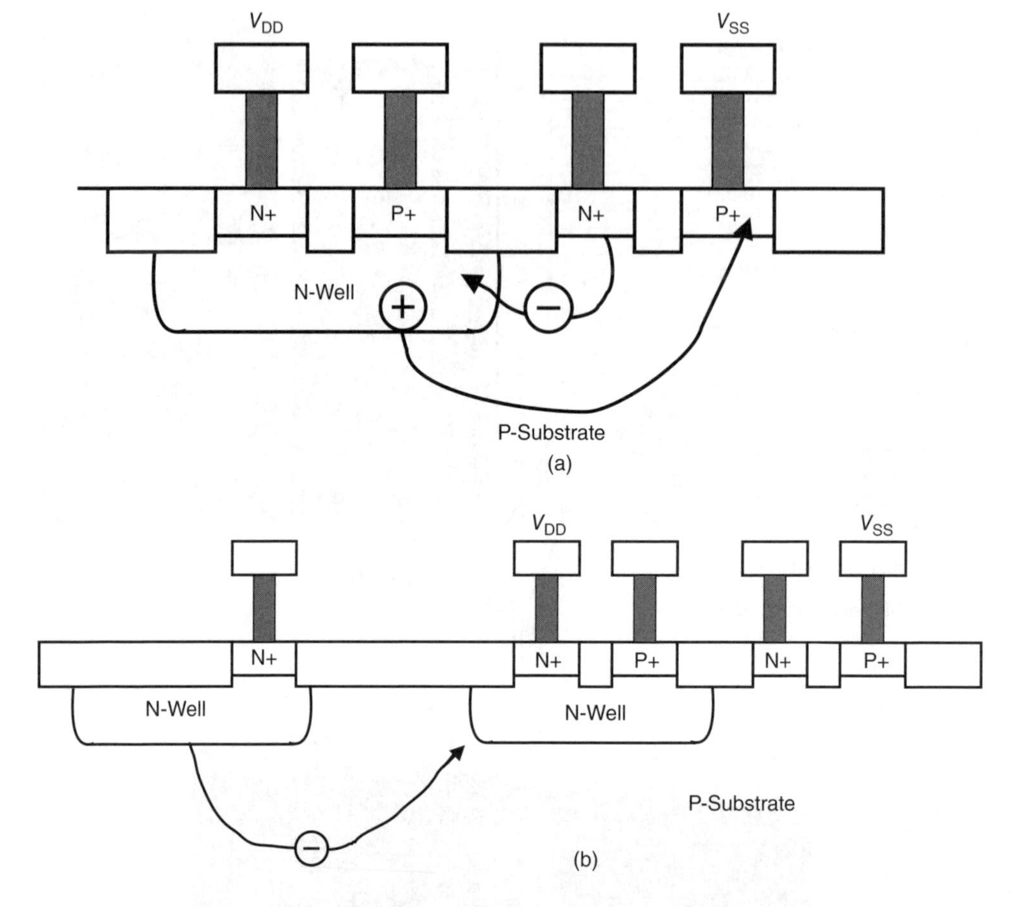

Figure 11.12 (a) Internal latchup current paths. (b) External latchup current paths

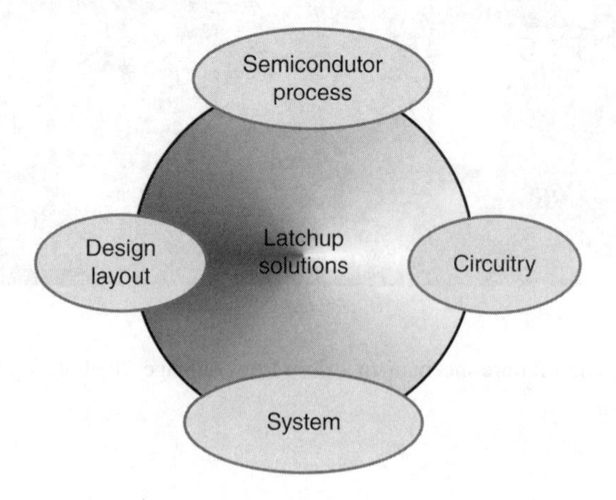

Figure 11.13 Classes of latchup protection solutions

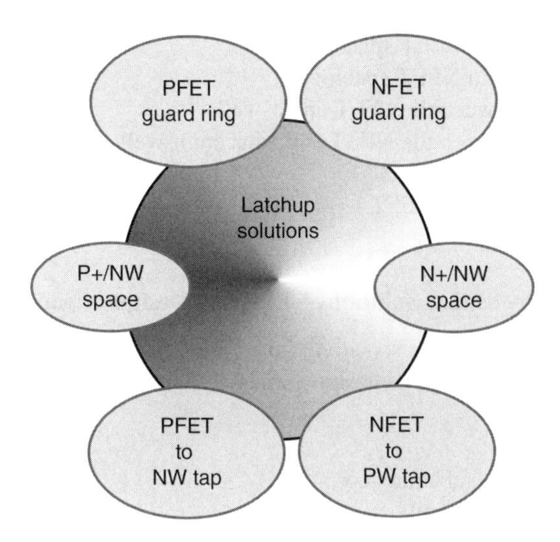

Figure 11.14 Latchup design layout solutions – internal

11.10.1 Latchup Protection Solutions – Semiconductor Process

Semiconductor processes can be developed to provide immunity to latchup events. The process solutions can include the following:

- Retrograde n-well design [1, 2]
- Retrograde p-well design
- Triple well isolation [2]
- Heavily doped buried layer (HDBL) [33]
- Buried Grid (BGR) [33]
- Shallow trench isolation [2]
- Deep trench isolation [2]
- Silicon on insulator buried oxide layer.

11.10.2 Latchup Protection Solutions – Design Layout

Design layout of circuitry can be used to minimize CMOS latchup sensitivity. Design variables can be separated into intra- and intercircuit latchup design [2].

11.10.2.1 Latchup Protection Solutions – Internal Design Layout

To decrease a semiconductor design sensitivity to latchup, physical spacing and structure can be used (Figure 11.14). The design variables to improve the latchup immunity are as follows:

- Increase the spacing between the p+ diffusion and n-well edge.
- Increase the spacing between the n+ diffusion and adjacent n-well.

- Decrease the n-well tap to PFET spacing.
- Decrease the p-well tap to NFET spacing.
- Place the n-well tap between the PFET and n-well.
- Place the p-well tap between the NFET and adjacent n-well.
- Place guard rings around the PFET.
- Place guard rings around the NFET.

11.10.2.2 Latchup Protection Solutions – External Design Layout

To decrease semiconductor design sensitivity to external latchup, physical spacing and structure can be used (Figure 11.15). The design variables to improve the latchup immunity are as follows:

- Increase the width of the substrate tap.
- Increase the width of the n-well tap.
- Place guard rings around the PFET.
- Place guard rings around the NFET.
- Place guard ring external to complete circuit.

11.10.3 Latchup Protection Solutions – Circuit Design

To decrease semiconductor design sensitivity to latchup, circuit design solutions exist (Figure 11.16) [2]. Circuit solutions for latchup immunity can include the following:

- TOS shutdown networks
- Electrical overvoltage (EOV) networks
- Electrical overcurrent (EOC) networks
- Sequence-independent power supply system
- Active clamp circuits

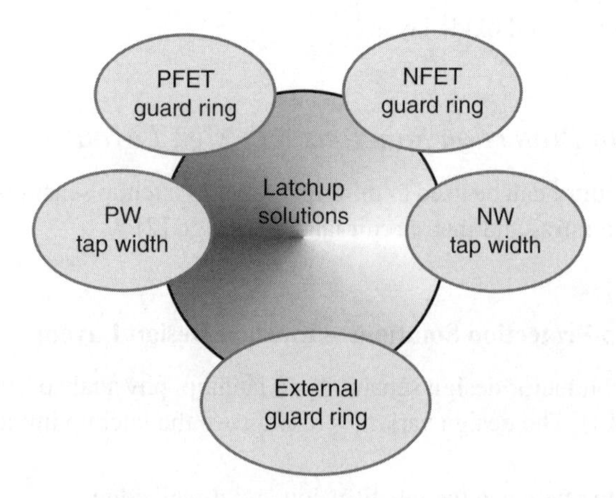

Figure 11.15 Latchup design layout solutions – external

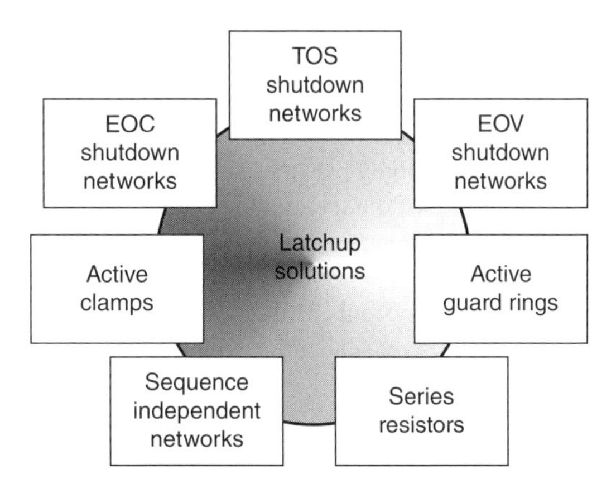

Figure 11.16 Latchup circuit design solutions

- Active guard ring structures
- Resistor elements in series with the power supply.

11.10.4 Latchup Protection Solutions – System Level Design

To decrease semiconductor design sensitivity to latchup, system level solutions exist (Figure 11.17). System level solutions for latchup immunity can include the following [2, 5]:

- Printed circuit board low ground resistance
- Power-up and power-down procedures
- Thermal shutdown networks
- EOV networks
- EOC networks
- Thermal circuit breakers
- Electrical overstress (EOS) protection devices.

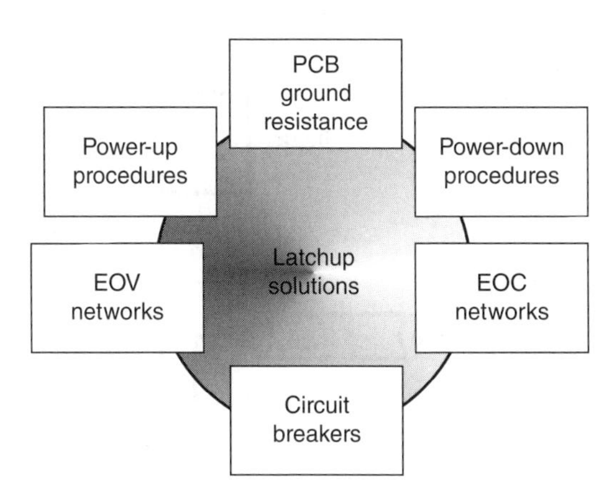

Figure 11.17 Latchup system level design solutions

11.11 Alternate Test Methods

Alternate test methods to study the latchup immunity of systems exist beyond electrical testing [2–5]. Optical visual techniques are used for latchup integrity to visualize latchup interior to a semiconductor chip. Electrical testing only provides signal pin and terminal characteristics but does not allow for an understanding of the location of latchup within a chip. Different failure analysis techniques include the following:

- CCD photoemission visualization systems [35]
- Picosecond current analysis (PICA) visualization tool [34]
- Liquid crystal techniques [3]
- SQUID imaging technique [3].

In the following sections, the PICA tool and CCD imaging tool will be used as examples.

11.11.1 Photoemission Techniques – PICA–TLP

A photoemission method for characterizing latchup is using a test system that scans the component to determine photoemissions [2, 3, 34]. An advanced test system known as the PICA is shown in Figure 11.18. Using a high current source, photon emissions can be detected. A high-quality photomultiplexer can determine the photon emission in space and time. A transmission line pulse (TLP) source can apply a pulse train where the photodetector can store and count the photon emissions. In this system, an animation movie can be produced that

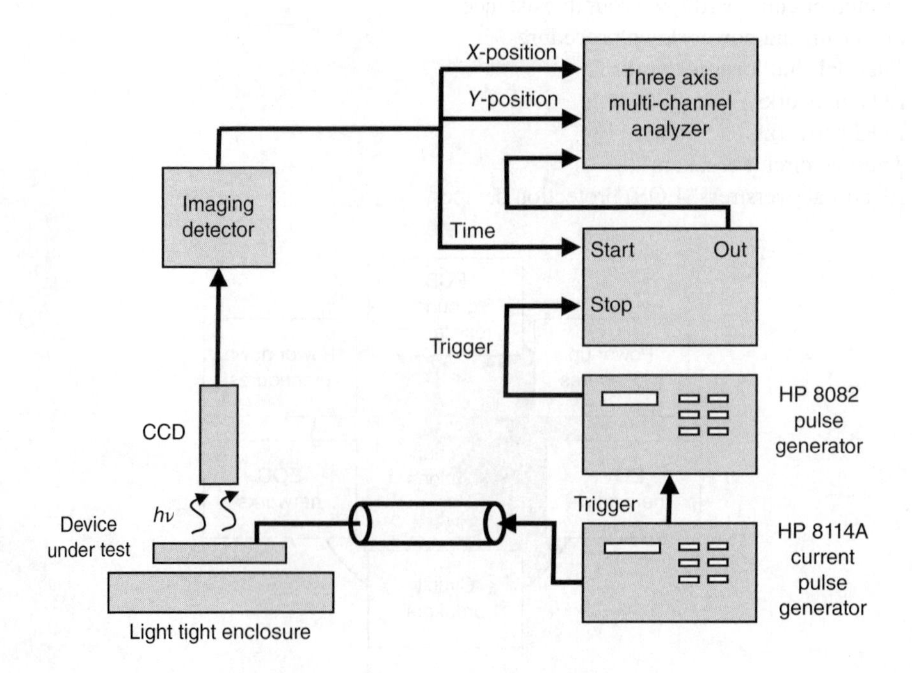

Figure 11.18 PICA – TLP test method

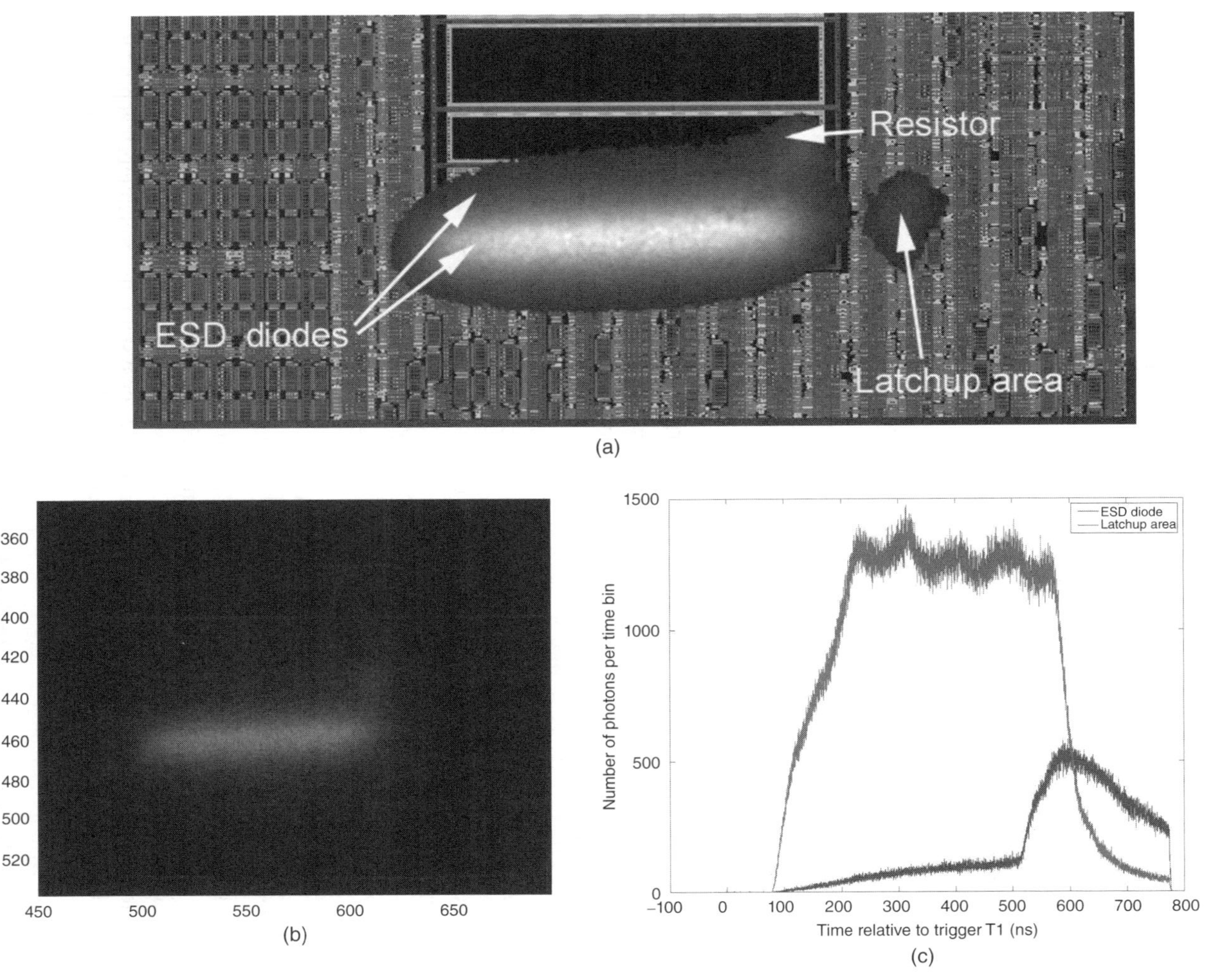

Figure 11.19 (a) Photon emissions overlaid on semiconductor chip design. (b) Photon emissions from ESD diode. (c) Photon emission count versus time

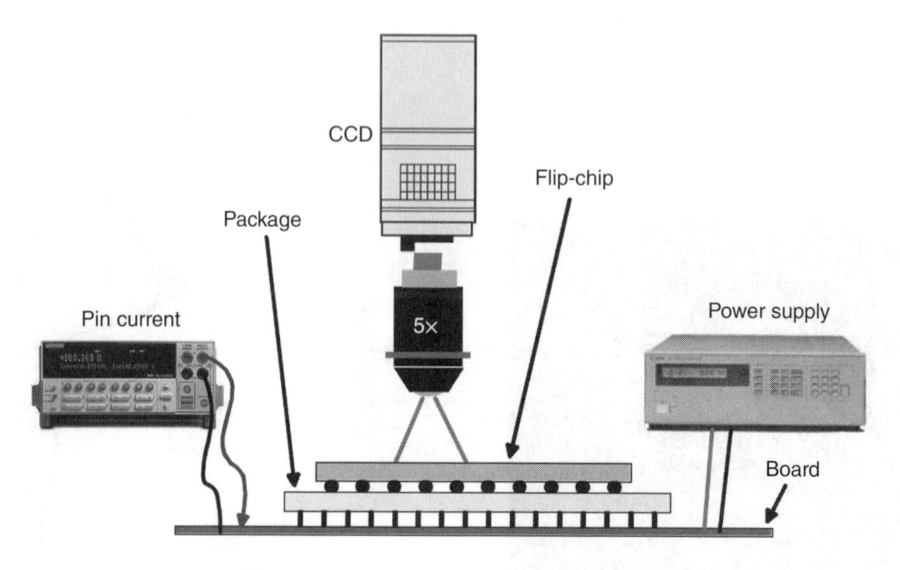

Figure 11.20 Photon emission test system

shows the occurrence of latchup in the test chip. The system can overlay the design layout and the photon emission allowing one to visualize where in the physical design is occurring. Figure 11.19 shows examples of the photon emissions from the test system.

11.11.2 Photoemission Techniques – CCD Method

A second methodology using photon emission can be a CCD system [35]. The CCD test system is shown in Figure 11.20. In this system, the photon emission can also be obtained at different times. Figure 11.21(a) and (b) are examples of the photon emission patterns observed from the test system.

11.12 Single Event Latchup (SEL) Test Methods

A single particle can lead to initiation of latchup in a CMOS technology. This is known as single event latchup (SEL). SEL can occur in space applications. Heavy ions can lead to a latchup event in silicon or gallium arsenide products (Figure 11.22).

A testing methodology is to test semiconductor chips or systems in accelerators that produce particles that may lead to CMOS latchup. A test methodology is illustrated in Figure 11.23.

11.13 Summary and Closing Comments

This chapter discussed CMOS latchup. CMOS latchup physics and theory are briefly discussed, followed by test methods and issues. CMOS DC and transient latchup are highlighted as well as "internal latchup" versus "external latchup." The chapter introduced

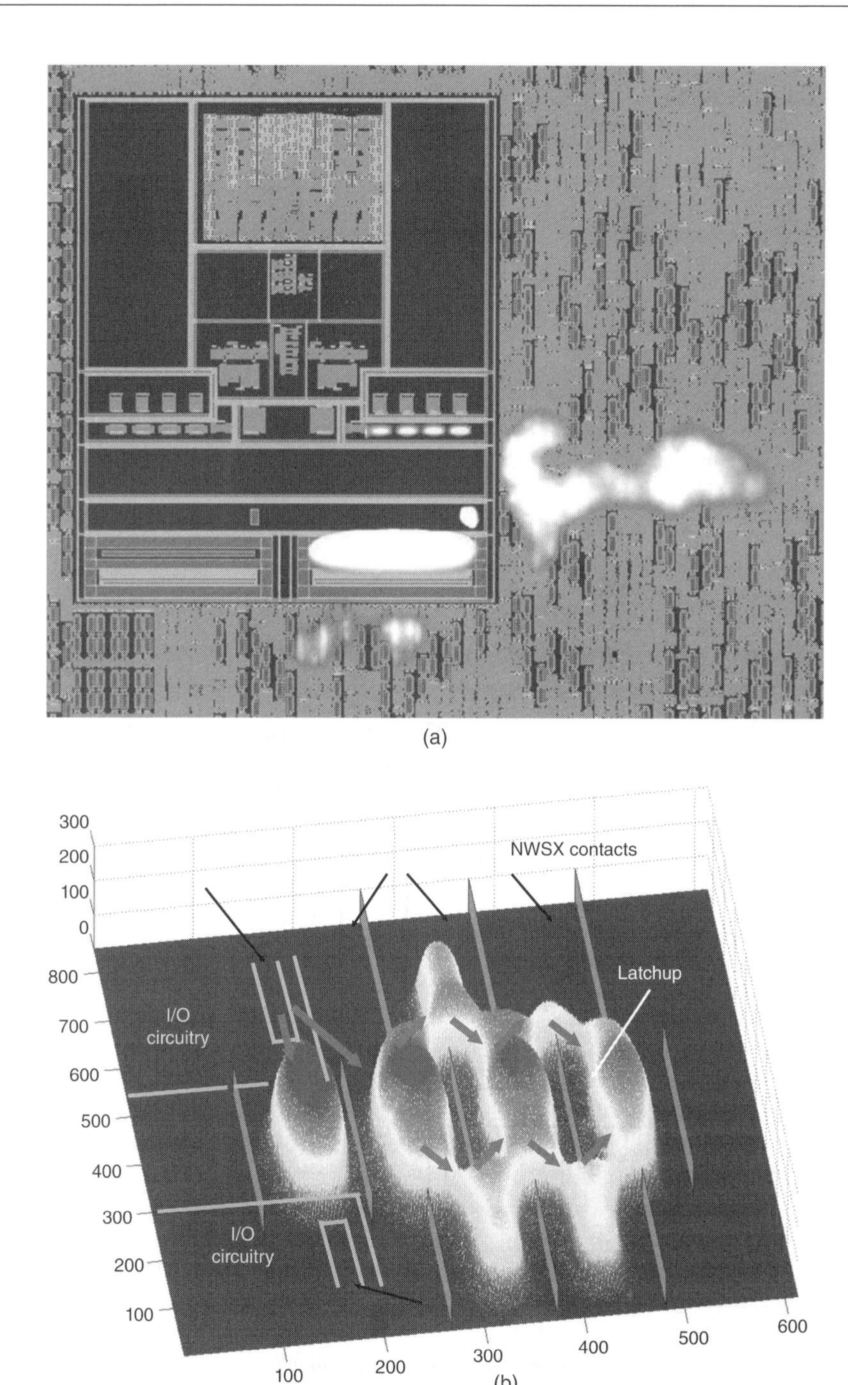

Figure 11.21 (a) Photon emission image. (b) Photon emission image (expanded)

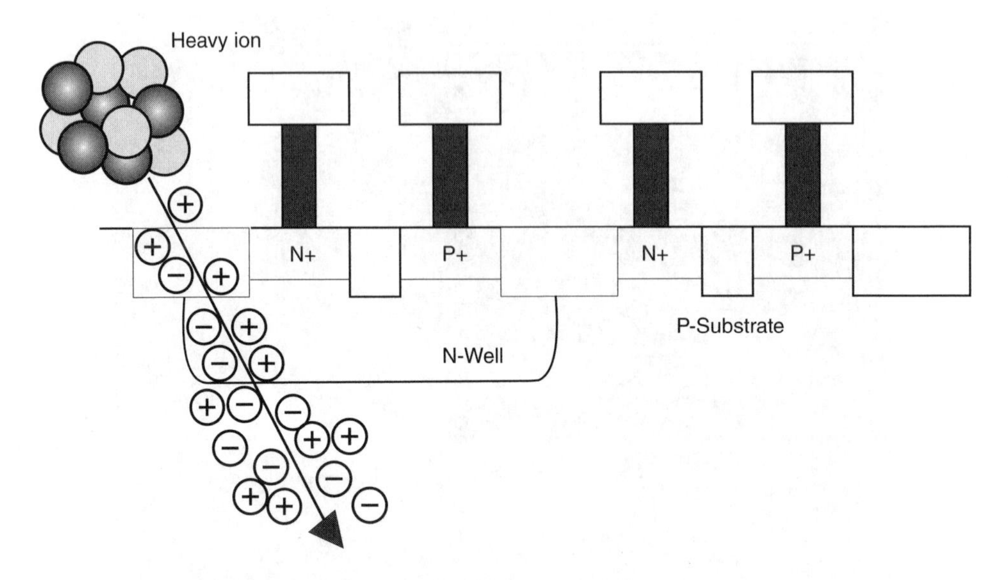

Figure 11.22 Single event latchup (SEL)

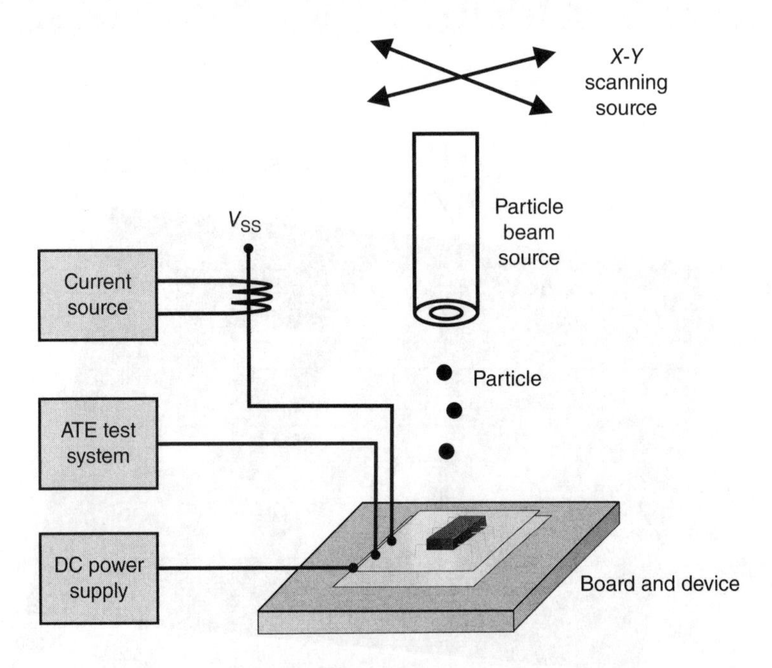

Figure 11.23 Testing technique

latchup benchmark test structures and test procedures to quantify the latchup robustness of a technology.

Chapter 12 looks at the broader issue of EOS. The topic of EOS will be approached providing distinctions of EOS versus ESD phenomena and specific EOS tests that are presently used for systems and for semiconductor chips. EOS failure mechanisms and solutions will be quantified. With the growing increase of system level issues, EOS has been receiving a renewed interest from both semiconductor chip developers and system designers.

Problems

1. What is latchup?
2. Draw a cross section of a CMOS process, and the process of the carrier transport that initiates latchup.
3. Draw a circuit schematic on the cross section of Problem 2 that are fundamental to the latchup process.
4. Draw a circuit schematic between the power supply V_{DD} and substrate V_{SS} highlighting the four fundamental elements involved in the latchup process.
5. Show from the layout how to reduce the likelihood of latchup from layout perspective.
6. Describe the distinction between direct current latchup and transient latchup.
7. Describe the difference between "internal latchup" and "external latchup." Show pictures to describe the distinctions.
8. How from the visual signature of failure analysis can you determine the failure was associated with latchup?
9. How from the electrical signature can one determine the failure was associated with latchup?
10. What is the distinction between long-pulse TLP and CMOS transient latchup testing?

References

1. R. Troutman. *Latchup in CMOS Technology: The Problem and the Cure*. New York: Kluwer Academic Publishers, 1985.
2. S. Voldman. *Latchup*. Chichester, England: John Wiley and Sons, Ltd., 2007.
3. S. Voldman. *ESD: Failure Mechanisms and Models*. Chichester, England: John Wiley and Sons, Ltd., 2009.
4. S. Voldman. *ESD Basics: From Semiconductor Manufacturing to Product Use*. Chichester, England: John Wiley and Sons, Ltd., 2012.
5. S. Voldman. *Electrical Overstress (EOS): Devices, Circuits, and Systems*. Chichester, England: John Wiley and Sons, Ltd., 2013.
6. J.L. Moll, M. Tannenbaum, J.M. Goldey, and N. Holonyak. P-N-P-N transistor switches. *Proceedings of the IRE*, Vol. 44, Sept 1956; 1174–1182.
7. A.K. Jonscher. P-N-P-N switching diodes. *Journal of Electronic and Control*, Vol. 3, (6), Sept 1957; 573–586.
8. F.E. Gentry, F.W. Gutzwiller, N. Holonyak, and E.E. von Zastrow. *Semiconductor Controlled Rectifiers: Principles and Applications of p-n-p-n Devices*. Englewood Cliffs, NJ: Prentice-Hall, Inc., 1964.
9. B.L. Gregory, and B.D. Shafer. Latchup in CMOS integrated circuits. *IEEE Transaction on Nuclear Science*, Vol. NS-20, Dec 1973; 293–299.
10. B.L. Gregory. Latchup studies in bulk silicon CMOS integrated circuits. Sandia Laboratory Report No. 73–3003, Albuquerque, New Mexico, January 20, 1973.
11. B.D. Shafer. CMOS latchup analysis and prevention. Sandia Laboratories Report SAND75-0371, Albuquerque, New Mexico, June 1975.

12. L.L. Sivo, F. Rosen, and L.C. Jeffers. Latchup screening of LSI devices. *IEEE Transaction on Nuclear Science*, Vol. NS-25, Dec 1978; 1534–1537.
13. D.B. Estreich, A. Ochoa, Jr.,, and R.W. Dutton. An analysis of latchup prevention in CMOS IC's using an epitaxial-buried layer process. *International Electron Device Meeting (IEDM) Technical Digest*, 1978; 230–234.
14. D.B. Estreich, and R.W. Dutton. Modeling latchup in CMOS integrated circuits. *1978 Asilomar Conference on Circuits, Systems, and Computer Digest*, Monterey, California, November 1978; 489–492.
15. D.B. Estreich, and R.W. Dutton. Latchup in CMOS integrated circuits. *1978 Government Microcircuit Applications Conference (GOMAC) Digest of Papers*, Monterey, California, November 1978; 110–111.
16. D.B. Estreich. Latchup and Radiation Integrated Circuit (LURIC): a test chip for CMOS latchup investigation. *Sandia Laboratories Report SAND78-1540*, Albuquerque, New Mexico, November 1978.
17. D.B. Estreich. The physics and modeling of latch-up and CMOS integrated circuits. *Technical Report No. G-201-9*, Integrated Circuits Laboratory, Stanford Electronic Laboratories, Stanford University, Stanford, California, November 1980.
18. P.V. Dressendorfer and A. Ochoa. An analysis of modes of operation of parasitic SCR's. *IEEE Transactions on Nuclear Science*, Vol. NS-28, (6), Dec 1981; 4288–4291.
19. A. Ochoa and P.V. Dressendorfer. A discussion of the role of distributed effects in latchup. *IEEE Transactions on Nuclear Science*, Vol. NS-28, (6), Dec 1981; 4292–4294.
20. A.W. Wieder, C. Werner, and J. Harter. Design model for bulk CMOS scaling enabling accurate latchup prediction. *International Electron Device Meeting (IEDM) Technical Digest,* 1981; 354–357.
21. C.C. Huang, M.D. Hartanft, N.F. Pu, C. Yue, C. Rahn, J. Schrankler, G.D. Kirchner, F.L. Hampton, and T.E. Hendrickson. Characterization of CMOS latchup. *International Electron Device Meeting (IEDM) Technical Digest,* 1982; 454–457.
22. R.R. Troutman and H.P. Zappe. Power-up triggering condition for latchup in bulk CMOS. *1982 Symposium on VLSI Technology*, Oiso, Japan. September 1982, 52–53.
23. 23.R.R. Troutman and H.P. Zappe. A transient analysis of latchup in bulk CMOS. *IEEE Transactions on Electron Devices*, Vol. ED-30, Feb 1983; 170–179.
24. R.D. Rung, and H. Momose. DC holding and dynamic triggering characteristics of bulk CMOS latchup, *IEEE Transaction on Electron Devices*, Vol. ED-30, Dec 1983; 1647–1655.
25. R.R. Troutman. Epitaxial layer enhancement of n-well guard rings for CMOS circuits. *IEEE Electron Device Letters*, Vol. EDL-4, Dec 1983; 438–440.
26. E. Hamdy, and A. Mohsen. Characterization and modeling of transient latchup in CHMOS technology. *International Electron Device Meeting (IEDM) Technical Digest,* 1983; 172–175.
27. R.R. Troutman and M.J. Hargrove. Transmission line model for latchup in CMOS circuits. *Symposium on VLSI Technology*, September 13–15, 1983; 56–59.
28. G.J. Hu. A better understanding of CMOS latchup. *IEEE Transactions on Electron Devices*, Vol. ED-31, Jan 1984; 62–67.
29. W. Craig. Latchup test structures and their characterization. IEEE VLSI Workshop on Test Structures, San Diego, February 20, 1984.
30. R.R. Troutman, and H.P. Zappe. Layout and bias considerations for preventing transiently triggered latchup in CMOS. *IEEE Transactions on Electron Devices*, Vol. ED-31, Mar 1984; 279–286.
31. S. Odanaka, M. Wakabayashi, and T. Ohzone. The dynamics of latchup turn-on behavior in scaled CMOS. *IEEE Transactions on Electron Devices*, Vol. ED-32, July 1985; 1334–1340.
32. M. Hargrove, S. Voldman, J. Brown, K. Duncan, and W. Craig. Latchup in CMOS technology. *Proceedings of the International Reliability Physics Symposium (IRPS)*, 1998; 269–278.
33. W. Morris. Latchup in CMOS. *Proceedings of the International Reliability Physics Symposium (IRPS)*, 2003; 76–84.
34. A. Weger, S. Voldman, F. Stellari, P. Song, P. Sanda, and M. McManus. A transmission line pulse (TLP) picosecond imaging circuit analysis (PICA) methodology for evaluation of ESD and latchup. *Proceedings of the International Reliability Physics Symposium (IRPS)*, 2003; 99–104.
35. F. Stellari, P. Song, M.K. McManus, A.J. Weger, K. Chatty, M. Muhammad, and P. Sanda. Study of critical factors determining latchup sensitivity in ICs using emission microscopy. *Proceeding of the International Symposium for Testing and Failure Analysis (ITSFA)*, 2003; 19–24.

36. Y. Huh, K.J. Min, P. Bendix, V. Axelrad, R. Narayan, J.-W. Chen, L.D. Johnson, and S. Voldman. Chip-level layout and bias considerations for preventing neighboring I/O cell interaction induced latchup and inter-power supply latchup in CMOS technologies. *Proceedings of the Electrical Overstress/Electrostatic Discharge (EOS/ESD) Symposium*, 2005; 100–107.
37. M. Hernandez. Electrical overstress, and transient latchup pulse generator system, circuit, and method. U.S. Patent No. 7,928,737, April 19, 2011.
38. JEDEC, JESD78D, IC Latch-Up Test. 2011.
39. Thermo Scientific. https://www.thermofisher.com. Thermo Scientific MK4^{TM} test system, 2016.

12

Electrical Overstress (EOS)

12.1 History

Electrical overstress (EOS) has been an issue with the coming of the electrical age, when electricity and electrical product were first introduced into the mainstream of society [1]. With the introduction of electrical power systems, the telephone, and electronics, inventions such as circuit breakers and fuses became the first type of EOS protection concepts to avoid overload of electronic systems [1–5].

EOS has been an issue in devices, circuit, and systems for electronics for many decades, as early as the 1970s and continues to be an issue till today [6–12]. Market segments from consumer, industrial, aerospace, military, and medical are all influenced by this issue. The experience of EOS failures are occurring at the device manufacturer, supplier, assembly, and the field. In the electronic industry, many products and applications are returned from the field due to "EOS" failure. To make progress in addressing the EOS issue, it is important to provide a framework for evaluation and analysis of EOS phenomena. As part of this framework, it is important to apply a vocabulary and definitions. It is key to apply both physical and mathematical definitions to quantify the EOS conditions. It is equally important to establish a methodology of failure analysis and testing. It is also critical to establish an awareness of the origins and sources of EOS concerns. In the end, to provide better EOS robust products, it is important to define design practices and procedures, as well as EOS control programs for manufacturing and production areas.

EOS sources exist from natural phenomena, power distribution, power electronics, and machinery [1, 5–12]. A significant natural phenomenon is lightning. Power distribution includes switches, cables, and other power electronics that can be a source of EOS. EOS sources also occur from design characteristics of devices, circuits, and systems. In the following sections, these issues are discussed.

Many of the EOS issues can occur from the design of the semiconductor component, the system, and its integration. Examples of EOS source design issues are as follows [1]:

- Semiconductor process – application mismatch
- Printed circuit board (PCB) inductance
- PCB resistance

ESD Testing: From Components to Systems, First Edition. Steven H. Voldman.
© 2017 John Wiley & Sons, Ltd. Published 2017 by John Wiley & Sons, Ltd.

- Latchup sensitivity [14]
- Safe operating area (SOA) power rating violation
- SOA voltage rating violation
- SOA current rating violation
- Transient SOA – *di*/*dt* and *dv*/*dt*.

Testing and test simulation of devices, components, and systems are an important part of the evaluation to EOS. EOS test simulation is valuable part of understanding EOS failures. EOS testing provides the following [1]:

- *Root Cause Analysis*: Determining the root cause of component and system failures
- *Replication of Failure*: Repeat the visual and electrical signature
- *Technology Hardness Evaluation:* Determining the robustness (or hardness) of a technology
- *Technology Benchmarking:* Comparative analysis of EOS or electrostatic discharge (ESD) hardness technology-to-technology
- *Component Reliability Qualification:* Qualification and release of components
- *System Qualification:* Qualification and release of systems.

One of the key concerns of EOS is the cost. There are different types of costs associated with EOS. In this section, the cost associated with field returns is discussed. In order to quantify the cost of EOS events on products, it is critical to categorize what percentage of the field returns are in fact EOS related.

Product field returns occur in all electronic components independent of the technology generation and period of time of evaluation. One of the key difficulties in the semiconductor industry is the ability to track, record, and maintain a database of these field failures.

A key question in the electronic industry is: what is the percentage of the field returns that is due to EOS?

In the mid-1980s, the military established an in-house program to track, record, and categorize field failures to answer this question [1, 21]. The U.S. military and the Reliability Analysis Center (RAC) in Rome, NY, jointly established the Field Failure Return Program (FFRP), with the objective of providing feedback to the semiconductor industry and determining the root cause of failure. By establishing the root cause of failure, the corrective action can be initiated. The FFRP goals were as follows [21]:

- Identify high failure rate or component problems.
- Identify their root causes of failure from failure analysis.
- Feedback the information to the supplier, industry, or government organization for corrective action.

In this early reliability study, data from 24 different systems were collected and reviewed. In this review, 1650 parts were evaluated, of which the part numbers were from actual field failures that were operational from 2 to 10 years. Table 12.1 shows the results of the field failure categories [1].

From this study, 46% of the field returns were associated with EOS. It was regarded from this study, a number of EOS issues were associated with poor system design, improper maintenance procedures, and improper operational procedures. In the second category, it was regarded that

Table 12.1 Field failure categories and percentage

Field failure category	Percentage of field failures (%)
Electrical overstress (EOS)	46
IC design, fabrication, and assembly	25
Retested without observed failure	17
Electrostatic discharge (ESD)	6
EOS or ESD	6

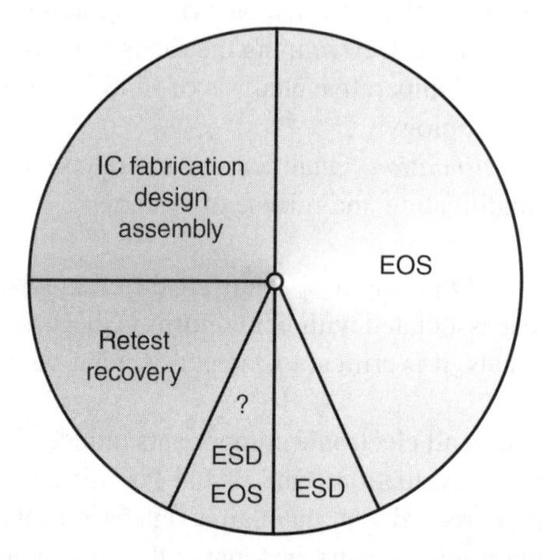

Figure 12.1 Field failure categories

these failures were from inherent flaws and latent defects. Of the field returns, only a few percent were related to ESD. Note, that in some cases, it was decided that it was not possible to determine if the failures were EOS or ESD (Figure 12.1).

The results of this study are not significantly distinct from other future studies. It is typically quoted that EOS is a high percentage of field failures, and a certain percentage cannot distinguish from EOS or ESD.

In a more recent study by C. Thienel for the automotive industry called "Avoiding Electrical Overstress for Automotive Semiconductors by New Connecting Concepts" attributed 6% of the failures to ESD and 94% to EOS [7]. A large percentage of the fails were "no defect found" and approximately 32% were EOS/ESD failures.

12.2 Scope

The scope of EOS test is for the testing, evaluation, and classification of components and micro- to nanoelectronic circuitry. The test is to quantify the sensitivity or susceptibility of these components to damage or degradation to EOS [1].

12.3 Purpose

The purpose of the EOS test is for establishment of a test methodology to evaluate the repeatability and reproducibility of components or microelectronic circuitry to a pulse event in order to classify or compare EOS levels of components [1].

12.4 Pulse Waveform

ESD event characteristic time response is associated with a specific process of charge accumulation and discharge. Hence, the characteristic time response is definable enough to establish an ESD standard associated with the specific process. Second, the time response of ESD events is fast processes. The time constant for ESD events range from subnanoseconds to hundreds of nanoseconds [1].

In contrast, EOS events do not have a characteristic time response. They can have short time response or long (note: today, it is popular to separate the "ESD events" as distinct from "EOS events" which is what will be followed in this text). EOS processes are typically slower, and distinguishable from ESD events by having longer characteristic times. The time constant for EOS events range from submicroseconds to seconds (Figures 12.2–12.4).

12.5 Equivalent Circuit

In EOS, there is no defined pulse width nor pulse shape. As a result, there is no equivalent circuit. What is known is that the time constant is long.

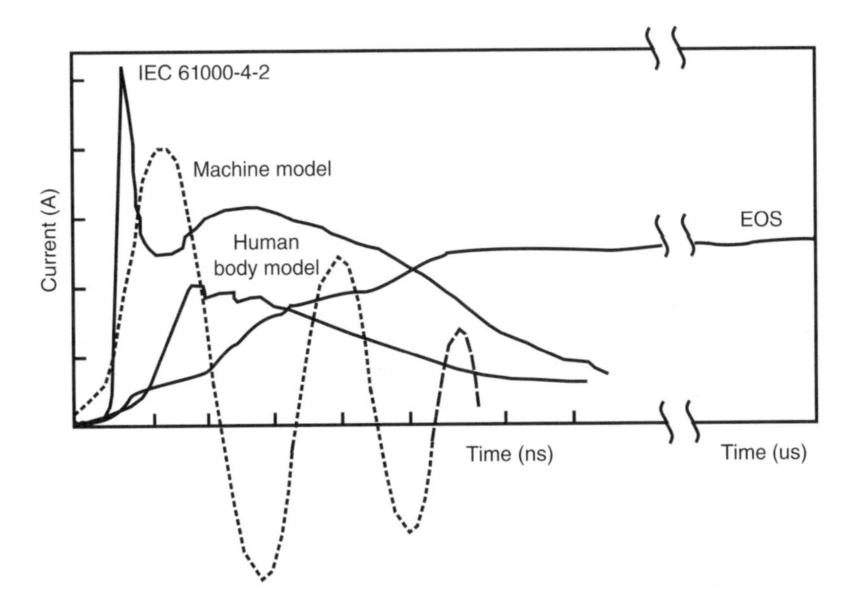

Figure 12.2 EOS versus ESD waveforms

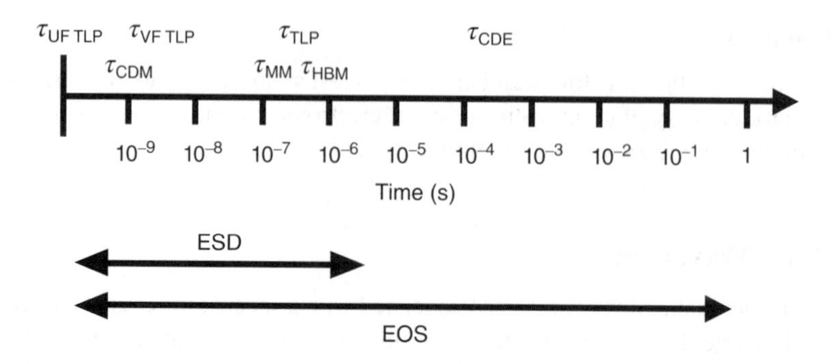

Figure 12.3 EOS and ESD event time constant spectrum

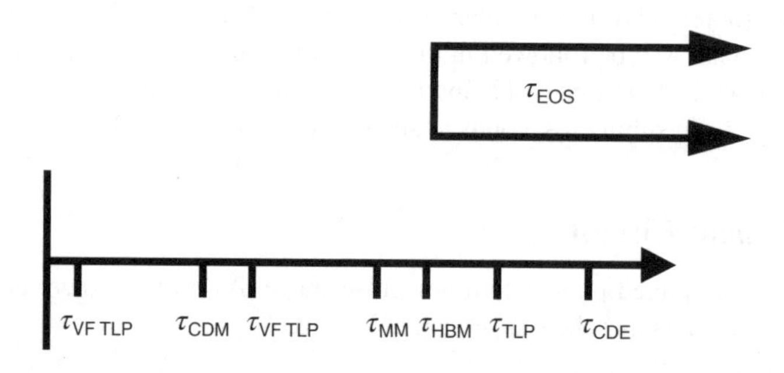

Figure 12.4 EOS and ESD event time constant hierarchy

Today, there is discussion of a future EOS-defined pulse and equivalent circuit, which is a "long duration pulse transmission line pulse (LD-TLP)" of pulse width 500 ns. Given that this is an accepted pulse waveform, then the equivalent circuit would be a low loss transmission line.

12.6 Test Equipment

In EOS, there is neither defined pulse width nor pulse shape. As a result, there is no equivalent circuit. What is known is that the time constant is long. Using square pulse testing or oscillatory pulse events, EOS phenomena can be quantified.

12.7 Test Procedure and Sequence

In EOS, a test procedure and sequence can be defined to evaluate the EOS robustness. Figure 12.5 illustrates an example of a test procedure and sequence.

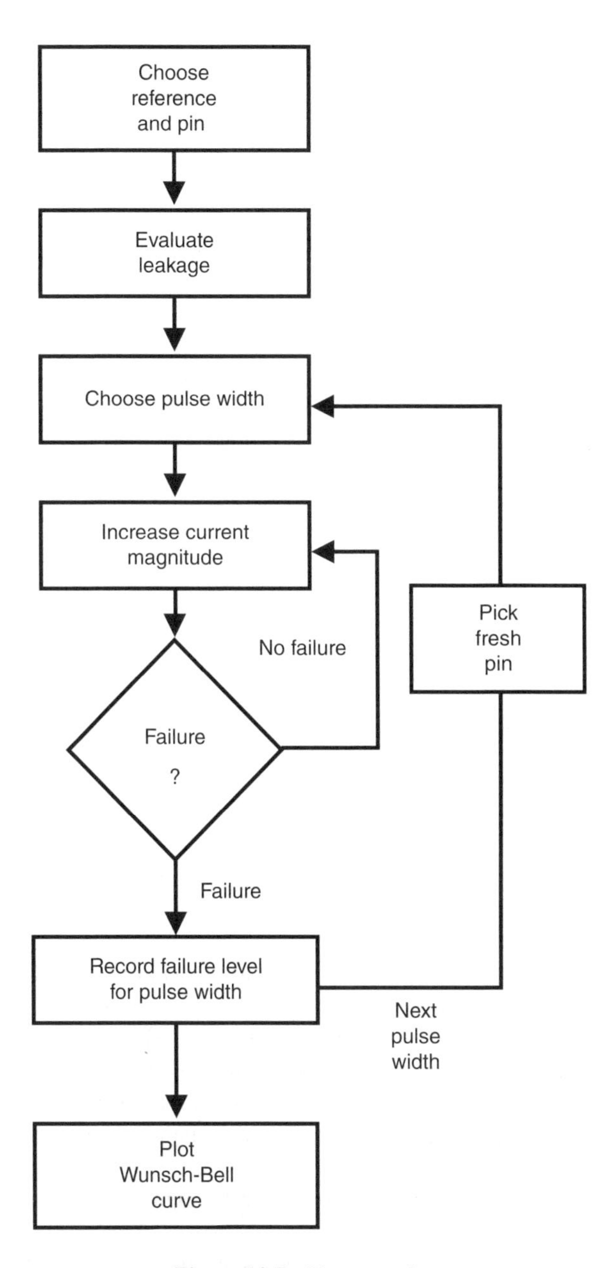

Figure 12.5 Test procedure

12.8 Failure Mechanisms

EOS failure mechanisms can occur from electrical overvoltage (EOV), electrical overcurrent (EOC), and electrical overpower (EOP) [1, 6, 13–71]. These conditions can lead to melted packages, blown single component capacitors and resistors, ruptured packages, blown bond wires, cracked dielectrics, fused and melted metal layers, and molten silicon.

Failure analysis of components and systems involves investigation, science, and experience. As a failure analysis matures, a failure analyst's experience will allow a faster resolution of the root cause and solution. A failure analyst is like a forensic investigator who much finds the problem and resolves the case for each product failure. This is important in product development of semiconductor components where the answer will be required prior to undergoing the next "design pass." Time is limited between design passes to allow shorter cycles of the design for release and volume ramping into production [1].

The steps of the failure analysis process are as follows:

- *Information gathering*
- *Failure verification*
- *Failure site identification and localization*
- *Root cause determination*
- *Feedback of root cause*
- *Corrective action*
- *Documentation reports*
- *Statistical analysis, record retention, and control.*

12.8.1 Information Gathering

In the information gathering process for evaluation of EOS events, there are many elements to acquire and assemble to get a coherent view of the failure:

- *Failure facility or field location:* Where did the failure occur? Did it occur in manufacturing, production, or the field? In what stage of the process did this occur? What tools did the product undergo in manufacturing? What is the yield history of the different tools? In semiconductor manufacturing and assembly, there are yield analysis experts who know how to determine the path of the hardware and where yield losses occur. Yield experts will also have inspection history, and qualification of the tools, stations, and environment history.
- *Failure determination method:* How was the failure discovered? Is it important to determine root cause of how it was found? Thermal stress? Electrical measurements? Visual signature? Electrical characteristics?
- *Visual signature:* From visual inspection, is there any sign of the failure or defect? This is possible on the PCB, plastic molding (in the forms of discoloration, molten package, holes, and bubbles in the package).
- *Electrical signature:* From the electrical signature, what is the type of electrical signature? Short? Open? I_{DD} leakage? I–V characteristic shift? V_{DD}–V_{SS} resistance?
- *Product description and datasheet:* Knowledge of the product description, function, and electrical parameters is important for the evaluation of the product. It is important to understand the functional blocks in the physical design.

- *Comparative study:* It is valuable to evaluate examples of good product and compare the electrical response of good versus bad part numbers. For visualization of the electrical characteristics, it is a good practice to have the data overlaid, or together in a report.
- *Database review:* It is valuable to evaluate and review the product history in terms of temporal variation and statistical parameters. It is important to evaluate the site-to-site, wafer-to-wafer, lot-to-lot, and foundry-to-foundry differences within a manufacturing environment or assembly facility. This may provide a clue to the source of the problem.
- *Electrical test simulation:* A means to verify and resimulate the event is important to see if the failure mechanism can be repeated and show the same electrical signature or visual signature. For example, a key question is to determine whether the event is ESD or EOS. ESD tests (e.g., human body model (HBM), machine model (MM), charged device model (CDM), IEC 61000-4-2, HMM) can be performed to try to replicate the electrical and visual signatures. To determine if it is latchup, JEDEC latchup testing can be completed, where the current is increased to failure. For EOS, a series of electrical simulations can be performed for different current magnitudes and pulse widths.

12.8.2 Failure Verification

The goal of failure verification is to determine whether one can reproduce the failure. This can be completed by product test and retest. Retesting is important, since some failures recover, or never really occurred. A percentage of field returns at times never repeat the failure signature. Comparative testing of good/bad product can also be performed for the verification process. This can be done by overstress test simulation on untested or "good" parts.

Both nondestructive and destructive testing methods exist for failure verification [1]. Nondestructive testing methods are as follows [1]:

- Optical inspection
- Acoustic microscope
- X-ray inspection
- Pin-to-pin testing
- Pin-to-rail testing
- Parametric functional testing.

12.8.3 Failure Site Identification and Localization

For the identification of the failure site within the system, this can be done by either visual external inspection or internal inspection. For visual inspection, one can look for the following visual damage signatures [1]:

- Package lead damage
- Foreign material
- Cracks
- Package discoloration
- Corrosion.

For internal inspection, the following visual damage signatures are as follows:

- Melted metallurgy
- Cracked interlevel dielectrics
- Molten silicon.

12.8.4 Root Cause Determination

Root cause of the failure can be determined by evaluation of the manufacturing and production center records and processes, the electrical and/or visual signatures, evaluation of the circuit schematics and design layout, and test simulation. Once the root cause is determined, it is important to review the results and draw definitive conclusions.

12.8.5 Feedback of Root Cause

Once the root cause is determined, it is important to provide feedback to the manufacturing, production, design team, reliability, and quality teams. Results and conclusions should be reviewed with the complete team to close the issue and establish corrective actions.

12.8.6 Corrective Actions

Corrective actions can be taken to avoid repeating the event after its root cause is determined. The corrective action can be the chip design (e.g., design layout and/or circuit), manufacturing tooling, or production processes. It is also important in the case of a corporation with a high "reuse" of circuit blocks, to prevent EOS-sensitive or ESD sensitive blocks from being used in other products until the corrective actions are completed.

12.8.7 Documentation Reports

Documentation of the results in a report is important for the EOS control program management. To have a good EOS control program, documents may be a source of learning for future design reviews, checklists, procedures, and audits.

12.8.8 Statistical Analysis, Record Retention, and Control

Statistical analysis and record retention are important for an EOS control program and to maintain an EOS Safe Protected Area.

A common question that arises is: whether from the failure analysis can one determine if the root cause of the failure is EOS or ESD? Resolving this question is dependent on the visual failure signature.

There are certain categories of failures that ESD does not typically cause and EOS events do cause (Figure 12.6). Failures that are typically caused by EOS phenomena but not ESD are as follows [1]:

- PCB damage
- Package molding damage

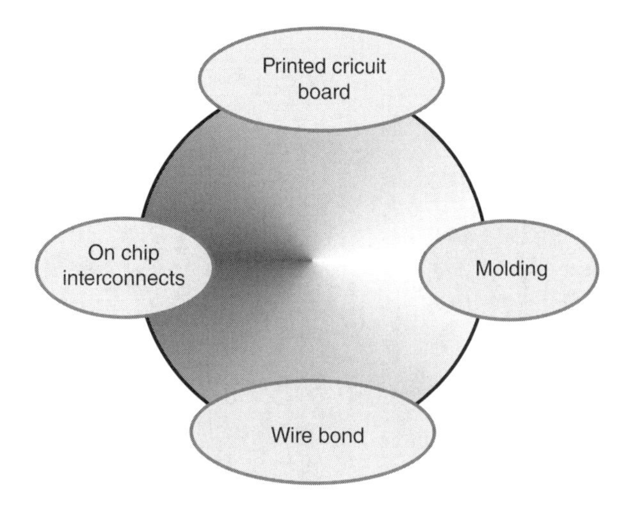

Figure 12.6 EOS failure mechanisms

- Package pin damage
- Wire bond damage.

Today, PCBs, packaging, and wire bonds do not typically fail due to an ESD event caused by HBM, MM, or CDM events. Figure 12.7 is an example of a wire bond failure [1].

Figure 12.8 is an example of package pin damage due to EOS. From the package damage and package pin, it is clear that the region underwent significant currents leading to failure [1].

There are also failure types associated with long time constant pulses. Phenomena affiliated with long pulse width significantly longer than the thermal diffusion time constant of the

Figure 12.7 EOS failure mechanism – wire bond

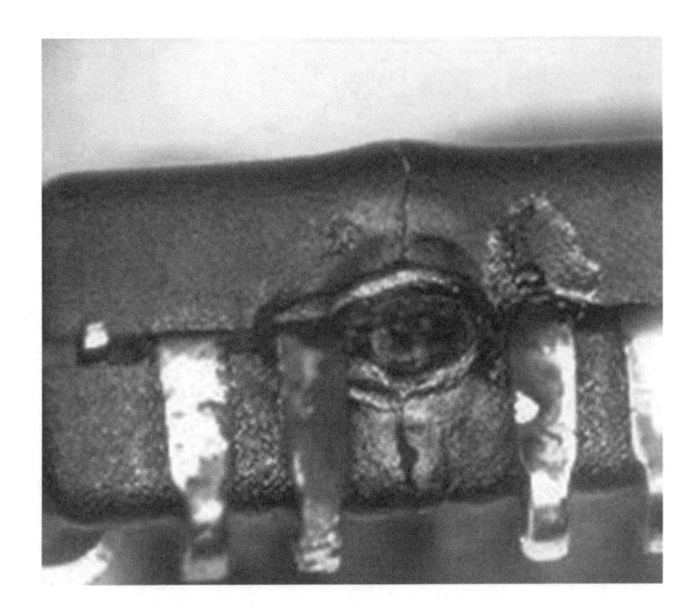

Figure 12.8 EOS failure mechanism – packaging

structure are prone to thermal transport, heating, and electromigration [47, 50, 56–59, 71]. Hence, processes that involve material transport, thermal stress, and mechanical stress are more affiliated with EOS. The resulting packaging ablation is associated with both thermal and mechanical stress [1].

EOS events can shorten the reliability lifetime of components, leading to an early wearout and inducing shifts in the reliability "bathtub" curve of a component. As discussed in the previous sections, this will be a bigger issue as advanced components are scaled.

Where it is hard to distinguish between EOS and ESD is in the failure signature of a component. Both EOS and ESD can introduce "on-chip" failure mechanisms whose visual failure signatures are similar. Both EOS and ESD can create the following failures [1]:

- Dielectric breakdown
- Interlevel dielectric cracking
- Metal failures
- Molten semiconductor damage.

At times, failure analysts distinguish these events based on physical size. Typically, EOS events create a larger failure damage, more distributed through the semiconductor chip; but this is dependent on the ESD event type. Figure 12.9 is EOS of the wire interconnects bond pad.

12.9 Electrical Overstress (EOS) Protection Circuit Solutions

Today, EOS is still an issue in today's electronic systems [6–10]. To address EOS in systems, EOS protection devices are added to PCB, cards, and systems [29–31, 45, 53, 54, 60, 72–83].

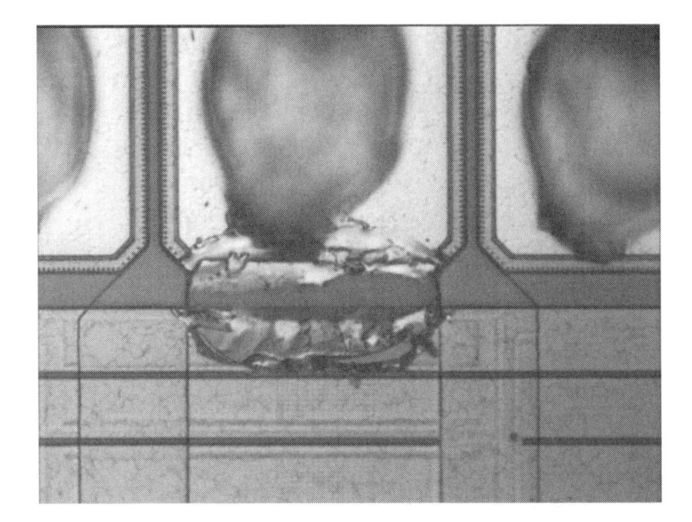

Figure 12.9 EOS failure mechanism – bond pads

In this chapter, the focus will be on providing an understanding of the different classifications and types of EOS protection devices used today. EOS protection devices are supported by a large variety of technologies. Although material and operation may differ between the EOS protection devices, their electrical characteristics can be classified into a few fundamental groups. EOS protection networks can be identified as a voltage suppression device or as a current-limiting device. The voltage suppression device limits the voltage observed on the signal pins or power rails of a component, preventing EOV. The current-limiting device prevents a high current from reaching sensitive nodes, avoiding EOC [1].

Voltage suppression devices can also be subdivided into two major classifications [6]. Figure 12.10(a) illustrates examples of voltage suppression categories. Voltage suppression devices can be segmented into devices that remain with a positive differential resistance and those that undergo a negative resistance region. For positive differential resistance, these devices can be referred to as "voltage clamp" devices where dI/dV remains positive for all states; for the second group, there exists a region where dI/dV is negative. The first group can be classified as "voltage clamp devices," whereas the second group can be referred to as an "S-type I–V characteristic device," or as a "snapback device." In the classification of voltage suppression devices, the second classification can be associated with the directionality; a voltage suppression device can be "unidirectional" or "bidirectional." Figure 12.10(b) illustrates an example of an EOV device solution integrated with a differential circuit.

To prevent EOS associated with EOV, voltage suppression devices are mounted on PCB, or integrated into systems. The choice of EOS device to use in an application is dependent on the electrical characteristics, cost, and size. The electrical characteristics that are of interest are the breakdown voltage and the forward conduction.

Types of voltage suppression devices used EOS are as follows [1]:

- Transient voltage suppression (TVS) diodes
- Thyristor devices

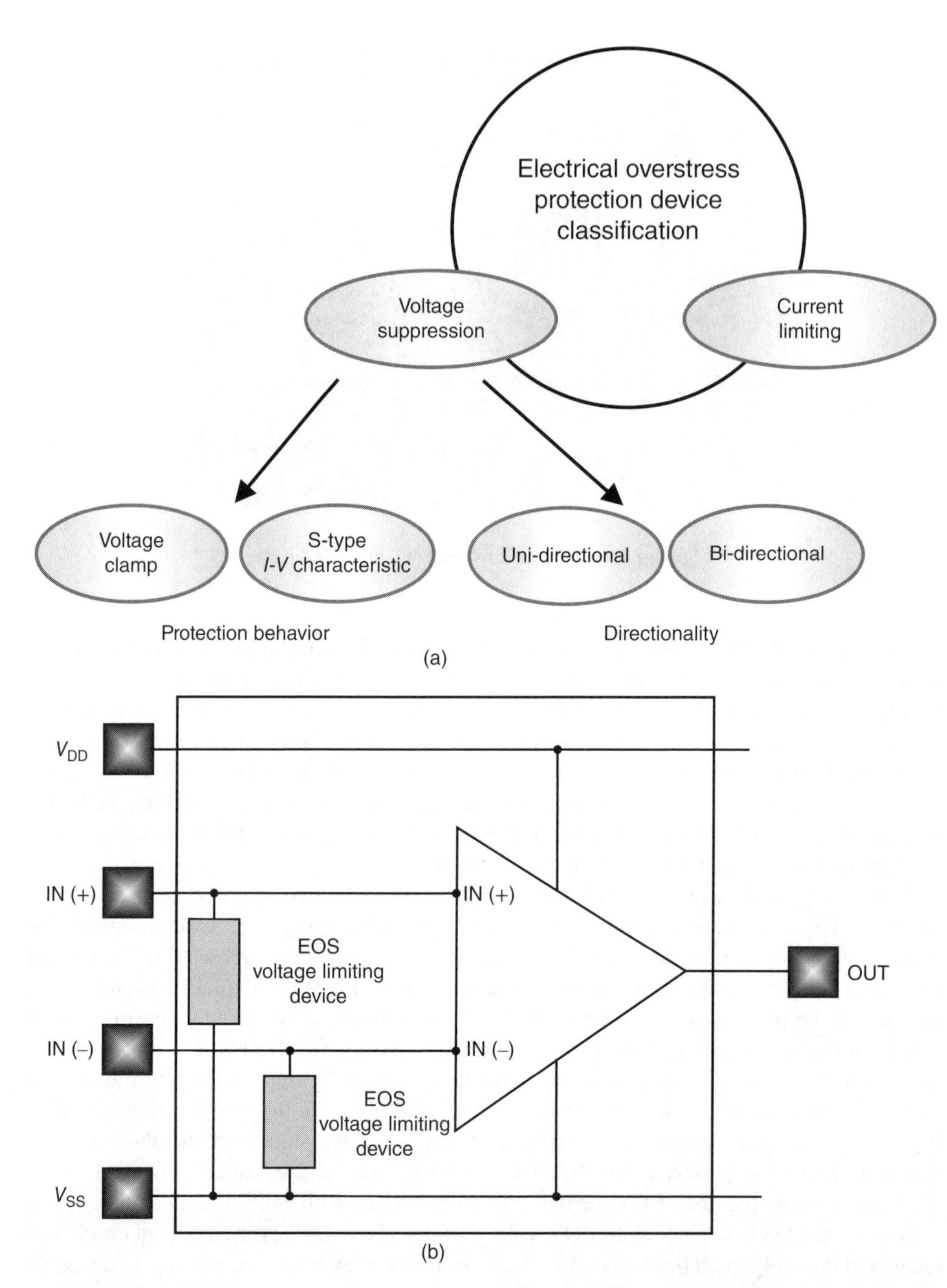

Figure 12.10 (a) Electrical overvoltage (EOV) protection classification. (b) Electrical overvoltage (EOV) protection network

- Varistor devices
- Polymer voltage suppression (PVS) devices
- Gas discharge tube (GDT) devices.

Current-limiting devices can be used in a series configuration for EOS protection. EOS current-limiting devices can be as follows [1]:

- Resistors
- Resetting fuses
- Nonresetting fuses
- eFUSE
- Positive temperature coefficient (PTC) devices
- Circuit breakers.

The choice of the current-limiting EOS protection device is a function of the cost, size, rated current, time response, I^2t value, rated voltage, voltage drops, and application requirements.

Figure 12.11 illustrates an example of an EOC protection device solution integrated with a differential circuit. For EOC solutions, series resistor elements or other networks are suitable to limit the current flowing into the inputs of sensitive circuitry.

Diodes are a commonly used unidirectional EOS protection device. Diodes provide a forward conduction state and reverse blocking state. For EOS single component, diodes are mounted on PCB by soldering in the leads through vias or surface mount. Diodes are commonly used within components to provide ESD protection for internal circuitry [1].

Electrical characteristics of a diode structure include forward turn-on, the on-resistance, and reverse breakdown voltage. At high current conditions, in the forward conduction state,

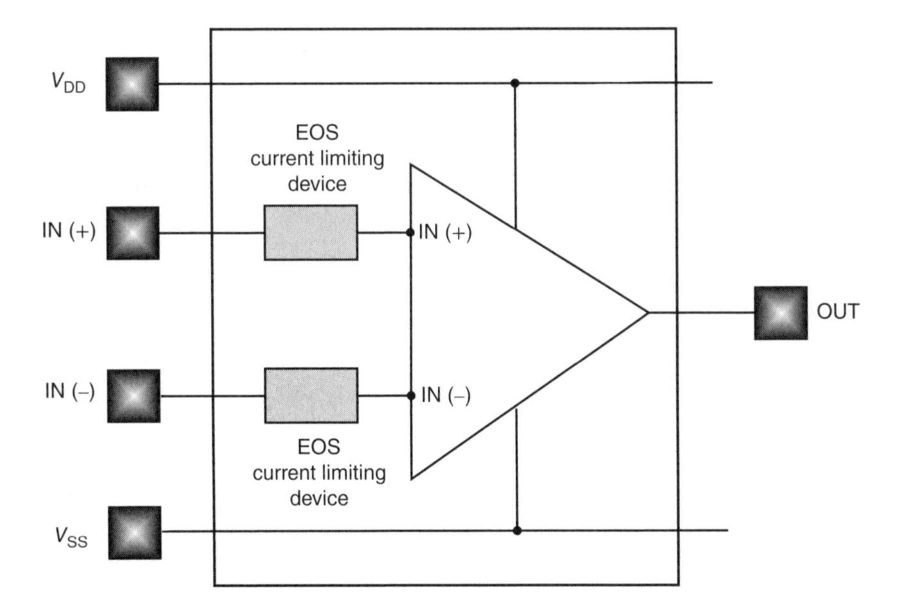

Figure 12.11 Electrical overcurrent (EOC) protection

thermal breakdown occurs as well as thermal failure of the diode structure. If the EOS stress current exceeds the thermal breakdown of the diode structure, destructive failure of the element can lead to system degradation or an EOS failure.

Diodes are unidirectional type EOS structure but can be utilized in a forward or reverse breakdown mode of operation for a voltage-limiting EOS solution [6]. The power-to-failure of a diode structure is higher in the forward mode of operation compared to its reverse breakdown mode of operation. In a forward conduction mode, the power distributes through all sections of the element (e.g., anode, cathode, and metallurgical junction). In the reverse conduction mode, power is contained only in the metallurgical junction region.

Schottky diodes are a commonly used EOS protection device (Figure 12.12). Schottky diodes have a forward conduction state and reverse blocking state. Schottky diodes have a

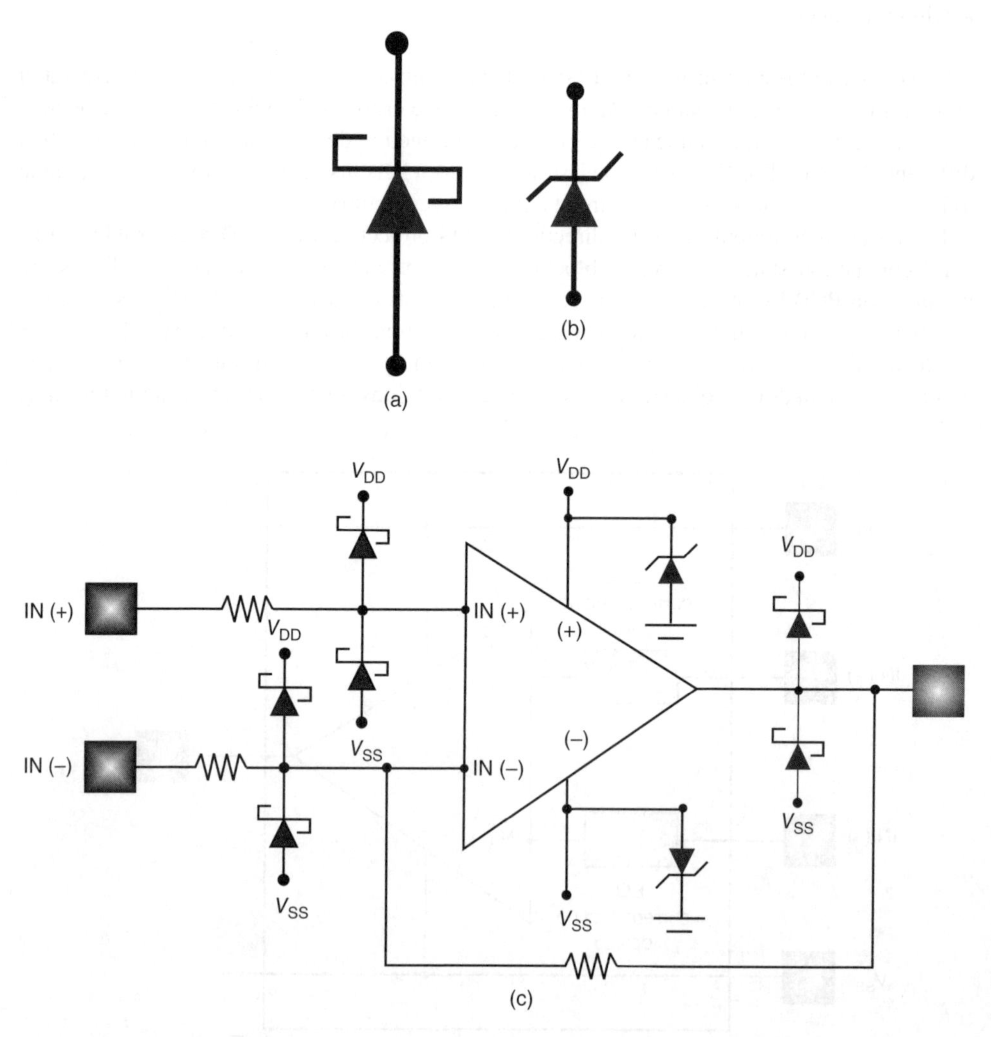

Figure 12.12 (a) EOS protection – Schottky diode. (b) EOS protection – Zener diode. (c) EOS protection – operational amplifier with Schottky and Zener protection

lower forward turn-on (e.g., 0.35 V) compared to standard silicon p–n junction (e.g., 0.7 V). For EOS single component, Schottky diodes are mounted on PCB by soldering in the leads through vias or surface mount. Schottky diodes are not commonly used within components to provide ESD protection due to lack of availability. Schottky diodes are unidirectional type EOS structure but can be utilized in a forward or reverse breakdown mode of operation for a voltage-limiting EOS solution.

Zener diodes are used as a unidirectional EOS protection device [1]. Zener diodes are typically used as a voltage clamping EOS protection device, and typically used in the breakdown state. Zener diodes have a well-defined voltage breakdown value.

Zener diodes are used for ESD protection for high voltage and power applications. Zener diodes are not used for ESD protection for low-voltage complementary metal-oxide semiconductor (CMOS) applications. For EOS single component, Zener diodes are mounted on PCB by soldering in the leads through vias or surface mount.

Zener diodes are unidirectional type EOS structure but can be utilized primarily in reverse breakdown mode of operation for a voltage-limiting EOS solution.

Zener diodes are used as a unidirectional EOS protection device [1]. Zener diodes are typically used as a voltage clamping EOS protection device and typically used in the breakdown state. Zener diodes have a well-defined voltage breakdown value. Figure 8.11 is the symbol for the Zener diode.

Schottky and Zener diodes can be integrated into a given application. Figure 12.12(c) shows an example of an EOS protection scheme that utilizes both Schottky and Zener diode elements.

An EOS protection device used for high voltages is varistor [1]. A varistor is also known as a voltage-dependent resistor (VDR). A varistor is a portmanteau – combining the word for variable and resistor. In reality, the varistor element behaves like a diode, forming a nonlinear current–voltage (*I–V* characteristic). The element has the characteristics of being bidirectional voltage clamp (note: it does not form an S-type *I–V* characteristic or undergo a negative resistance regime). At low voltages, the device has a high series resistance, serving as an "off-state." At higher voltages, the device "turns-on" and has a low resistance.

The metal oxide varistor (MOV) device is the most common varistor composition. Figure 12.13(a) is the circuit schematic symbol for an MOV device. Zinc oxide, combined with other metal oxides, is integrated between two metal electrodes. Other metal oxides integrated include bismuth, cobalt, and manganese. The operation of the MOV device is based on conduction through ZnO grains; current flows "diode-like" through the grain structures creating a low current flow at low voltages. At higher voltages, the current flow is dominated by a combination of thermionic emissions and tunneling. This behavior forms a diode-like *I–V* characteristic providing the high-resistance/low-voltage state, and the low-resistance/high-voltage state. From the *I–V* characteristic, a "clamping voltage" can be defined (e.g., analogous to a diode turn-on voltage). The characteristic that influences the on-resistance and the turn-on voltage is a function of the ZnO grain structure, the film thickness, the physical size, and the other metal oxides integrated into the structure.

The advantage of the MOV structure is it has a high trigger voltage, making it suitable for EOS protection in power electronics (e.g., 120–700 V applications). The disadvantage of these elements is that it has high capacitance and high on-resistance. A second disadvantage is that the high trigger voltage does not make it suitable for advanced high-speed or low-voltage electronics. A third disadvantage is the variability of the device response (e.g., on resistance and clamping voltage) in the MOV device characteristics. Key device parameters of varistor

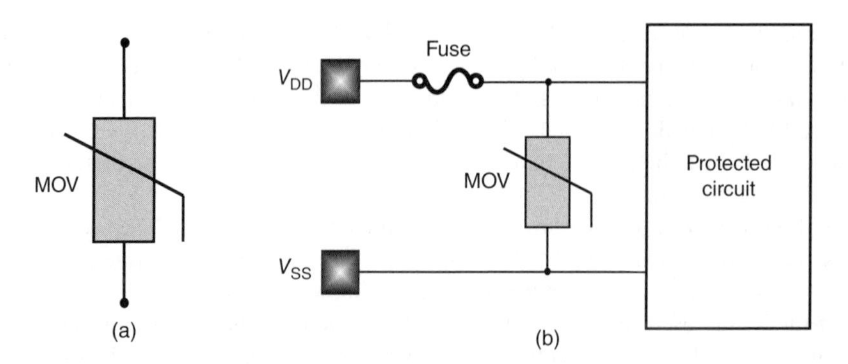

Figure 12.13 (a) EOS protection – metal oxide varistor (MOV). (b) EOS protection – metal oxide varistor (MOV) for power supply

are the energy rating, operating voltage, response time, maximum current, and breakdown voltages.

Figure 12.13(b) is an example of usage of an MOV for power supply protection. The MOV element can be placed between the power supply and ground to avoid EOS to the V_{DD} power supply.

GDT devices can be used to avoid EOS in systems (Figure 12.14(a)) [1]. GDT are bidirectional, allowing for protection of both positive and negative EOS events. GDT elements are suitable for surge protection, high-voltage electronic "crowbars," to lightning. GDT devices are not suitable for EOS protection of low-voltage electronics due to the high trigger voltages (unless used as a first stage followed by other low-voltage secondary EOS solutions).

GDT utilize electrical discharge in gases. An applied voltage initiates the device by ionizing the electrical gas, followed by electrical glow discharge and an electrical arc. With creation of an electrical arc, the GDT device becomes a low-resistance shunt for EOS protection. These gas-filled tubes can contain hydrogen, deuterium, and noble gases (e.g., helium, neon, argon, krypton, and xenon). GDT devices can vary their electrical characteristics by choices of the gas type, pressure, electrode design, and spacings [1].

Figure 12.14(b) shows an example of a ramp impulse voltage versus time for a GDT device. GDT devices undergo three states: (1) electrical breakdown, (2) glow discharge, and (3) electrical arc. The electrical breakdown is a high-voltage, low-current state prior to triggering of the GDT device. A glow discharge region forms a second state that incorporates a low-current, high-voltage state. Lastly, after full ionization of the gas, a low-voltage, high-current state occurs with a low "on-resistance."

GDT devices have high trigger voltages suitable for LDMOS power electronic applications to HV LDMOS (e.g., 120 V), and UHV LDMOS applications (e.g., 600–700 V). These devices are used in a number of high-voltage switch devices, such as ignitrons, krytrons, and thyratrons. One of the disadvantages of the GDT devices is the slow turn-on times typically in the microseconds. An example of some of the electrical characteristics can exhibit DC breakdown from 75 to 600 V, with a single surge response of 40 kA in 10–20 μs, or multiple surges of magnitude of 20 kA.

In time, when a voltage disturbance reaches the GDT sparkover voltage, the GDT will switch into a low impedance state, also known as the "arc mode." With this low resistance

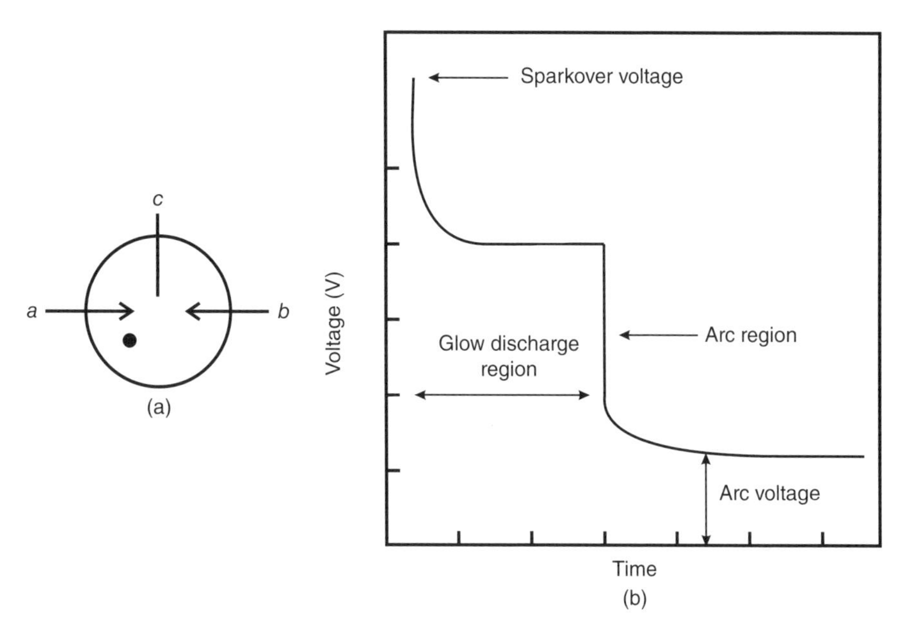

Figure 12.14 (a) EOS protection – gas discharge tube (GDT). (b) EOS protection – gas discharge tube (GDT) *I–t* plot

state, the GDT device will discharge the EOS event to the ground, avoiding EOS failure of the component.

At voltages below the sparkover voltage, the GDT remains in a high impedance off-state. As the voltage increases, the GDT enters the "glow voltage region"; this glow region is where ionization starts to occur within the GDT. As the current increases, avalanche multiplication occurs, leading to further ionization of the gas within the GDT. This is followed by avalanche breakdown and the low impedance state. The voltage condition developed across the GDT during this state is called the "arc voltage."

The transition time between the glow and arc regions is dependent on the impulse current, electrode shape (e.g., electrode curvature), electrode spacing, gas composition, gas pressure, and emission coatings.

For EOC, the electrical circuit breaker is used in industrial, commercial, and residential electrical systems. Electrical circuit breakers are typically not found in semiconductor chips, or small systems due to the physical size, weight, cost, and time response. Circuit breakers can be used to protect household appliances and large-scale switchgear high-voltage circuits.

The circuit breaker is an electrical switch designed for the purpose of EOC events, short circuits, or fault detection. Circuit breakers are typically "tripped" by the high current event and can be manually reset. The concept of the circuit breaker was invented by Charles Grafton Page in 1836 [1]. A pilot device senses a fault current and operates a trip opening mechanism. The trip solenoid releases a latch. Some high-voltage circuit breakers are self-contained with current transformers, protection relays, and an internal control power source. With detection of an electrical fault, the circuit breaker contacts open to interrupt the circuit; some mechanically stored energy contained within the breaker is used to separate the contacts. The circuit breaker

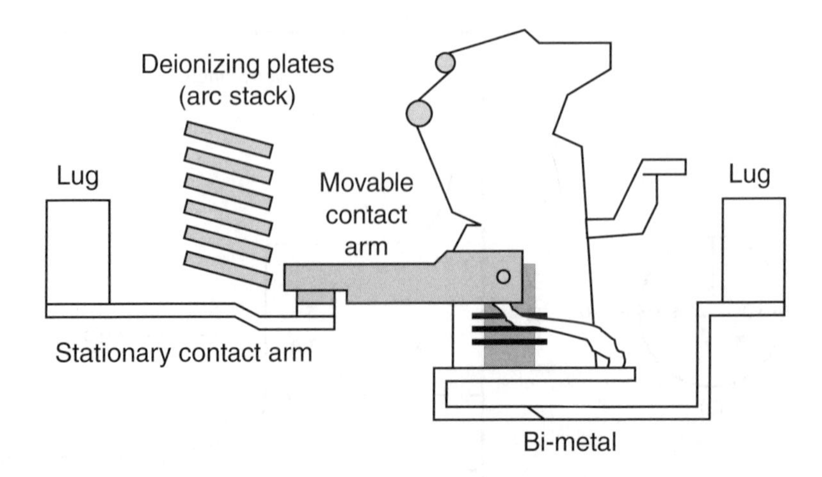

Figure 12.15 EOS protection – thermal circuit breakers

can be reset manually after the event is over. Circuit breakers are nondestructive, as opposed to fuses that may only have a single use (Figure 12.15). Nowadays, both fuses and circuit breakers can be integrated into a common system.

A class of circuit breakers is the thermal-magnetic circuit breaker. Thermal-magnetic circuit breakers are used to avoid "short-circuit" currents. Thermal-magnetic circuit breakers are sensitive to temperature.

Figure 12.15 shows an example of a thermal-magnetic circuit breaker. Thermal-magnetic circuit breakers contain a bimetal switch and an electromagnet. The bimetal switch provides overcurrent protection. During current overload, the bimetal switch heats up, leading to bending of the element.

The electromagnet responds to short-circuit currents. The electromagnetic is a wire coil and an iron core. As a high current goes through the coil, induced magnetic field attracts an armature of the thermal-magnetic circuit breaker. When the armature extends toward the electromagnet due to the magnetic force, the armature rotates the trip bar. With the tripping of the trip bar, the current path is "open," and circuit breaker prevents the EOC.

Power controllers are used for low-power and low-voltage applications [1]. Power controllers are typically low-voltage, high-efficiency products that can carry amperes of current per channel. Buck-converters contain overcurrent protection logic and networks; overcurrent functions protect the switching converter from an output short by monitoring current flow in the application. Overcurrent fault counters prevent turn-on of circuitry until the fault has passed, and the system is on its next switching cycle. Flags are triggered in the counters after the overcurrent fault condition (OFC). When the overcurrent condition flag is reset, the circuitry is allowed to reinitiate. This "soft start" eliminates the inrush current during the start and restart.

Figure 12.16 shows the integration of the fault, soft start, and control logic at the outputs of the overcurrent and overvoltage comparators. The soft start function outputs a ramp reference for both the voltage and current loops.

Hence, in power applications, it is possible to integrate EOV and EOC within a component design. Many analog and power applications also contain thermal protection networks to avoid thermal runaway and EOS damage.

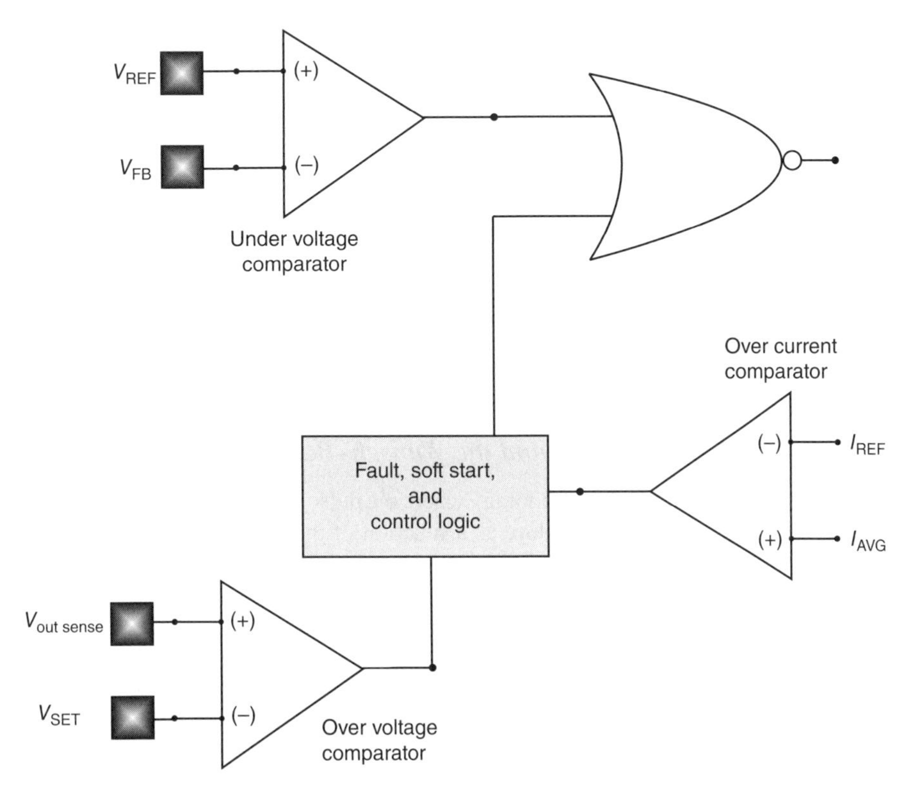

Figure 12.16 EOS protection – soft start EOC and EOV protection circuitry

12.10 Electrical Overstress (EOS) Testing – TLP Method and EOS

Transmission line pulse (TLP) testing methodology can be utilized to quantify EOS robustness in both components and systems. Today's TLP methodology is a "two-pin" test to determine the TLP *I–V* characteristic. TLP measurements can provide information of the characteristics of a single device or a signal pin. TLP measurements can provide the following current and voltage metrics:

- Trigger voltage (V_{t1})
- Holding voltage (V_h)
- Sustaining voltage (V_{sus})
- On-resistance (R_{on})
- Second trigger voltage (V_{t2})
- Second breakdown current (I_{t2}).

TLP testing is used till today for device characterization and qualification of components in a technology. The choice of the TLP pulse width was chosen for equivalency for the HBM energy (e.g., integrated energy of the applied pulse). For EOS, this methodology can be extended to longer pulse width for evaluation of long pulse events.

12.10.1 Electrical Overstress (EOS) Testing – Long Duration Transmission Line Pulse (LD-TLP) Method

The methodology of TLP testing can be utilized for components and systems for EOS. For evaluation of EOS, the TLP methodology can be modified as follows:

- *Long pulse*
- *Variable pulse width*
- *Component ground reference*
- *PCB ground reference.*

12.10.2 Electrical Overstress (EOS) Testing – Transmission Line Pulse (TLP) Method, EOS, and the Wunsch–Bell Model

Modification of the present TLP methodology where the pulse width is varied from short pulses to long pulse events, the power-to-failure as a function of the pulse event can be quantified. By establishing a power-to-failure versus pulse width plot, a "universal curve" is formed that fully characterizes the device under test for all pulse lengths. Historically, this was established methodology for evaluation of the power-to-failure for wide bandwidth pulse events (e.g., electromagnetic pulse, EMP). The power-to-failure versus pulse width plot is known as the Wunsch–Bell plot, as discussed in the previous chapters. With the evaluation of the Wunsch–Bell curve, the full spectrum from short pulse to long pulse events can be understood (Figures 12.17 and 12.18) [1].

12.10.3 Electrical Overstress (EOS) Testing – Limitations of the Transmission Line Pulse (TLP) Method for the Evaluation of EOS for Systems

For evaluations of components and systems, there are some limitations to the TLP method.

A first issue is that in a system environment, an EOS event can propagate through multiple current paths. TLP analysis of components is successful when applying a single device or a

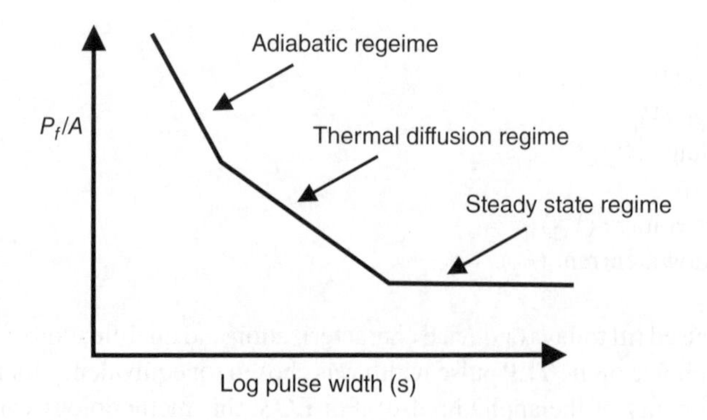

Figure 12.17 Wunsch–Bell power to failure plot

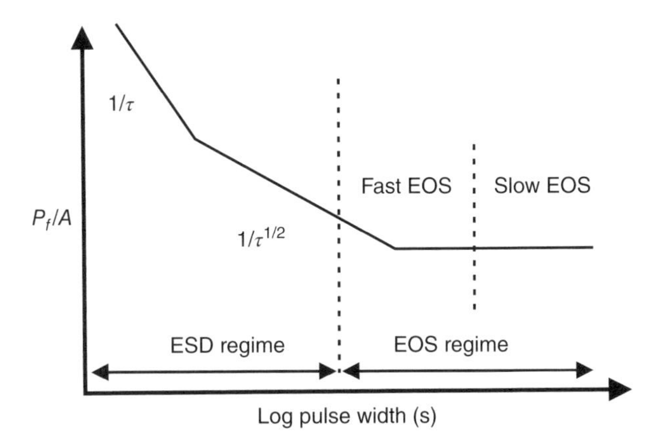

Figure 12.18 Slow and fast EOS domain in the Wunsch–Bell curve

single signal pin. In a full component or a system, the TLP current can flow through different current paths. A second issue is the bridging between the single device and full system level response. Due to the nonlinear nature of the semiconductor devices, the superposition of currents is not always valid in bridging from the single device TLP *I–V* characteristic to the full component or system level.

12.10.4 Electrical Overstress (EOS) Testing – Electromagnetic Pulse (EMP)

A source of an EOS event is the EMP event. The EMP event is defined as a high amplitude, single pulse, short duration, and broadband pulse of electromagnetic energy. EMP is the electromagnetic effect resulting from the detonation of a nuclear device. Other definitions of EMP at high altitudes are high-altitude electromagnetic pulse (HEMP), HAEMP, and HNEMP.

EMP test methods include MIL-STD-461; this provides radiated (RS 105) and conducted (CS 116) test methods and test levels for determining a device's immunity to EMP. The RS 105 radiated test methodology addresses the risk of radiated exposure to an EMP event; this test is applicable to equipment installed in exposed and partially exposed environments on aircraft, surface ships, submarines, and ground vehicles. The CS116 coupling test method addresses the effects of EMP coupling onto interconnecting wiring harnesses; this test is to ensure the equipment's immunity to damped sinusoidal transients induced on the equipment's cables. Testing is generally applicable to all applications with limited applicability to submarine equipment.

The MIL-STD-188-125 establishes minimum requirements and design objectives for HEMP hardening of fixed, ground-based facilities that perform critical, time-urgent command, control, communications, computer, and intelligence (C4I) missions. Similar to the approach described in MIL-STD-461, this standard provides both radiated and conducted test methods and test levels.

12.11 Electrical Overstress (EOS) Testing – DC and Transient Latchup Testing

CMOS latchup can occur from EOS when regenerative feedback occurs within a semiconductor chip circuitry [14]. From 1979 to 2000, CMOS latchup occurred within a given circuit [14]. Due to scaling of technologies, CMOS latchup occurred between the following:

- I/O PFET to I/O NFET
- I/O to ESD protection circuit
- I/O to ESD power clamp circuit
- I/O to I/O
- I/O to adjacent circuit element (e.g., decoupling capacitors)
- I/O to core circuitry
- Core to core circuitry.

CMOS latchup is evaluated in both the DC and transient state. The integrated circuit latchup standard most commonly used is the JEDEC JESD 78:2010 standard for integrated circuit latchup test, as discussed in Chapter 11. Many of the EOS pulse-defined events should be used to determine if CMOS latchup can occur. For example, cable discharge events (CDE) can lead to CMOS latchup. In addition, latchup is typically evaluated on a component level. A semiconductor component can survive the JESD78 latchup IC specification but fail on a PCB or card. Latchup testing should be evaluated on both the semiconductor chip level and the populated PCB and system level to avoid card failures due to CMOS latchup.

12.12 Summary and Closing Comments

This chapter looked at the broader issue of EOS. The topic of EOS was approached providing distinctions of EOS versus ESD phenomena, and EOS versus CMOS latchup. The chapter discussed specific EOS tests that are presently used for systems and for semiconductor chips, as well as growing interest in "long pulse transmission line pulse (TLP)" testing method that is having increased popularity to address EOS phenomena. EOS failure mechanisms and solutions were quantified. With the growing increase of system level issues, EOS has been seeing a renewed interest for both semiconductor chip developers and system designers.

Chapter 13 addresses electromagnetic compatibility (EMC). In this chapter, scanning methods for evaluation of EMC sensitivity are discussed. New state-of-the-art EMC scanning techniques are the highlights of the chapter.

Problems

1. How can we quantify EOS robustness when the variety of EOS events are significant?
2. For EOS robustness qualification, it is possible to have a set of different criteria and tests that are required to pass in order to qualify a component. Define a set of tests that can be performed in order to pass "EOS robustness qualification."
3. On a Wunsch–Bell plot, highlight the regions that are regarded as the regime where EOS events occur.

4. Using a Wunsch–Bell plot, define a set of tests that can be performed to cover quantification of the EOS robustness of a test structure or component.
5. Long pulse transmission line pulse (LP-TLP) is a test to cover the time regime of EOS events. What are the advantages? What are potential problems?
6. EOS events can be defined as a "slow EOS" regime. How do these regimes correspond to the Wunsch–Bell model (e.g., adiabatic, thermal diffusion, and steady state regimes)?
7. EOS events can be defined as "slow EOS" regime and "fast EOS regime." How do these regimes correspond to the Dwyer–Cambell–Franklin model?
8. EOS can include electromagnetic pulse (EMP) events. How does one quantify robustness from an EMP event? What is the frequency that one evaluates for EMP?

References

1. S. Voldman. *Electrical Overstress (EOS): Devices, Circuits, and Systems*. Chichester, England: John Wiley and Sons, Ltd., 2013.
2. T.A. Edison. Fuse block. U.S. Patent No. 438,305, October 14, 1890.
3. C.G. Page. Improvement in induction-coil apparatus and in circuit breakers. U.S. Patent No. 76,654. April 14, 1868.
4. A. Wright, and P.G. Newbury. *Electric Fuses 3rd Edition*, Institute of Electrical Engineers, 2004; 2–10.
5. W. Schossig. Introduction to History of Selective Protection. *PAC Magazine*, 2007; 70–74.
6. V. Kraz. Origins of EOS in manufacturing environment. *Proceedings of the Electrical Overstress/Electrostatic Discharge (EOS/ESD) Symposium*, 2009; 44–48.
7. C. Thienel. Avoiding electrical overstress for automotive semiconductors by new connecting concepts, *SAE International Journal of Passenger Cars – Electronic Electrical Systems*, Vol. 2, (1), Apr 2009; 101–102.
8. ESD Association Workshop, Electrical overstress (EOS): many failures and few solutions. *Proceedings of the Electrical Overstress / Electrostatic Discharge (EOS/ESD) Symposium*, 2009; 433.
9. C. Thienel. *Electrical Overstress for Automotive Semiconductors*, Stuttgart, Germany: Euroforum, 2010.
10. K.T. Kaschani, and R. Gaertner. The impact of electrical overstress on the design, handling and application of integrated circuits. *Proceedings of the Electrical Overstress/Electrostatic Discharge (EOS/ESD) Symposium*, 2011; 220–229.
11. K.P. Yan, and C.Y. Wong. Poor grounding – major contributor to. *Proceedings of the Electrical Overstress/Electrostatic Discharge (EOS/ESD) Symposium*, 2012; 215–220.
12. ESD Association, Workshop A1. EOS versus ESD: definition, field failures, and case studies. *Proceedings of the Electrical Overstress/Electrostatic Discharge (EOS/ESD) Symposium*, 2012; 423.
13. S. Voldman. *ESD: Physics and Devices*. Chichester, England: John Wiley and Sons, Ltd., 2004.
14. S. Voldman. *Latchup*. Chichester, England: John Wiley and Sons, Ltd., 2007.
15. S. Voldman. *ESD Basics: From Semiconductor Manufacturing to Product Use*. Chichester, England: John Wiley and Sons, Ltd., 2012.
16. S. Voldman. *ESD: Design and Synthesis*. Chichester, England: John Wiley and Sons, Ltd., 2011.
17. J.S. Smith. Analysis of electrical overstress failures. *Proceedings of the International Reliability Physics Symposium (IRPS)*, 1973; 105–107.
18. L.J. Gallace, and H.L. Pujol. The evaluation of CMOS static charge protection networks and failure mechanisms associated with overstress conditions as related to device life. *Proceedings of the International Reliability Physics Symposium (IRPS)*, 1977; 149–157.
19. H. Domingos. Electro-thermal overstress failure in microelectronics, Specification for microcircuits electrical overstress tolerance. *Technical Report RADC TR-73-87, Rome Air Development Center (RADC),* Air Force Systems Command, Griffis Air Force Base, N.Y. 1973.
20. J.S. Smith. Electrical overstress failure analysis in microcircuits. *Proceedings of the International Reliability Physics Symposium (IRPS)*, 1978; 41–46.
21. R.J. Antinone. Specification for microcircuits electrical overstress tolerance. RADC Contract F30602-76-C-0308, The BDM Corporation, *RADC TR-78-28*, 1978.
22. D.T. McCullough, C.H. Lane, and R.A. Blore. Reliability for EOS-screened gold doped 4002 CMOS devices. *Proceedings of the Electrical Overstress/Electrostatic Discharge (EOS/ESD) Symposium*, 1979; 41–45.

23. D.H. Rutherford, and J.F. Perkins. Effects of electrical overstress on digital bipolar microcircuits and analysis techniques for failure site location. *Proceedings of the Electrical Overstress/Electrostatic Discharge (EOS/ESD) Symposium*, 1979; 64–77.
24. T. Uetsuki, and S. Mitani. Failure analysis of microcircuits subjected to electrical overstress. *Proceedings of the Electrical Overstress/Electrostatic Discharge (EOS/ESD) Symposium*, 1979; 88–96.
25. A. Baruah, and P.P. Budenstein. An electrothermal model for current filamentation in second breakdown of silicon on sapphire diodes. *Proceedings of the Electrical Overstress/Electrostatic Discharge (EOS/ESD) Symposium*, 1979; 126–131.
26. N. Kusnezov and J.S. Smith. Modeling of electrical overstress in silicon devices. *Proceedings of the Electrical Overstress/Electrostatic Discharge (EOS/ESD) Symposium*, 1979; 133–139.
27. C.J. Petrizio. Electrical overstress versus device geometry. *Proceedings of the Electrical Overstress/Electrostatic Discharge (EOS/ESD) Symposium,* 1979; 183–187.
28. J.A. Madison. The analysis and elimination of EOS induced secondary failure mechanisms. *Proceedings of the Electrical Overstress/Electrostatic Discharge (EOS/ESD) Symposium*, 1979; 205–209.
29. H.R. Philipp, and L.M. Levinson. Transient protection with ZnO varistors: technical considerations. *Proceedings of the Electrical Overstress/Electrostatic Discharge (EOS/ESD) Symposium*, 1980; 26–34.
30. D.C. Hopkins. Protective level comparisons for voltage transient suppressors. *Proceedings of the Electrical Overstress/Electrostatic Discharge (EOS/ESD) Symposium*, 1980; 35–43.
31. A. Bazarian. Gas tube surge arresters for control of transient voltages. *Proceedings of the Electrical Overstress/Electrostatic Discharge (EOS/ESD) Symposium*, 1980; 44–53.
32. E.L. Horgan. Analytical assessment of electrical overstress effects on electronic systems. *Proceedings of the Electrical Overstress/Electrostatic Discharge (EOS/ESD) Symposium*, 1980; 140–148.
33. J.B. Smyth Jr.,, V.A.J. van Lint, and A.R. Hart. Solar cell electrical overstress analysis. *Proceedings of the Electrical Overstress/Electrostatic Discharge (EOS/ESD) Symposium*, 1980; 149–153.
34. D.L. Durgin. An overview of the sources and effects of electrical overstress. *Proceedings of the Electrical Overstress/Electrostatic Discharge (EOS/ESD) Symposium*, 1980; 154–160.
35. R.F. Hess. Test waveforms and techniques to assess the threat to electronic devices of lightning-induced transients. *Proceedings of the Electrical Overstress/Electrostatic Discharge (EOS/ESD) Symposium*, 1980; 161–167.
36. K.E. Crouch. Lightning protection design for a photovoltaic concentrator. *Proceedings of the Electrical Overstress/Electrostatic Discharge (EOS/ESD) Symposium*, 1980; 167–175.
37. D.R. Kressler. Surge tests on plug-in transformers. *Proceedings of the Electrical Overstress/Electrostatic Discharge (EOS/ESD) Symposium*, 1980; 176–183.
38. R.J. Antinone. Microelectronic electrical overstress tolerance testing and qualification. *Proceedings of the Electrical Overstress/Electrostatic Discharge (EOS/ESD) Symposium*, 1980; 184–188.
39. R.J. Karaskiewicz, P.A. Young, and D.R. Alexander. Electrical overstress investigations in modern integrated circuit technologies. *Proceedings of the Electrical Overstress/Electrostatic Discharge (EOS/ESD) Symposium*, 1981; 114–119.
40. D.G. Pierce, and D.L. Durgin. An overview of EOS effects on semiconductor devices. *Proceedings of the Electrical Overstress/Electrostatic Discharge (EOS/ESD) Symposium*, 1981; 120–131.
41. N. Kusnezov, and J. Smith. Modeling of EOS in silicon devices. *Proceedings of the Electrical Overstress/Electrostatic Discharge (EOS/ESD) Symposium*, 1981; 132–138.
42. D.C. Wunsch. An overview of EOS effects on passive components. *Proceedings of the Electrical Overstress/Electrostatic Discharge (EOS/ESD) Symposium*, 1981; 167–173.
43. D.M. Tasca. Pulse power response and damage characteristics of capacitors. *Proceedings of the Electrical Overstress/Electrostatic Discharge (EOS/ESD) Symposium*, 1981; 174–191.
44. R.A. Hays. EOS threshold determination of electro-explosive devices. *Proceedings of the Electrical Overstress/Electrostatic Discharge (EOS/ESD) Symposium*, 1981; 202–207.
45. O.M. Clark. Lightning protection for computer lines. *Proceedings of the Electrical Overstress/Electrostatic Discharge (EOS/ESD) Symposium*, 1981; 212–218.
46. R.L. Pease, J. Barnum, W. Vuliet, T. Wrobel, and V. Van Lint. EOS damage in silicon solar cells. *Proceedings of the Electrical Overstress/Electrostatic Discharge (EOS/ESD) Symposium*, 1981; 229–235.
47. E.L. Horgan, O.E. Adams, and W.H. Rowan. Limitations of modeling electrical overstress failure in semiconductor devices. *Proceedings of the Electrical Overstress/Electrostatic Discharge (EOS/ESD) Symposium*, 1982; 19–33.
48. R.A. Hays. Electrical overstress threshold testing. *Proceedings of the Electrical Overstress/Electrostatic Discharge (EOS/ESD) Symposium,* 1982; 34–40.

49. D.L. Durgin, R.M. Pelzl, W.H. Thompson, and R.C. Walker. A survey of EOS/ESD data sources. *Proceedings of the Electrical Overstress/Electrostatic Discharge (EOS/ESD) Symposium*, 1982; 49–55.
50. H. Volmerange. An improved EOS conduction model of semiconductor devices. *Proceedings of the Electrical Overstress/Electrostatic Discharge (EOS/ESD) Symposium*, 1982; 62–70.
51. P.H. Noel, and D.H. Dreibelbis. EOS or ESD: can failure analysis tell the difference? *Proceedings of the Electrical Overstress/Electrostatic Discharge (EOS/ESD) Symposium*, 1983; 154–157.
52. T.F. Brennan. Invisible EOS/ESD damage: how to find it. *Proceedings of the Electrical Overstress/Electrostatic Discharge (EOS/ESD) Symposium*, 1983; 158–167.
53. J.E. May, and S.R. Korn. Metal oxide varistors for transient protection of 3 to 5-V integrated circuits. *Proceedings of the Electrical Overstress/Electrostatic Discharge (EOS/ESD) Symposium*, 1983; 168–176.
54. G. Forster. Protection of components against electrical overstress (EOS) and transients in monitors. *Proceedings of the Electrical Overstress/Electrostatic Discharge (EOS/ESD) Symposium*, 1984; 136–143.
55. D.G. Pierce, J. Perillat, and W.L. Shiley. An evaluation of EOS failure models. *Proceedings of the Electrical Overstress/Electrostatic Discharge (EOS/ESD) Symposium*, 1984; 144–156.
56. B.C. Roberts. Determination of threshold energies and damage mechanisms in semiconductor devices subjected to voltage transients. *Proceedings of the Electrical Overstress/Electrostatic Discharge (EOS/ESD) Symposium*, 1984; 157–164.
57. D.G. Pierce. Electro-thermomigration as an electrical overstress failure mechanism. *Proceedings of the Electrical Overstress/Electrostatic Discharge (EOS/ESD) Symposium*, 1985; 67–76.
58. P.S. Neelakantaswamy, T.K. Sarkar, and I.R. Turkman. Residual fatigues in microelectronic devices due to thermoelastic strain caused by repetitive electrical overstressing: a model for latent failures. *Proceedings of the Electrical Overstress/Electrostatic Discharge (EOS/ESD) Symposium*, 1985; 77–83.
59. P. Ryl, J. Brossier, R.M. Pelzl, and W.H. Cordova. A comparison of discrete semiconductor electrical overstress damage models to experimental measurement. *Proceedings of the Electrical Overstress/Electrostatic Discharge (EOS/ESD) Symposium*, 1985; 100–102.
60. R.N. Shaw, and R.D. Enoch. An experimental investigation of ESD-induced damage to integrated circuits on printed circuit boards. *Proceedings of the Electrical Overstress/Electrostatic Discharge (EOS/ESD) Symposium*, 1985; 132–140.
61. T.G. Mahn. Liability issues associated with electrical overstress in computer hardware, design and manufacture. *Proceedings of the Electrical Overstress/Electrostatic Discharge (EOS/ESD) Symposium*, 1986; 1–11.
62. C.E. Stephens, and C.T. Amos. A study of EOS in microcircuits using the infra-red microscope. *Proceedings of the Electrical Overstress/Electrostatic Discharge (EOS/ESD) Symposium*, 1986; 219–223.
63. K. Siemsen. EOS test limits for manufacturing equipment. *Proceedings of the Electrical Overstress/Electrostatic Discharge (EOS/ESD) Symposium*, 1987; 168–173.
64. E. Horgan. Advanced semiconductor device EOS test and modeling methods. *Proceedings of the Electrical Overstress/Electrostatic Discharge (EOS/ESD) Symposium*, 1987; 174–178.
65. T. Green, and W. Denson. A review of EOS/ESD field failures in military equipment. *Proceedings of the Electrical Overstress/Electrostatic Discharge (EOS/ESD) Symposium*, 1988; 7–14.
66. J.J. Burke. The effect of lightning on the utility distribution system. *Proceedings of the Electrical Overstress/Electrostatic Discharge (EOS/ESD) Symposium*, 1990; 10–18.
67. J.R. Harford. Powerline disturbances – a primer. *Proceedings of the Electrical Overstress/Electrostatic Discharge (EOS/ESD) Symposium*, 1990; 19–26.
68. C. Diaz, S. Kang, C. Duvvury, and L. Wagner. Electrical overstress (EOS) power profiles: a guideline to qualify EOS hardness of semiconductor devices. *Proceedings of the Electrical Overstress/Electrostatic Discharge (EOS/ESD) Symposium*, 1992; 88–94.
69. W. Greason, and K. Chum. The integrity of gate oxide related to latent failures under EOS/ESD conditions. *Proceedings of the Electrical Overstress/Electrostatic Discharge (EOS/ESD) Symposium*, 1992; 106–111.
70. R. Wagner, C. Hawkins, and J. Soden. Extent and cost of EOS/ESD damage in an IC manufacturing process. *Proceedings of the Electrical Overstress/Electrostatic Discharge (EOS/ESD) Symposium*, 1993; 49–56.
71. S. Kiefer, R. Milburn, and K. Racley. EOS induced polysilicon migration in VLSI gate array. *Proceedings of the Electrical Overstress/Electrostatic Discharge (EOS/ESD) Symposium*, 1993; 123–128.
72. D. Pommeranke. Transient fields of ESD, *Proceedings of the Electrical Overstress / Electrostatic Discharge (EOS/ESD) Symposium*, 1994; 150–159.
73. M. Maytum, K. Rutgers, D. Unterweger, Lightning surge voltage limiting and survival properties of telecommunication thyristor-based protectors. *Proceedings of the Electrical Overstress/Electrostatic Discharge (EOS/ESD) Symposium*, 1994; 182–192.

74. D. Lin, and M.-C. Jon. Off-chip protection: shunting of ESD current by metal fingers on integrated circuits and printed circuit boards. *Proceedings of the Electrical Overstress/Electrostatic Discharge (EOS/ESD) Symposium*, 1994; 279–285.
75. S. Li, K. Lee, J. Hulog, S. Kazmi, S. Yin, and J. Pollock. Identification of electrical over stress failures from other package related failures using package delamination signatures. *Proceedings of the Electrical Overstress/Electrostatic Discharge (EOS/ESD) Symposium*, 1996; 95–100.
76. S. Ramaswamy, S. Kang, C. Duvvury, A. Amerasekera, and V. Reddy. EOS/ESD analysis of high density logic chips. *Proceedings of the Electrical Overstress/Electrostatic Discharge (EOS/ESD) Symposium*, 1996; 285–290.
77. G. Baumgartner, and J. Smith. EOS analysis of soldering iron tip voltage. *Proceedings of the Electrical Overstress/Electrostatic Discharge (EOS/ESD) Symposium*, 1998; 224–232.
78. W. Farwell, K. Hein, and D. Ching. EOS from soldering irons connected to faulty 120VAC receptacles. *Proceedings of the Electrical Overstress/Electrostatic Discharge (EOS/ESD) Symposium*, 2005; 238–244.
79. A. Wallash, H. Zhu, R. Torres, T. Hughbanks, and V. Kraz. A new electrical overstress (EOS) test for magnetic recording heads. *Proceedings of the Electrical Overstress/Electrostatic Discharge (EOS/ESD) Symposium*, 2006; 131–135.
80. R. Ashton, and L. Lescouzeres. Characterization of off chip ESD protection devices. *Proceedings of the Electrical Overstress/Electrostatic Discharge (EOS/ESD) Symposium*, 2008; 21–29.
81. A. Tazzoli, V. Peretti, E. Autizi, and G. Meneghesso. EOS/ESD sensitivity of functional RF-MEMs switches. *Proceedings of the Electrical Overstress/Electrostatic Discharge (EOS/ESD) Symposium*, 2008; 272–280.
82. R. Ashton. Types of electrical overstress protection. AND9009D, On Semiconductor Application Note, http://www.onsemi.com, 2011; 1–13.
83. M. Hernandez. Electrical overstress, and transient latchup pulse generator system, circuit, and method. U.S. Patent No. 7,928,737, April 19, 2011.

13

Electromagnetic Compatibility (EMC)

13.1 History

Electromagnetic compatibility (EMC) has been a concern from the beginning of power plants, power distribution, telephones, and electrical appliances [1–54]. Power disturbances from lightning to switching electronics increased the issue of EMC. With the introduction of telephone, electrical noise suppression, noise coupling, and transient signals became a large issue. This leads to a need for new devices and techniques for noise suppression.

With the introduction of radio technology, new EMC issues occurred. With broadband wireless telegraphs, interference from multiple sources occurred. The introduction of the vacuum tube oscillator made narrow-band radio communications possible and greatly reduced instances of unintentional electromagnetic interference (EMI) between radio transmitters. Narrow-band radio transmitters enabled the transmission of voice communication resulting in the proliferation of commercial radio stations.

During World War II, radio frequency (RF) and microwave technology development accelerated. At MIT, in the "Rad Lab" radar and microwave devices were invented for radar development. Radio communications with both intentional and unintentional RF interference played a major role in the war. It was clear the importance of measuring, analyzing, and preventing EMC problems. This led to the development of procedures and standards for electromagnetic compatibility [2–7].

During WWII, the development of the atomic bomb led to another level of concern for military equipment during an atomic bomb ignition. A large electromagnetic pulse (EMP) event from an atomic bomb ignition led to the concern of the ability of military equipments to survive a harsh electromagnetic environment. It was this concern that accelerated the studying of the reliability and electrical overstress (EOS) of electronics in the late 1950s and 1960s. Driven primarily by the needs of the Cold War and the military, a new engineering specialization occurred for diagnosing, solving, or preventing EMC issues.

ESD Testing: From Components to Systems, First Edition. Steven H. Voldman.
© 2017 John Wiley & Sons, Ltd. Published 2017 by John Wiley & Sons, Ltd.

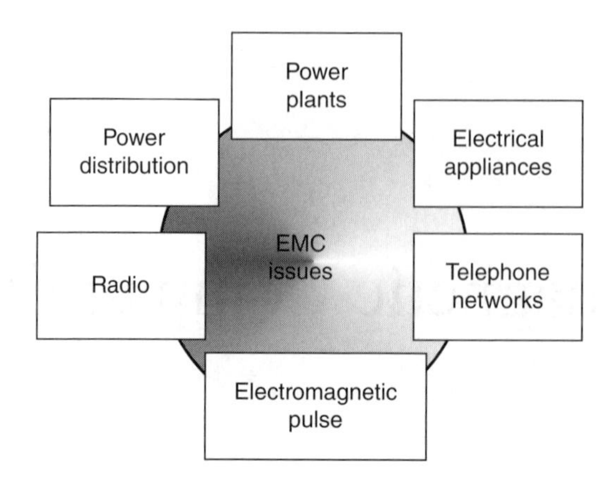

Figure 13.1 Electromagnetic compatibility issues

In the 1960s and the 1970s, EMC became a large issue as the electronics were developed for single components to integrated circuit (IC) technology (Figure 13.1). The initiation of the International Reliability Physics Symposium (IRPS) and the Electrical Overstress/Electrostatic Discharge (EOS/ESD) Symposium occurred in the late 1970s. The early publications were associated with addressing concerns in EOS, ESD, EMP, and EMC.

13.2 Purpose

The purpose of EMC test is for the establishment of a test methodology to evaluate the repeatability and reproducibility of systems in order to classify or compare sensitivity or susceptibility levels of systems.

13.3 Scope

The scope of the EMC tests is for the testing, evaluating, and classification of systems. The test is to quantify the sensitivity or susceptibility of these components to damage, degradation, or disturbs to electromagnetic events [6–16].

- Printed circuit board (PCB)
- Populated PCB.

13.4 Pulse Waveform

For EMC evaluation, the pulse waveform can vary depending on the issue being studied. EMC evaluation can introduce the following pulse waveforms (Figure 13.2):

- RF oscillating source
- Resonance pulse

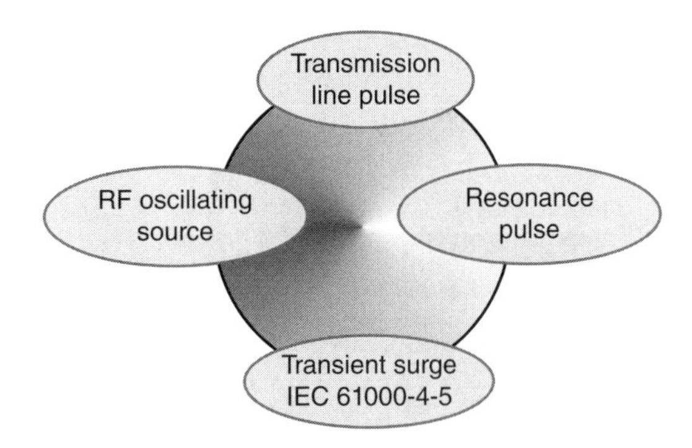

Figure 13.2 Electromagnetic compatibility pulse waveforms

- Transient surge
- Transmission line pulse (TLP) – square pulse.

RF oscillating sources have been used as early as the 1970s for the evaluation of the survivability of microwave electronics from an RF source. Resonance sources address the resonant coupling of an incoming electromagnetic source to a component, PCB, or system. Transient surge evaluation, such as the IEC 61000-4-5 pulse event, addresses the response of electronic coupling to a transient source. Additional surges such as combination wave, telecom wave, or ring oscillator wave can also be applied for EMC evaluation. Lastly, a square pulse event is used similarly to the TLP event for addressing electromagnetic coupling from a square pulse [45–50].

13.5 Equivalent Circuit

In the case of the EMC, the equivalent circuit is dependent on the type of source. The sources can be an RF oscillating source, single resonant frequency oscillating source, transient pulse to rectangular pulse events.

13.6 Test Equipment

Commercial test systems exist for EMC testing that can test a broad range of IEC immunity standards. The commercial test systems range from rack-mounted systems to new scanning methodologies. The following sections provide some examples of both the closed systems versus the scanning-based methodologies.

13.6.1 Commercial Test System

Commercial test systems exist for EMC testing that can test a broad range of IEC immunity standards. Figure 13.3 shows a commercial product known as the EMCPro Plus™ EMC Test System [55]. The EMCPro Plus™ EMC Test System tests up to seven global EMC standards for susceptibility to EMI and EMC with a single instrument.

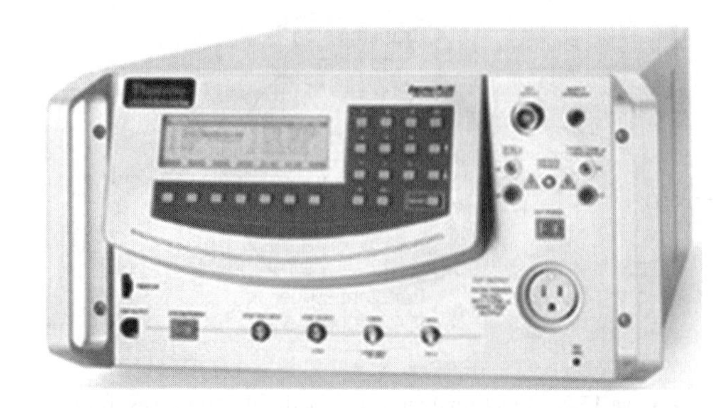

Figure 13.3 Commercial EMC test system – EMCPro Plus™ EMC (Reproduced by permission of Thermo Scientific Fisher)

The Thermo Scientific™ EMCPro PLUS™ EMC Test System is an advanced, multicapability, low-cost EMC immunity test system addressing a broad range of test capabilities and can be configured for testing to one or any combination of IEC basic immunity standards, such as the previously discussed IEC 61000-4-5. Additional function of this test system is as follows [55]:

- Conducts surge testing with combination, telecom, and ring waves to IEC 61000-4-5 limit (6 kV)
- Performs electrical fast transient (EFT) testing to IEC 61000-4-4 limit (4 kV)
- Monitors surge voltage and current at the output terminals.

The test system addresses common surge requirements, such as combination wave, telecom wave, and ring oscillator wave requirements.

13.6.2 Scanning Systems

In recent years, scanning systems for EMC have been introduced. New method and test systems have been developed to analyze EMC that are either in development or are in commercial products.

EMC/ESD scanning is a method developed to determine the susceptibility or upsets of systems, circuits, and components to ESD or other EMC events without causing hard failures [45–50, 53]. In order to provide visualization of the EMC/ESD sensitivity, a localized source and means of scanning over the system is important to evaluate the local system or circuit response.

An ESD/EMC scanning system has been developed that produces a localized E-field or H-field, scans the equipment under test (EUT), and monitors the product for disturbs or upsets. This test can scan exterior of products, subassemblies, and boards. The same test can be applied to components within the system (e.g., semiconductor ICs). In both cases, a mapping can be produced that overlays the system to determine entry points of electromagnetic noise and system sensitivity.

Figure 13.4(a) shows an example of a high-level diagram for a scanning system. The test system comprises the following [31, 32, 45–50, 53]:

- Test table or fixture
- Pulse or waveform source
- X–Y scan control system
- Scanning armature
- Signal processing equipment and software.

In the figure, the example of the pulse source shown is a TLP source used to generate a local EMP event. The scanning armature sweeps across the EUT and is driven by the motor driver electronics. The data are collected and stored in the computer.

Figure 13.4(b) shows a photograph of a small commercial scanning system. Figure 13.4(b) shows the armature for pulsing the event. The armature position is initiated by the supporting mechanical system [45–50].

Figure 13.5 is a photo of the Amber Precision Instrument full scanning system [31, 32]. The EUT is placed under the armature. A mechanical arm sweeps the source over the EUT driven by the motor drive circuitry. The pulse armature is connected to a pulse source that initiates the EMC event. The system also has voltage sources for powering the EUT during testing. The computer collects the information from the EMC test.

13.7 Test Procedures

In the following section, test procedures for ESD/EMC scanning system are discussed.

13.7.1 ESD/EMC Scanning Test Procedure and Method

Figure 13.6 illustrates a flowchart of the ESD/EMC scanning system test procedure. The procedure starts by applying a stimulus to the EUT. If no “upset” occurs, the armature moves to the next test location. If an upset does occur at that location, the system reduces the test level to the lowest value. At the lowest pulse level if there is an upset, then the system moves to the next test point. If there is no upset at the lowest level, the system increases the test level, until an upset does occur. Once the upset level occurs, then the armature moves to the next test point and repeats the process at the next location. This process continues until the entire area of the EUT is completed. In this fashion, a plot of all the sensitive areas is obtained [31, 32, 45–50].

13.8 Failure Mechanisms

For EMC testing, the failure of the system can be destructive or nondestructive. For the destructive failure, the failure can be the following:

- Component signal input parametric shift
- Component signal input permanent damage.

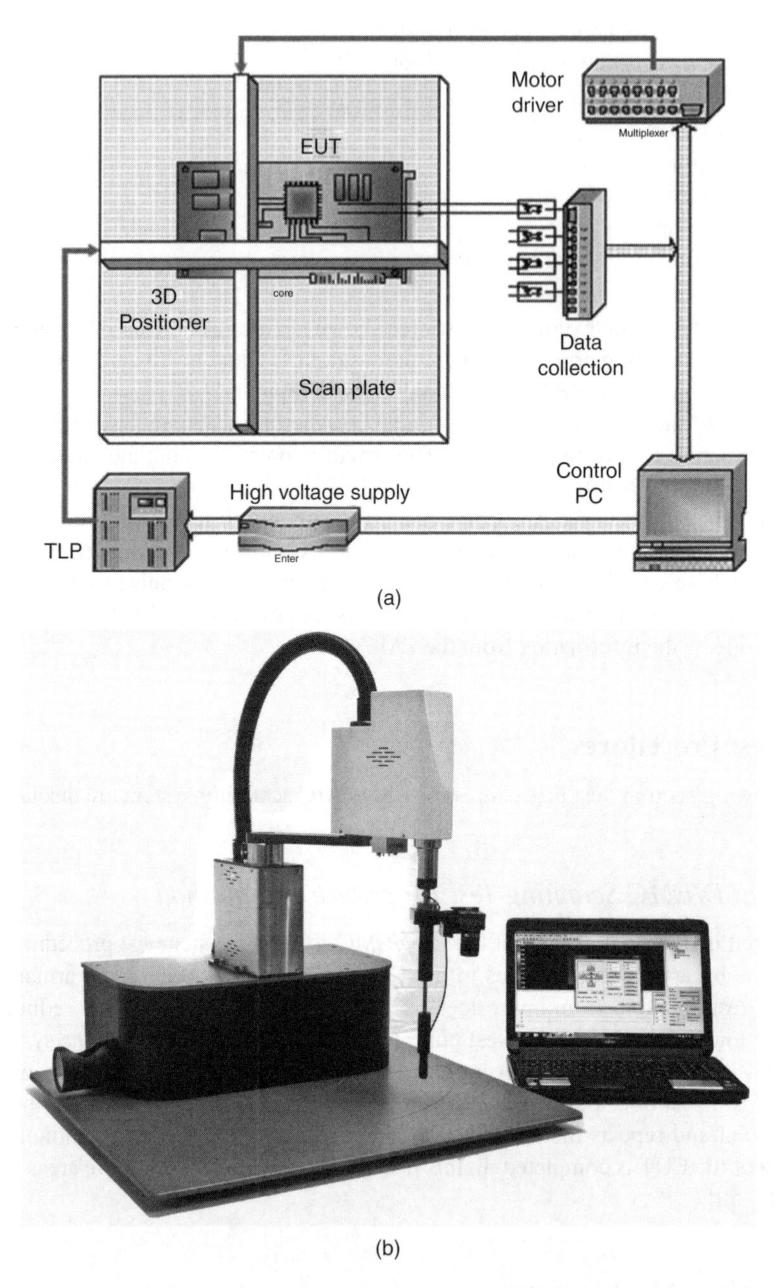

Figure 13.4 (a) EMC/ESD high-level diagram (Reproduced with permission of Amber Precision Instruments). (b) EMC/ESD scanning system – small system (Reproduced with permission of Amber Precision Instruments)

Figure 13.5 EMC/ESD scanning system – large system (Reproduced with permission of Amber Precision Instruments)

For the nondestructive failure, the failure can be the following:

- System disturbs
- System interrupts
- System timing delay.

EMC failures can be nondestructive event. System disturbs, interrupts, or time delays can lead to a system failure.

13.9 ESD/EMC Current Paths

EMC failures can be destructive or nondestructive events. EMC failures can occur in the component's internal circuitry, the component packaging, or in the PCB.

EMC current that flows into a component can be destructive if the current level exceeds the failure level of the bond pad, interconnects, ESD network, or internal circuitry. In the case where the current flows through the circuitry, and no destructive ESD current occurs, the return current flow through the chip substrate can lead to a disturb, interrupt, or timing delay in the component.

EMC current that flows into a PCB trace can be destructive if the current level exceeds the failure level of the PCB trace or materials. The return current flow through the PCB can also

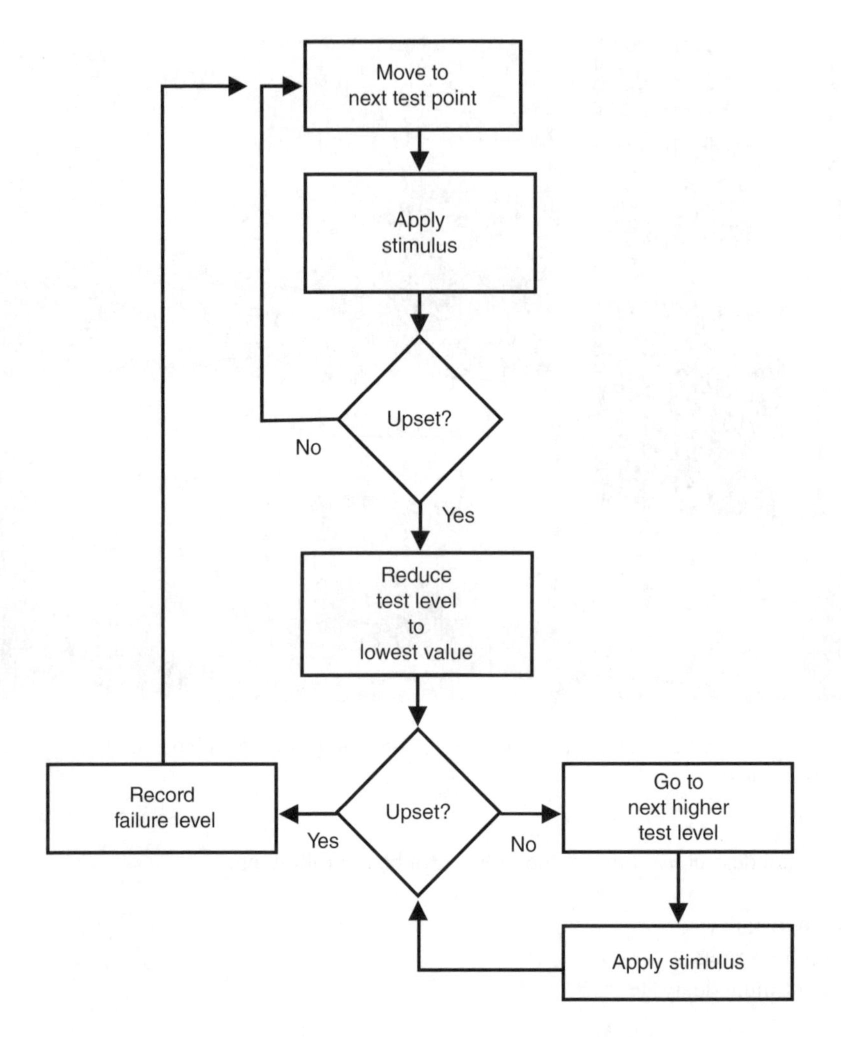

Figure 13.6 ESD/EMC scanning test procedure

lead to a system disturb, interrupt, or timing delay. Given that the return current of a digital component flows under an analog component, system disturbs can occur.

Figure 13.7 is an example of an EMC failure associated with the return current of a digital component flows under an analog component mounted on a common PCB [6, 7, 33].

13.10 EMC Solutions

PCB design can influence the sensitivity to EMI and provide good EMC characteristics. Today, PCB design is done with electronic design automation (EDA) tools. The EDA PCB design process undergoes the following steps [7, 31, 32]:

- *Circuit schematic:* Schematic capture is achieved using an EDA.
- *Card dimension:* Card dimension and template size are determined for the application. This is based on the component sizes and heat sinks.
- *Number of design levels:* The number of the design layers is determined based on designer's designation or requirements. Decisions include the following:
  - Power plane
  - Ground plane
  - Signal plane
- *Trace line impedance:* Decisions are made of the dielectric layer thickness copper routing thickness and trace width.
- *Component placement:* The placement of the components on the PCB is decided based on power distribution and physical dimensions.
- *Signal trace routing:* Signal lines are created using a "place and route" function.
- *File generation:* A Gerber file is generated for the final design.

In the design of the PCB, a number of the design steps can influence the EMI and the EMC characteristics [7, 31, 32]:

- *Card dimension*
- *Number of design levels*
- *Power plane design*
- *Ground plane design*
- *Signal plane design*
- *Trace line impedance*
- *Copper routing thickness*
- *Trace width*
- *Component placement*
- *Signal trace routing.*

Within this subject, guidelines can be created to minimize the systems' sensitivities.

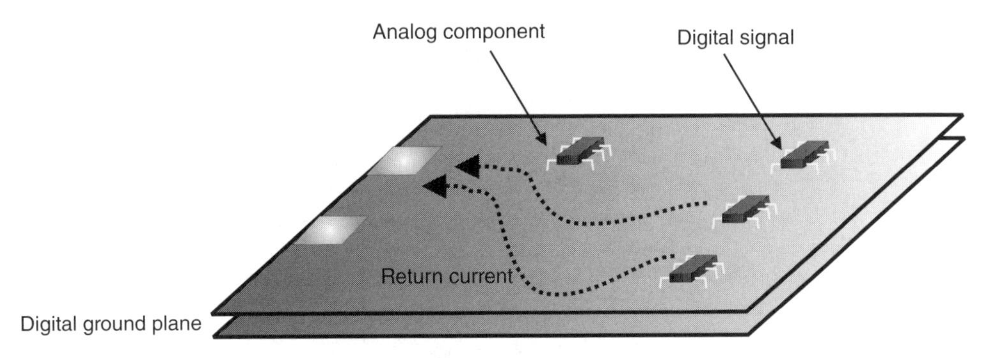

Figure 13.7 EMC current paths – the return current

13.11 Alternative Test Methods

Alternative test methods exist for a variety of different EMC issues. In this section, alternative methods are discussed, as well as the concept of susceptibility and vulnerability of electronics to EMC.

13.11.1 Scanning Methodologies

For evaluation of system level susceptibility and vulnerability, new methodologies have been developed that provide scanning of populated PCB [46–57]. The different types of PCB scanning include the following methods (Figure 13.8):

- ESD/EMC immunity scanning method [45–50]
- EMI emissions scanning method
- RF immunity scanning method [48]
- Resonance scanning method [46]
- Current spreading scanning method [53]
- Current reconstruction scanning method [53].

13.11.2 Testing – Susceptibility and Vulnerability

For the understanding of system failure, these scanning methods provide information about a system's susceptibility and vulnerability (Figure 13.9). In this section, the distinction is highlighted [53].

Susceptibility is a system's sensitivity to "upset." Susceptibility may be a nondestructive event. Electrical currents that flow on the PCB generate electromagnetic (EM) signals that can disrupt operation of a system. The electrical currents can flow into the semiconductor components through the signal pins, power pins, as well as the power plane and ground plane With the presence of transient voltage suppression (TVS) devices on the PCB, EOS event current can be shunted to a power or ground plane [53]. Prior to the EOS protect devices, the current will generate EM emissions along the PCB trace. In addition, a "residual current" can

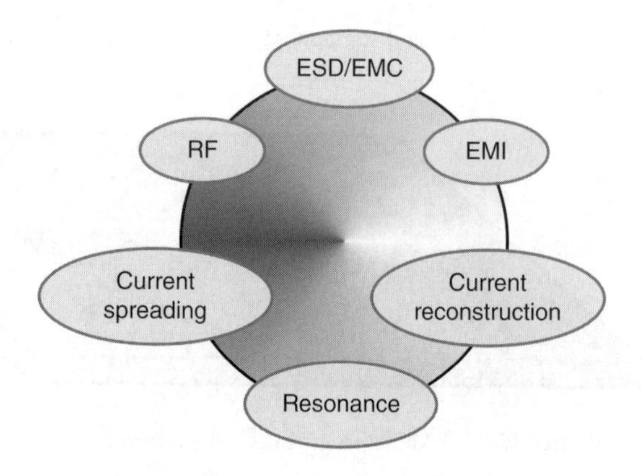

Figure 13.8 Scanning methodologies

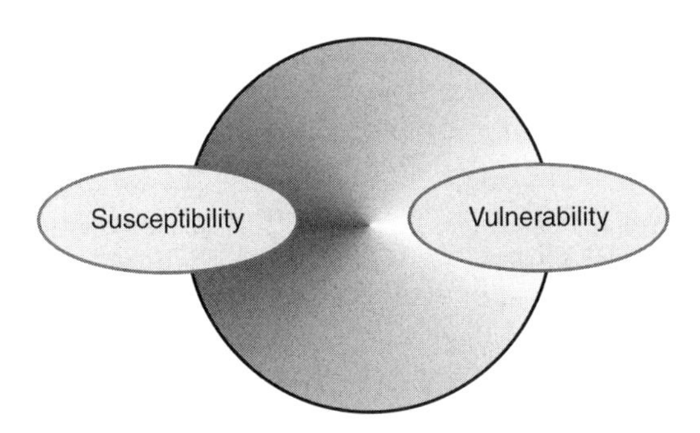

Figure 13.9 Susceptibility and vulnerability

also flow into the semiconductor components beyond the EOS protection networks [30, 31, 45, 54].

Scanning methods will provide a mapping of the susceptibility of regions of a PCB to electromagnetic emissions. As will be discussed, susceptibility is a measure of the PCB or components to the EM events. But, not all methods demonstrate actual "events" that enter the PCB traces, or components. Hence, a new method provides both susceptibility and vulnerability to EOS events [53].

13.11.3 EMC/ESD Scanning – Semiconductor Component and Populated Printed Circuit Board

EMC/ESD scanning can be performed on the semiconductor component and on a populated PCB. EMC scanning can be performed on an unpopulated or a full populated PCB.

13.12 EMC/ESD Product Evaluation – IC Prequalification

EMC/ESD scanning can be used for product evaluation. As an IC prequalification, EMC/ESD can be used as part of a qualification process. This can be completed in the design phase, or in a reliability evaluation.

13.13 EMC/ESD Scanning Detection – Upset Evaluation

EMC/ESD scanning methodology can be used for upset evaluation. By applying an EMI event, evaluation of the susceptibility of a populated PCB can be initiated and determine the upset sensitivity across a system.

13.13.1 ESD/EMC Scanning Stimulus

Figure 13.10 illustrates an ESD/EMC scanning stimulus. Scanning systems can include EMI scanning, local H-field scanning, fixed resonant RF sources, and variable frequency scanning methods.

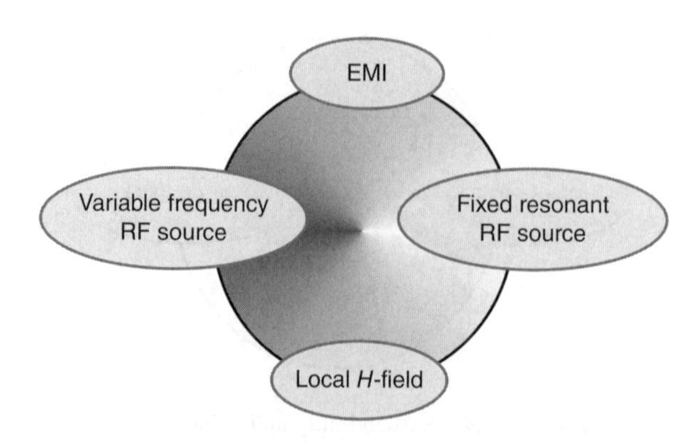

Figure 13.10 ESD/EMC scanning stimulus

13.14 EMC/ESD Product Qualification Process

EMC/ESD qualification process flowchart is shown in Figure 13.11. The qualification process should include the ESD device qualification for human body model (HBM), machine model (MM), and charged device model (CDM). If the component passes the standard ESD qualification, then evaluation of the device susceptibility scanning using standard test boards is completed. If it passes the susceptibility scanning process, the next step is the full system EMC compliance testing. If the system EMC compliance testing does not pass, then the susceptibility scanning at the system level is required to identify the problem areas [31, 32].

13.14.1 EMC/ESD Reproducibility

In order for a test method to achieve the status as a standard test method, it is important that the test method is both repeatable and reproducible. Figure 13.12 shows an example of the susceptibility mapping from the same component under the same test. From the mapping, it can be observed that the same regions of the test show comparable susceptibility [31, 32].

13.14.2 EMC/ESD Failure Threshold Mapping and Histogram

The results can be displayed in a failure threshold mapping and in a histogram. Figure 13.13(a) shows the EMC/ESD failure threshold mapping, and Figure 13.13(b) shows an example of the display in histogram form. Figure 13.13(a) shows the output failure threshold mapping that are sensitive regions. The test system clearly identifies a two-dimensional mapping of the areas of concern. The second plot displays a histogram of the voltage magnitude across the device.

13.14.3 ESD Immunity Test – IC Level

The scanning methodology can be used as a comparative ESD immunity tool. The scanning tool can be used to identify which pins are sensitive to noise. The failure levels of components

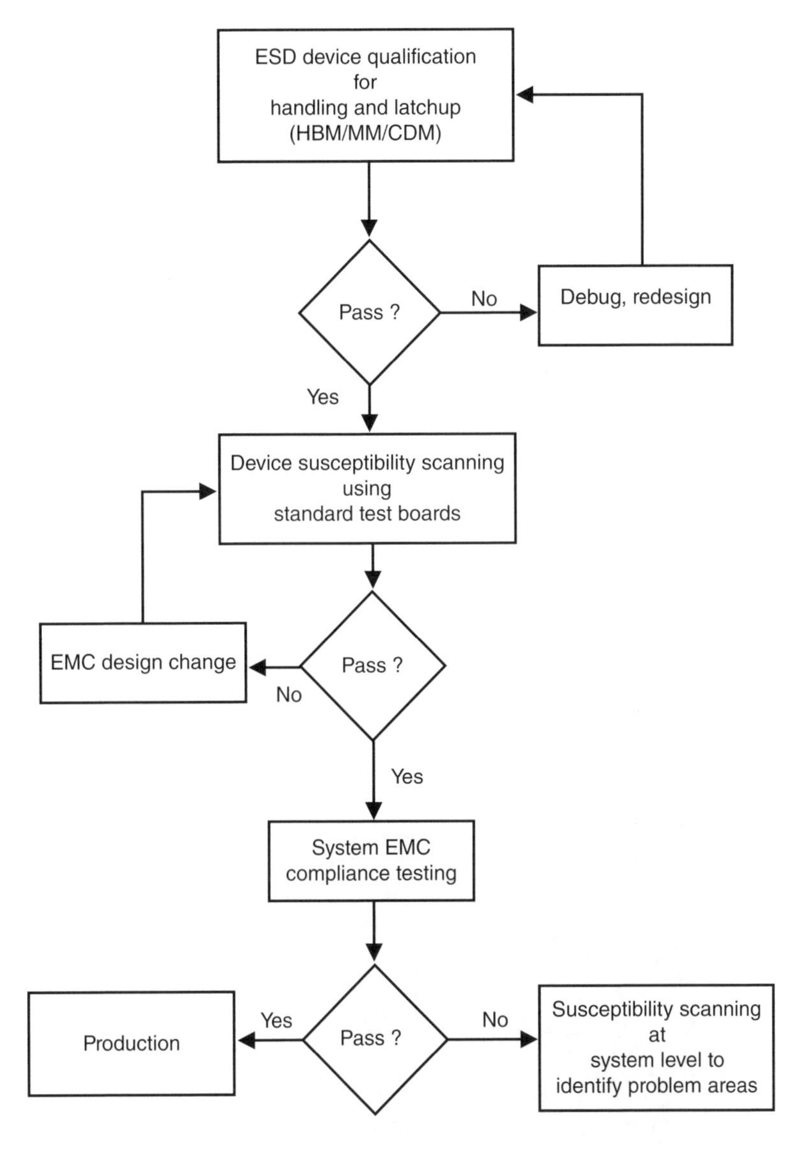

Figure 13.11 ESD/EMC qualification process

can be evaluated by determining the failure frequency. A mapping of the test is shown in Figure 13.14(a). Figure 13.14(b) and (c) is an example of the pin response and failure levels.

After the scanning process, the system produces a mapping of the susceptibility overlaid on the physical image of the board or components. Figure 13.14(d) shows a product sensitivity mapping of two products on the same system. It can be observed that the image mapping on the left (vendor A) is more sensitive compared to the image mapping on the right (vendor B). Although both fulfill the same electrical functional specification, their sensitivity to EMI is not equivalent.

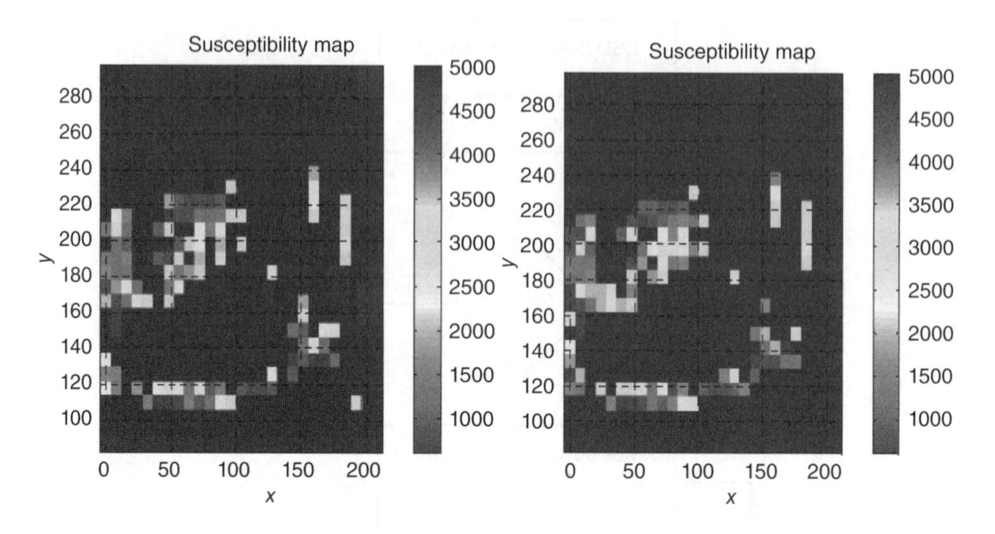

Figure 13.12 ESD/EMC comparison of reproducibility

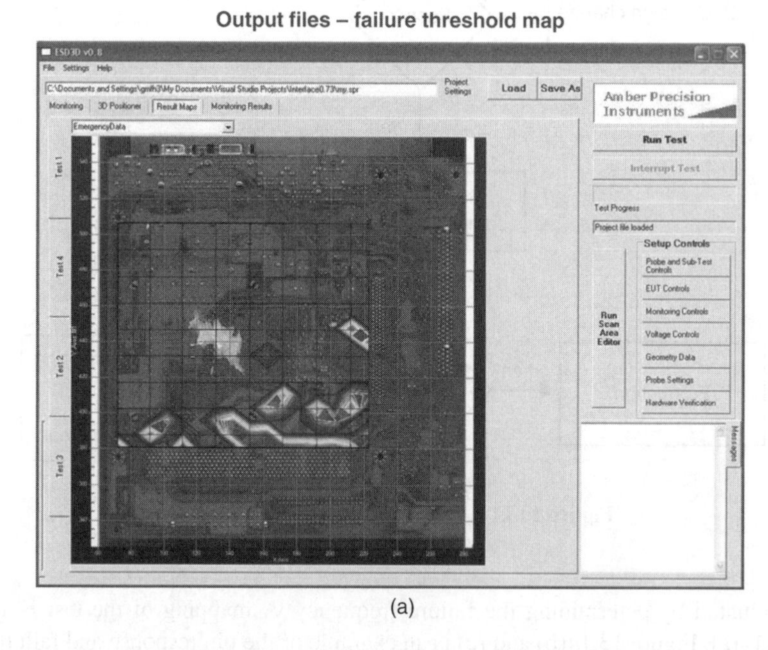

Figure 13.13 (a) ESD/EMC failure threshold mapping (Reproduced with permission of Amber Precision Instruments). (b) ESD/EMC failure threshold histogram (Reproduced with permission of Amber Precision Instruments)

Output files – failure threshold histogram

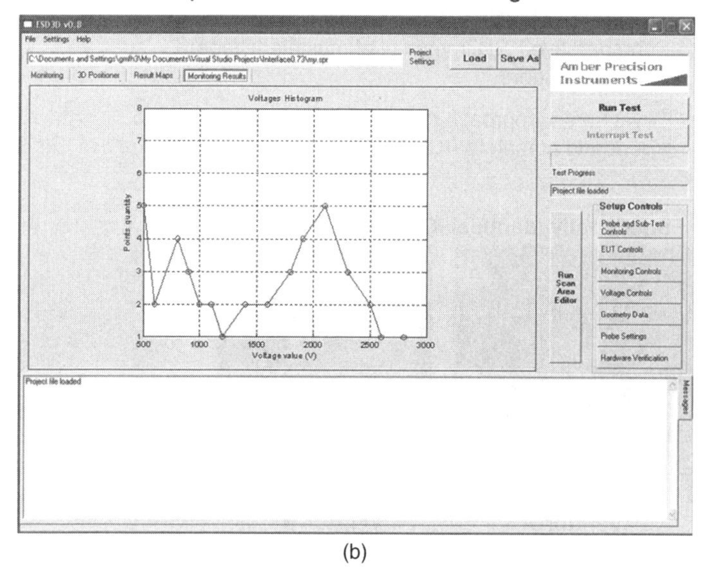

(b)

Figure 13.13 *(continued)*

13.14.4 ESD Immunity Test – ATE Stage

ESD immunity test can be initiated at the automatic test equipment (ATE) stage. At the ATE stage, evaluation of the ESD immunity can be performed with a manual scan process. The manual process can apply pulse events, or noise on the sensitive component that is mounted on the board. System tests can be performed that verify "system disturb" errors instead of hard failure of components. Figure 13.15 shows an example of the error message associated with the failure of the signal. The ATE test system verified that the scan failure was reproduced on multiple parts.

13.15 Alternative ESD/EMC Scanning Methods

A methodology is needed for both component and system level manufacturers to test for susceptibility of both the components and the systems and have a means to correlate the relationship. A susceptibility test that can correlate system level upsets and the location of the upset is key to building better systems.

Today, component ESD models, which are unpowered, cannot provide enough information to estimate the performance at a "powered" state of the system.

13.15.1 Alternative ESD/EMC Scanning Methods – Printed Circuit Board

In a PCB, motherboards, and the components itself, the ability to determine where the susceptibility problem occurs is very important. With the complexity of computer chips, and the

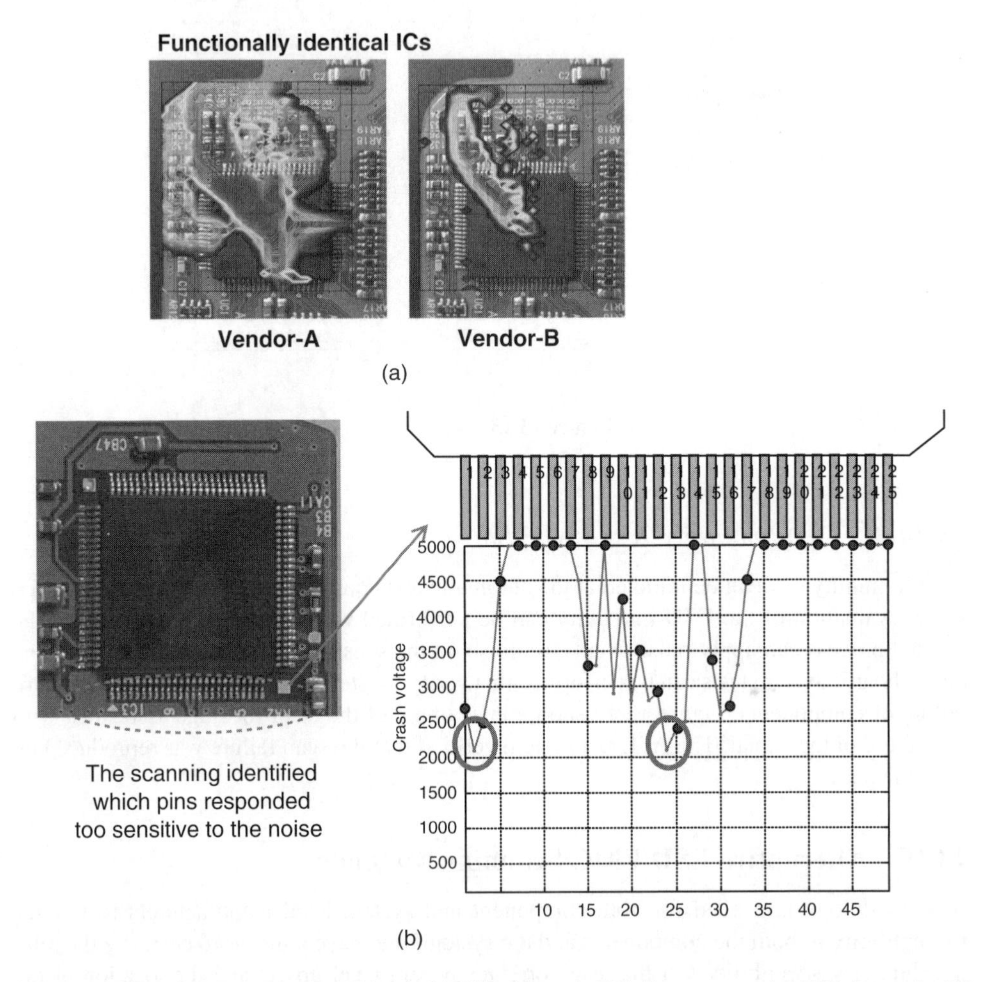

Figure 13.14 (a) ESD immunity IC level. (b) ESD immunity IC pin response. (c) ESD immunity IC pin failure levels. (d) ESD/EMC image comparison (vendors A and B)

spatial extent of the system board, it is very important to determine exactly where the problems occur in order to improve and fix the problems. One of the biggest problems today is that there is no visualization capability of the location of the susceptibility concern. As a result, EMC susceptibility improvements are evaluated by relying on both trial-and-error experimentation and EMC experience [45–50].

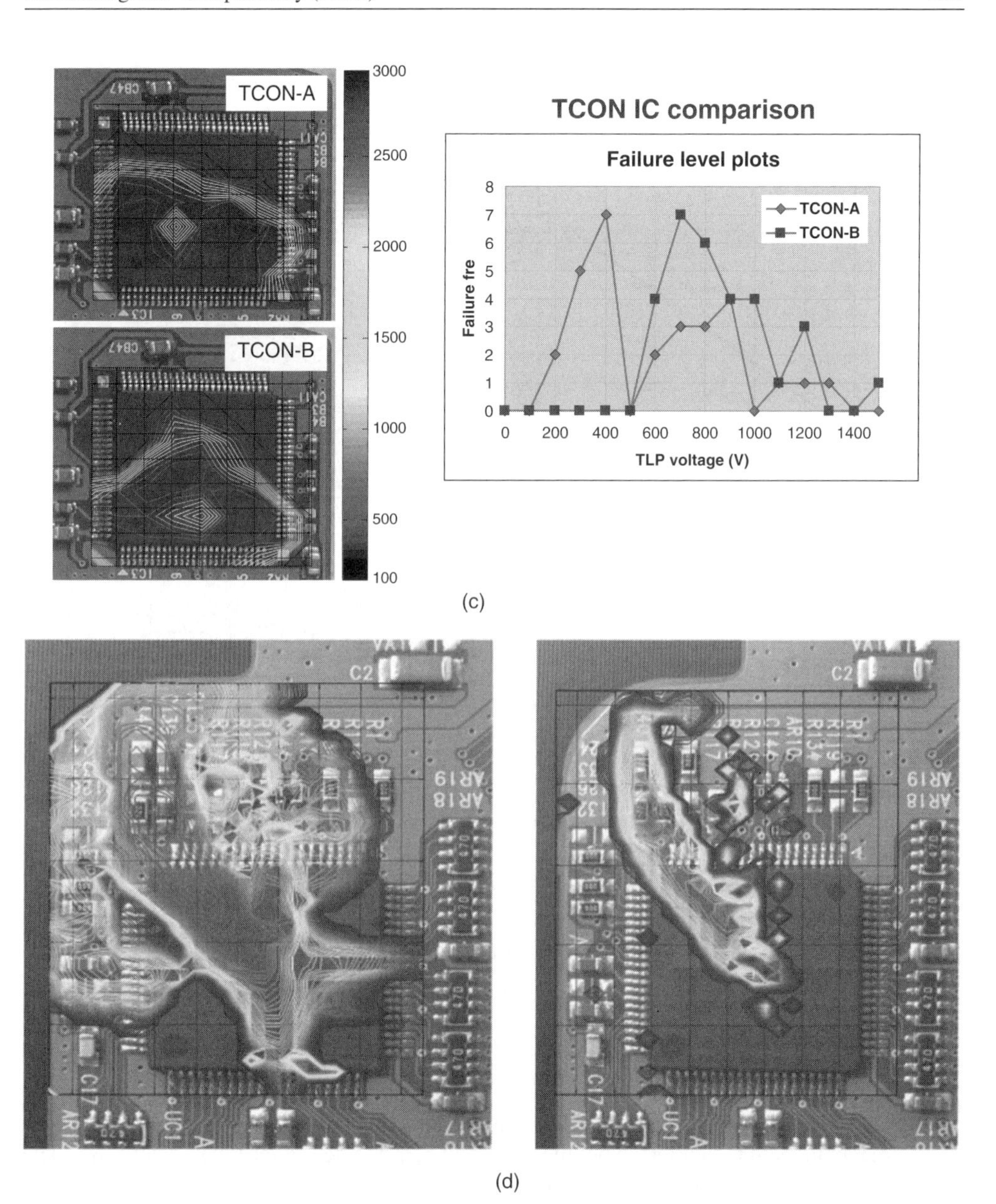

Figure 13.14 *(continued)*

This ESD/EMC scanning system can apply different stimulus sources [45–50, 56]. The pulse test can be an RF source, EFT, transmission line, or ESD gun. A TLP source can be used to provide a TLP or very fast transmission line pulse (VF-TLP) event. For simulation of system-like events, the human metal model (HMM) or IEC 61000-4-2 test can be applied.

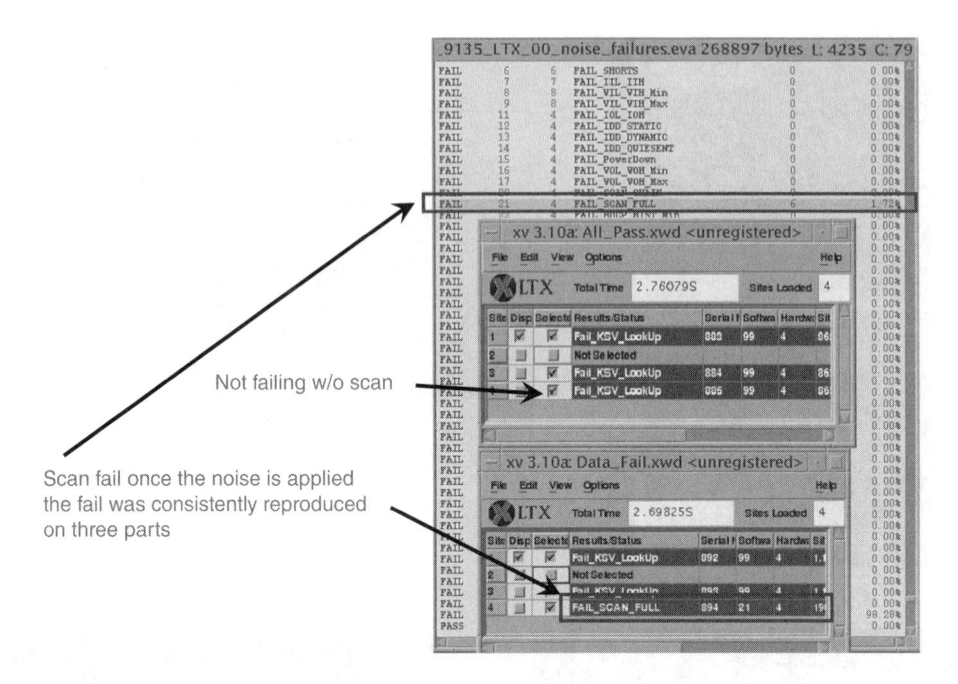

Figure 13.15 ESD immunity at ATE stage

Figure 13.16 is an example of the scanning results from a motherboard. In the image mapping, there are regions of sensitivity. With this mapping, it provides guidance for potential redesign of the motherboard [56].

With this EMC/ESD scan system, components and system boards can be evaluated in the design phase, assembly, or for qualification. A convergence of system level and chip level ESD testing is occurring today with these new methodologies [56].

13.15.2 Electromagnetic Interference (EMI) Emission Scanning Methodology

In the ESD/EMC method discussed earlier, current was injected into the current probe to generate a magnetic field to couple the populated circuit board to determine the susceptibility. The probe was then swept across the entire sample to provide a mapping of susceptibility.

Another methodology is the EMI emission scanning method [45–50]. In this method, the PCB is powered, and operable, and the probe is used in a "listening mode" where it is mapping the electromagnetic emissions on the PCB sample. This method is valuable for evaluation of the regions of the PCB that are sources of electromagnetic emissions.

13.15.3 Radio Frequency (RF) Immunity Scanning Methodology

A method to determine the sensitivity and susceptibility of RF signals can be achieved using an RF immunity scanning method [48–50]. Using an RF source, an RF signal can be connected

Figure 13.16 Motherboard image scan (Reproduced with permission of Amber Precision Instruments)

to a scanning probe. The populated PCB or motherboard is then powered and cycled. The RF signal can be scanned over the populated PCB and system level upsets can be evaluated.

13.15.4 Resonance Scanning Methodology

A method to determine the sensitivity and susceptibility of resonant frequency signals can be achieved using a resonance scanning method. Resonant frequencies of the PCB, antennae, and components can be an issue. Using an RF source, a signal may be applied to the scanning probe to evaluate the effect of a resonant signal across the PCB [46].

13.15.5 Current Spreading Scanning Methodology

In the prior methodologies, the scanning probe was established a signal across the populated PCBs to determine the relative susceptibility of different regions [45–50]. The scanning process provides insight into potential regions where a PCB and components can be sensitive to electromagnetic emissions, RF signals, and resonance states. The sensitivity mapping can assist the system developer to determine which component to choose by determining susceptibility of equivalent components. It can also determine potential weak spots in a motherboard design where EMI, RF, noise, and other signals may enter a system.

13.16 Current Reconstruction Methodology

For EOS, the influence on the system is a function of both the flow of current from the signal and power connections, and the sensitivity of the regions that they influence. Hence, susceptibility mapping does not provide the actual vulnerability of the system [53].

By injecting EOS events into the ports, and connections, the current will flow into the system. As the EOS event is injected into the PCB, the trace signal will generate E- and H-field emissions. Using the scanning probe, the emissions from the injected EOS current can be mapped and determine where the current flows in the network [53].

13.16.1 EOS and Residual Current

Figure 13.17 shows an example of a PCB with a component chip mounted. An EOS on-board protection device is shown on the PCB. The EOS current is injected into the system, where a large percentage of the pulse event flows through the EOS protection device. In the scanning system, the current flow can be mapped from the E- and H- fields generated using the scanning probe, as the signal or power pin is pulsed.

13.16.2 Printed Circuit Board (PCB) Trace Electromagnetic Emissions

Figure 13.18 demonstrates a populated PCB and the electromagnetic emissions coming from the signal traces. In the plot are incoming signals that are electrically connected to resistor elements in series with a chip. Measurements show the EM emissions from the pulse event prior to the resistors and after the chip. Results show that only a small percentage of the EM signal exits the semiconductor chip, showing the influence of the resistor in lowering the current entering the semiconductor chip during an ESD event.

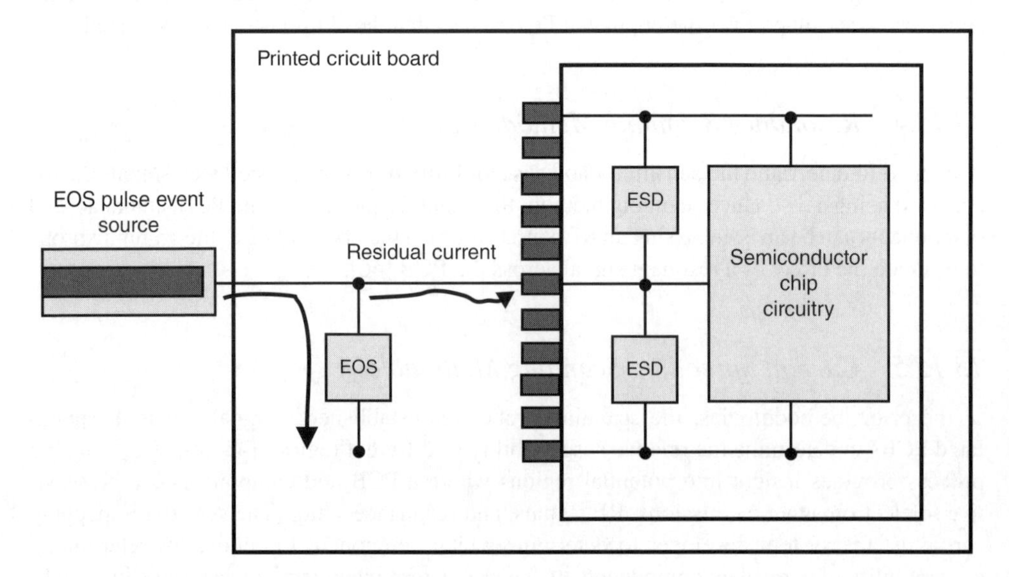

Figure 13.17 EOS and residual current

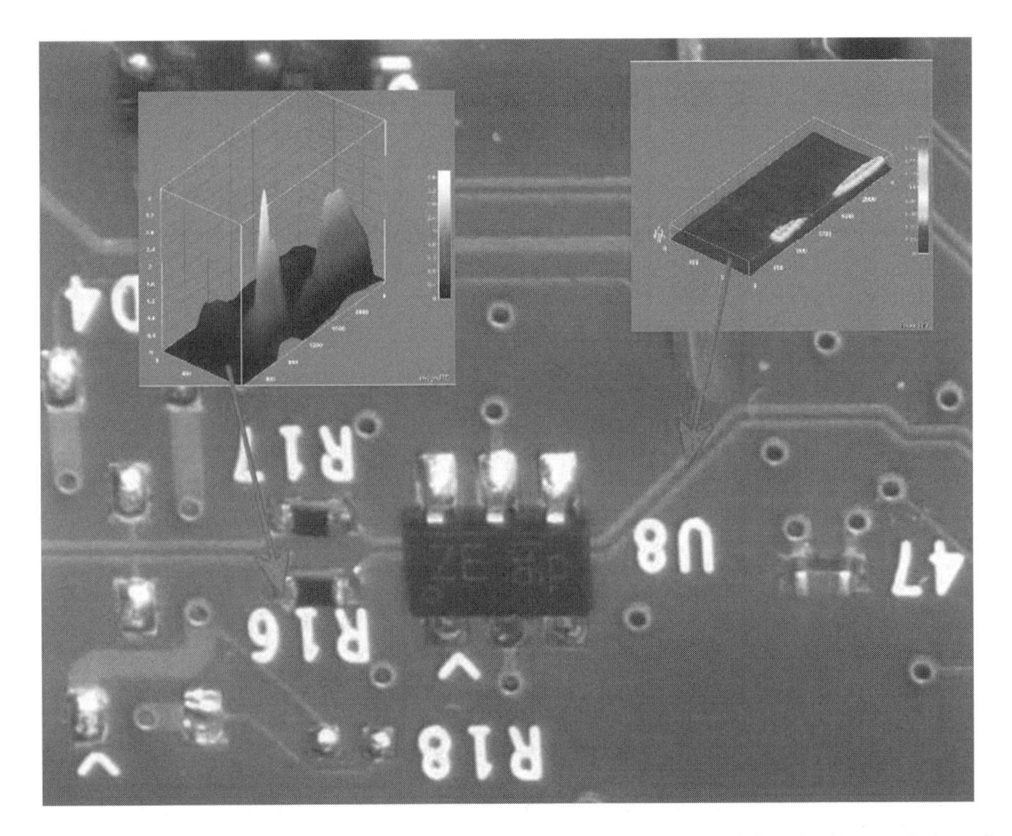

Figure 13.18 Printed circuit board trace electromagnetic emissions (Reproduced with permission of Pragma Design)

13.16.3 Test Procedure and Sequence

Figure 13.19 is a susceptibility scanning flowchart of the process. In the process, a pulse train is injected into the system. The device under test (DUT) is observed where the failure level is recorded. After this is achieved, the probe is shifted to a new location and the process is continued until a full mapping is achieved [31, 32, 53].

Figure 13.20 highlights a comparison of three signals produced from the current reconstruction methodology. Figure 13.21 shows a populated PCB using the scanning methodology.

13.17 Printed Circuit Board (PCB) Design EMC Solutions

There are many decisions in the design of PCBs that influence the EOS, EMI, and EMC sensitivities. The component selection and placement influences these issues. Some guidelines for printed circuit design for placement and component selection are as follows [6, 7, 30, 31]:

- *Component placement and ESD return current:* Components should avoid being placed over ESD return current ground traces.

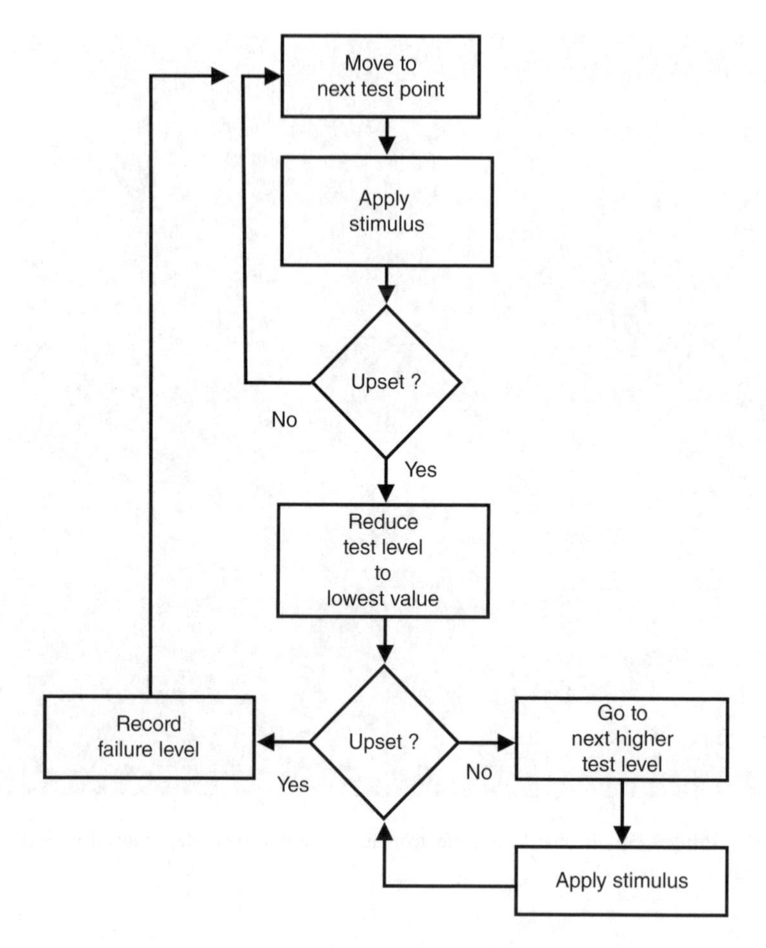

Figure 13.19 Current reconstruction – test procedure and sequence

- *Connector edge placement:* Connectors should be located on the same edge of a PCB.
- *Connector corner placement:* Connectors should be located on one corner of a PCB if possible.
- *Common connector:* All off-board signals from a single device should be routed through a common connector.
- *Connector to on-board I/O components spacing:* Components connected to I/O nets, connectors, and off-board components should be located within 2 cm away from connectors.
- *Connector and I/O to on-board non-I/O components spacing:* Components not connected to I/O nets or connectors should be located 2 cm away from I/O nets and connectors.
- *Digital component off-chip transition timing:* Digital components should be selected to have a maximum off-chip rise and fall times.
- *Clock and clock oscillator placement:* Clock drivers should be located within proximity of clock oscillators.

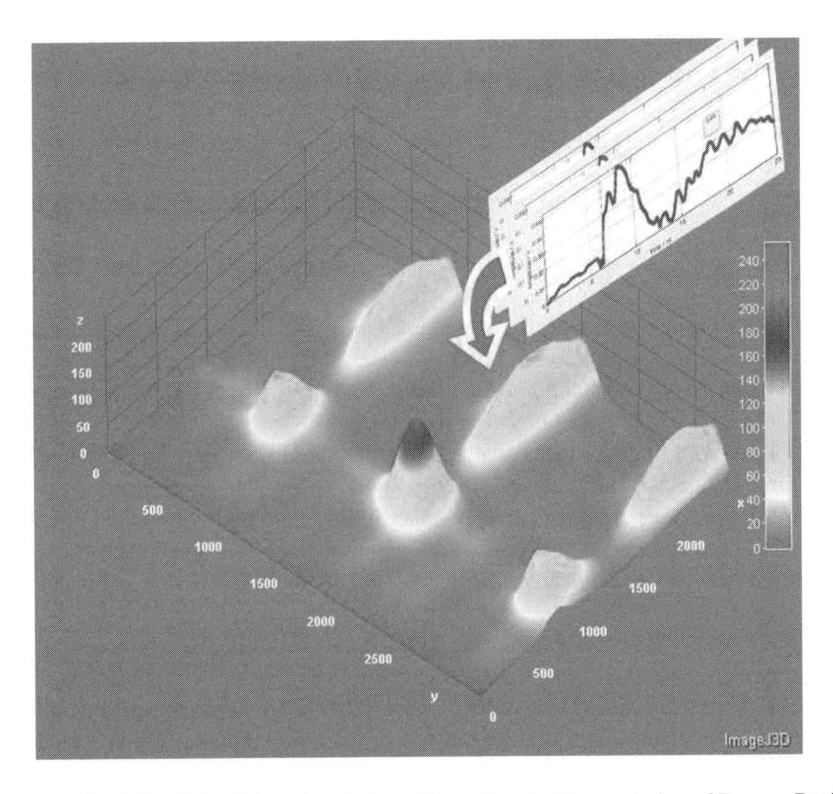

Figure 13.20 Printed circuit board emissions (Reproduced with permission of Pragma Design)

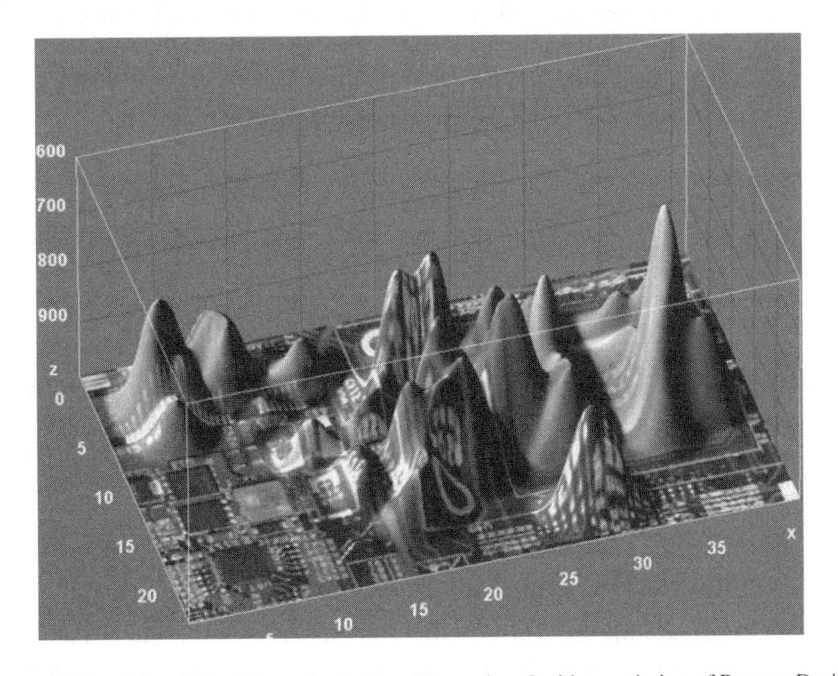

Figure 13.21 Printed circuit board emissions (Reproduced with permission of Pragma Design)

Trace routing and power/ground plane decisions are key to avoid EMI, and EMC concerns. Some guidelines for printed circuit design for signal traces are as follows [1]:

- *Power trace width:* All power supply line traces should be of suitable width for EOS robustness objectives.
- *Ground trace width:* All ground line traces should be of suitable width for EOS robustness objectives.
- *Signal trace width:* All signal line traces should be of suitable width for EOS robustness objectives.
- *Critical signal trace placement:* Critical signal traces should be placed between power and ground planes. This avoids susceptibility of noise, EMI, and other sources.
- *Critical signal trace placement and ESD return current:* Critical signal traces should avoid being placed over ESD return current ground traces.
- *Non-I/O trace placement:* Traces that are not I/O should not be located between an I/O connector and a device that receives or sends signals using that connector.
- *Non-I/O trace placement and I/O components:* Signals with high-speed contents should not be placed under components that contain I/O components.
- *Signal trace and power plane separation:* No trace should be used for connection to the power plane.
- *Signal trace and ground plane separation:* No trace should be used for connection to the ground plane.
- *I/O to connector trace length:* Trace lengths from I/O to connectors should be minimized.
- *High-speed digital trace length:* Trace lengths for high-speed digital signals should be minimized.
- *Clock trace length:* Trace lengths for clocks should be minimized.
- *High-speed trace to board edge:* High-speed traces should be routed at least $2X$ from the board edge, where X is the distance between the trace and its return current.
- *Differential signal trace pairs:* Differential signal trace pair should be routed together and maintain same distances from adjacent shapes, objects, and solid planes.

Some guidelines for printed circuit design for power planes are as follows:

- *Power plane and traces common layer:* Power planes and traces should be routed on the same layer.
- *Power plane – power return plane (ground) common layer*: All power planes that are referenced to the same power return plane should be routed on the same layer.

Figure 13.22 shows an example of a design methodology to optimize a PCB for EMC. The methods of the design include identifying (1) EMI sources, (2) critical paths, (3) antennas, and (4) coupling mechanisms. Using the scanning methodologies, these issues can be evaluated and quantified.

13.18 Summary and Closing Comments

This chapter addressed EMC as well as EMI as an introduction to many of the new state-of-the-art methods. In the present and the future, the direction of EMC and EMI analysis will

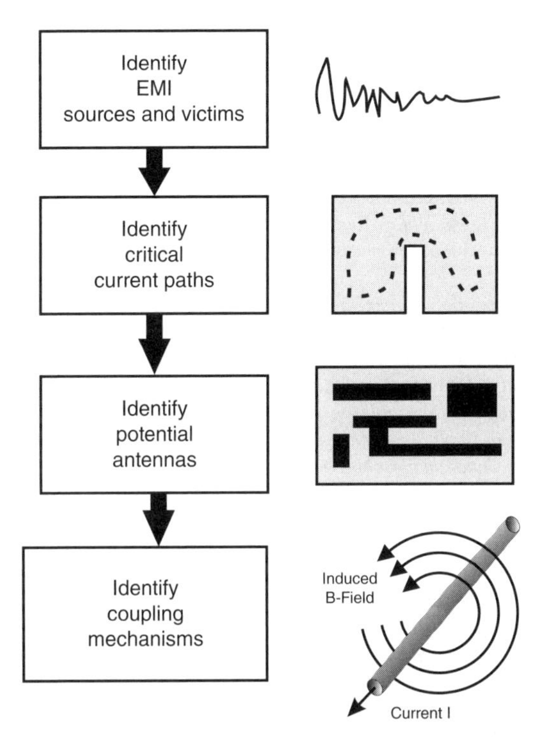

Figure 13.22 Printed circuit board design

take on the process of EMC scanning methods to pinpoint the location and susceptibility of components. It will also be used to evaluate competitive analysis and choosing vendors to provide the best designs and best systems in the future. In the coming years, new methods will be developed, and some methods will be shown to be advantageous over others. It will be interesting in the future how this field of interest will evolve.

Problems

1. Describe electromagnetic compatibility (EMC). How do you define it? How wide is the scope?
2. Describe electromagnetic interference (EMI). What technologies is this an issue?
3. Magnetic recording heads outside of the disk drive can fail from electromagnetic interference (EMI). Explain the process of failure.
4. Describe susceptibility.
5. Describe vulnerability. How does vulnerability differ from susceptibility?
6. Describe what the scanning process signal demonstrates.
7. Describe what the current reconstruction process demonstrates.
8. Explain the distinction of the ESD/EMC scanning from the current reconstruction process.
9. Describe how the TLP is used for the scanning processes versus the current reconstruction process.
10. How will these test methods change system engineering and quality of future products?

References

1. C.E. Jowett. *Electrostatics in the Electronic Environment*. New York: Halsted Press, 1976.
2. W.H. Lewis. *Handbook on Electromagnetic Compatibility*. New York: Academic Press, 1995.
3. R. Morrison, and W.H. Lewis. *Grounding and Shielding in Facilities*. New York: John Wiley and Sons Inc., 1990.
4. C.R. Paul. *Introduction to Electromagnetic Compatibility*. New York: John Wiley and Sons Inc., 2006.
5. R. Morrison, and W.H. Lewis. *Grounding and Shielding*. New York: John Wiley and Sons Inc., 2007.
6. H.W. Ott. *Electromagnetic Compatibility Engineering*. Hoboken, New Jersey: John Wiley and Sons Inc., 2009.
7. H.W. Ott. Controlling EMI by proper printed wiring board layout. *Sixth Symposium on EMC*, Zurich, Switzerland, 1985.
8. ANSI C63.4-1992. *Methods of Measurement of Radio-Noise Emissions from Low-Voltage Electrical and Electronic Equipment in the Range of 9 kHz to 40 GHz*, IEEE, July 17, 1992.
9. EN 61000-3-2. *Electromagnetic Compatibility (EMC) – Part 3–2: Limits-Limits for Harmonic Current Emissions (Equipment Input Current < 16 A Per Phase)*, CENELEC, 2006.
10. EN 61000-3-3. *Electromagnetic Compatibility (EMC) – Part 3–3: Limits-Limitation of Voltage Changes, Voltage Fluctuations and Flicker in Public Low-Voltage Supply Systems for Equipment with Rated Current < 16A Per Phase and Not Subject to Conditional Connection*, CENELEC, 2006.
11. EN 61000-4-2. *Electromagnetic Compatibility (EMC) – Part 4–2: Testing and Measurement Techniques – Electrostatic Discharge Immunity Test*, 2001.
12. MDS-201-0004. *Electromagnetic Compatibility Standards for Medical Devices*, U.S. Department of Health Education and Welfare, Food and Drug Administration, October 1, 1979.
13. MIL-STD-461E. *Requirements for the Control of Electromagnetic Interference Characteristics of Subsystems and Equipment*, August 20, 1999.
14. RTCA/DO-160E. *Environmental Conditions and Test Procedures for Airborne Equipment*, Radio Technical Commission for Aeronautics (RTCA), December 7, 2004.
15. SAE J551. *Performance Levels and Methods of Measurement of Electromagnetic Compatibility of Vehicles and Devices (60 Hz to 18 GHz)*, Society of Automotive Engineers, June 1996.
16. SAE J1113. *Electromagnetic Compatibility Measurement Procedure for Vehicle Component (Except Aircraft) (60 Hz to 18 GHz)*, Society of Automotive Engineers, June 1995.
17. A. Wall. Historical perspective of the FCC rules for digital devices and a look to the future. *IEEE International Symposium on Electromagnetic Compatibility*, August 9–13, 2004.
18. H.W. Denny. *Grounding For the Control of EMI*. Gainesville, VA: Don White Consultants, 1983.
19. W. Boxleitner. *Electrostatic Discharge and Electronic Equipment*. New York: IEEE Press, 1989.
20. D.D. Gerke, and W.D. Kimmel. Designing noise tolerance into microprocessor systems. *EMC Technology*, Mar/Apr 1986.
21. W.D. Kimmel, and D.D. Gerke, Three keys to ESD system design. *EMC Test and Design*, Sept 1993.
22. J.L.N. Violette, ESD case history – Immunizing a desktop business machine. *EMC Technology*, May/June 1986.
23. S.W. Wong. ESD design maturity test for a desktop digital system. *Evaluation Engineering*, Oct 1984.
24. S. Voldman. *ESD: Physics and Devices*. Chichester, England: John Wiley and Sons, Ltd., 2004.
25. S. Voldman. *ESD: Circuits and Devices*. Chichester, England: John Wiley and Sons, Ltd., 2005.
26. S. Voldman. *ESD: Circuits and Devices 2nd Edition*. Chichester, England: John Wiley and Sons, Ltd., 2015.
27. S. Voldman. *ESD: RF Circuits and Technology*. Chichester, England: John Wiley and Sons, Ltd., 2006.
28. S. Voldman. *Latchup*. Chichester, England: John Wiley and Sons, Ltd., 2007.
29. S. Voldman. *ESD: Circuits and Devices*. Beijing, China: Publishing House of Electronic Industry (PHEI), 2008.
30. S. Voldman. *ESD: Failure Mechanisms and Models*. Chichester, England: John Wiley and Sons, Ltd., 2009.
31. S. Voldman. *ESD Basics: From Semiconductor Manufacturing to Product Use*. Chichester, England: John Wiley and Sons, Ltd., 2012.
32. S. Voldman. *Electrical Overstress (EOS): Devices, Circuits, and Systems*. Chichester, England: John Wiley and Sons, Ltd., 2013.
33. S. Voldman. *ESD: Analog Design and Circuits*. Chichester, England: John Wiley and Sons, Ltd., 2014.
34. International Electro-technical Commission (IEC). *IEC 61000-4-2 Electromagnetic Compatibility (EMC): Testing and Measurement Techniques – Electrostatic Discharge Immunity Test*, 2001.
35. E. Grund, K. Muhonen, and N. Peachey. Delivering IEC 61000-4-2 current pulses through transmission lines at 100 and 330 ohm system impedances. *Proceedings of the Electrical Overstress/Electrostatic Discharge (EOS/ESD) Symposium*, 2008; 132–141.

36. IEC 61000-4-2. *Electromagnetic Compatibility (EMC) – Part 4–2: Testing and Measurement Techniques – Electrostatic Discharge Immunity Test*, 2008.
37. IEC 61000-4-5. *Electromagnetic Compatibility (EMC) – Part 4–5: Testing and Measurement Techniques – Surge Immunity Test*, 2000.
38. R. Chundru, D. Pommerenke, K. Wang, T. Van Doren, F.P. Centola, J.S. Huang. Characterization of human metal ESD reference discharge event and correlation of generator parameters to failure levels – Part I: reference event. *IEEE Transactions on Electromagnetic Compatibility*, Vol. 46, (4), Nov 2004; 498–504.
39. K. Wang, D. Pommerenke, R. Chundru, T. Van Doren, F.P. Centola, and J.S. Huang. Characterization of human metal ESD reference discharge event and correlation of generator parameters to failure levels – Part II: correlation of generator parameters to failure levels. *IEEE Transactions on Electromagnetic Compatibility*, Vol. 46, (4), Nov 2004; 505–511.
40. ESD Association. ESD-SP 5.6 – 2008. *ESD Association Standard Practice for the Protection of Electrostatic Discharge Sensitive Items – Electrostatic Discharge Sensitivity Testing – Human Metal Model (HMM) Testing Component Level*, Standard Practice (SP) document, 2008.
41. ANSI/ESD SP5.6 – 2009. *Electrostatic Discharge Sensitivity Testing – Human Metal Model (HMM) – Component Level*, 2009.
42. R. Ashton. Types of electrical overstress protection. AND9009D, On Semiconductor Application Note, http://www.onsemi.com. 2011; 1–13.
43. W. Schossig. Introduction to history of selective protection. PAC Magazine, 2007; 70–74.
44. R. Ashton, and L. Lescouzeres. Characterization of off-chip ESD protection devices. *Proceedings of the Electrical Overstress/Electrostatic Discharge (EOS/ESD) Symposium*, 2008; 21–29.
45. D.J. Pommerenke. System and method for testing the electromagnetic susceptibility of an electronic display unit. U.S. Patent No. 7,397,266, July 8, 2008.
46. D.J. Pommerenke, W. Huang, P. Shao, and J. Xiao. Resonance scanning system and method for test equipment for electromagnetic resonance. U.S. Patent No. 8,143,903, March 27, 2012.
47. K.J. Min, D.J. Pommerenke, G. Muchaidze, and B. Chickhradze. System and method for electrostatic discharge testing of devices under test. U.S. Patent No. 8,823,383, September 2, 2014.
48. D.J. Pommerenke, D. Dickey, J.J. DeBlanc, and V. Tsang. Method and apparatus for controlling electromagnetic radiation emissions from electronic device enclosures. U.S. Patent No. 6,274,807, August 14, 2001.
49. D.J. Pommerenke, D. Dickey, J.J. DeBlanc, V.T. Tam, and K.K. Tang. Method and apparatus for controlling electromagnetic radiation emissions from electronic device enclosures. U.S. Patent No. 6,620,999, September 16, 2003.
50. D.J. Pommerenke, D. Dickey, J.J. DeBlanc, and V. Tsang. Method and apparatus for controlling electromagnetic radiation emissions from electronic device enclosures. U.S. Patent No. 6,825,411, November 30, 2004.
51. P. Tamminen. System level EMC/ESD design – challenges and opportunities. Invited Seminar, International ESD Workshop, Tutzing, Germany, 2012.
52. T. Steinecke, M. Bischoff, F. Brandl et al. Generic IC EMC test specification. Asia-Pacific Symposium on Electromagnetic Compatibility, 2012, 5–8.
53. J. Dunnihoo. System ESD robustness: scanning/analysis. Texas ESD Association Meeting, January 29, 2013.
54. C. Duvvury, and H. Gossner. *System Level ESD Co-Design*. Chichester, England: John Wiley and Sons, Ltd., 2015.
55. Thermo Scientific. http://www.thermofisher.com. Thermo Scientific™ EMCPro Plus™ EMC Test System, 2016.
56. Amber Precision Instruments. http://www.amberpi.com, 2016.
57. Pragma Design. https://www.pragma-design.com, 2016.

Appendix A

Glossary of Terms

Automotive Electronic Council (AEC) A council that addresses electronic issues in the automotive industry.

Air Ionizers An electronic or nuclear device that generates ions from air to be used for dissipation of static charge typically used in manufacturing and assembly environments.

American National Standards Institute (ANSI) The ANSI is a private nonprofit organization that oversees the development of voluntary consensus standards for products, services, processes, systems, and personnel in the United States.

Antistatic Material or coating that prevents static buildup on worksurfaces or materials. An antistatic agent is a compound used for treatment of materials or their surfaces in order to reduce or eliminate buildup of static electricity generally caused by the triboelectric effect.

Audits Business processes review to verify conformance and compliance to ESD procedures and standards.

Cable Discharge Event (CDE) An electrostatic discharge (ESD) event from a cable source.

Cassette Model (CM) A test method whose source is a capacitor network with a 10 pF capacitor. This is also known as the Small Charge Model (SCM) and the "Nintendo model.

Charged Board Event (CBE) A test method for evaluation of the charging of a packaged semiconductor chip mounted on a board, followed by a grounding process. The semiconductor chip is mounted on a board during this test procedure. The board is placed on an insulator during this test.

Charged Device Model (CDM) A test method for evaluation of the charging of a packaged semiconductor chip, followed by a grounding a pin. The semiconductor chip is not socketed but placed on an insulator during the test.

Conductor A material that allows free flow of electrons. Example of conductors includes metal materials such as copper and aluminum. A material whose conductivity that exceeds insulators and semiconductors.

Current Reconstruction Method A test method for evaluation of the current in a signal line reconstructed from the measured electromagnetic (EM) signal.

ESD Testing: From Components to Systems, First Edition. Steven H. Voldman.
© 2017 John Wiley & Sons, Ltd. Published 2017 by John Wiley & Sons, Ltd.

Device Under Test (DUT) A component or device that is being tested in a test system.

Electromagnetic (EM) A field or signal that contains both electric and magnetic fields.

Electromagnetic Compatibility (EMC) A branch of electrical sciences which studies the unintentional generation, propagation, and reception of EM energy. EMC must address both the susceptibility of systems to EM interference and the propagation of EM noise.

Electromagnetic Interference (EMI) An EM disturbance that affects an electrical circuit due to either EM induction or EM radiation emitted from an external source.

Electromagnetic Pulse (EMP) A large EM burst event typically resulting from high energy or nuclear explosions. The resulting rapidly changing electric fields and magnetic fields may couple with electrical/electronic systems to produce damaging current and voltage surges.

Electrical Overcurrent (EOC) An electrical event of either overcurrent that leads to electrical component or electronic system damage and failure.

Electrical Overstress (EOS) An electrical event of either overvoltage or overcurrent that leads to electrical component or electronic system damage and failure.

Electrical Overpower (EOP) An electrical event of overpower that leads to electrical component or electronic system damage and failure.

Electrical Overstress (EOV) An electrical event of overvoltage that leads to electrical component or electronic system damage and failure.

ESD Protective Area (EPA) An area of a room or building where ESD protection precautions and measures were taken.

Electrostatic Discharge (ESD) ESD is a subclass of electrical overstress and may cause immediate device failure, permanent parameter shifts, and latent damage causing increased degradation rate.

Electrostatic Discharge (ESD) Gun An ESD source that contains a metal tip and RC network inside the test source.

Electrostatic Shielding Shielding used in electronic systems to prevent the entry or penetration of EM noise.

Electrostatic Susceptibility The sensitivity of a system to EM interference.

Equipment Under Test (EUT) Equipment or system that is being tested in a test system.

Equipotential A surface where all points on the surface are at the same electrical potential.

Equipotential Bonding A process where two objects are "bonded" whose electrostatic potential is the same to avoid ESD to occur.

ESD Control Program A corporate program or process for addressing ESD issues in manufacturing and handling in a corporation.

Field-Induced Charging Charging process initiated on an object after placement within an electric field. This is also known as Charging by Induction.

Horizontal Coupling Plane (HCP) A surface that is typically a metal where coupling is made through the EM field in a horizontal position.

Human Body Model (HBM) A test method whose source is an RC network with a 100 pF capacitor and 1500 Ω series resistor.

Human Metal Model (HMM) A test method that applies an IEC 61000-4-2 pulse to a semiconductor chip; only external pins exposed to system level ports are tested. Source can be an ESD gun that satisfies the IEC 61000-4-2 standard.

Inductive Charging A charging process that uses an EM field to transfer energy between two objects.

Insulator A material whose conductivity is less than a conductor and a semiconductor (less than 10^{-8} siemens per centimeter). Insulators are used to prevent flow of electrical current.

Integrated Circuit An electrical circuit constructed from semiconductor processing where different electrical components are integrated on the same substrate or wafer.

Ionization A method to generate ions from atoms. Ionization techniques include both electrical and nuclear sources.

Joint Electron Device Engineering Council (JEDEC) The JEDEC Solid State Technology Association, formerly known as the Joint Electron Device Engineering Council (JEDEC), is an independent semiconductor engineering trade organization and standardization body.

Japan Electronic and Information Technology Association (JEITA) An association in Japan that addresses standardization body.

Latchup A process electrical failure occurs in a semiconductor component or power system where a parasitic pnpn (also known as a silicon-controlled rectifier, thyristor, or Shockley diode) undergoes a high-current/low-voltage state. Latchup can lead to thermal failure and system destruction.

Latent Failure Mechanism A failure mechanism where the damage created deviates from the untested or virgin device or system. A latent failure can be a yield or reliability issue.

Machine Model (MM) A test method whose source is a capacitor network with a 200 pF capacitor.

Pin Under Test (PUT) A pin in a printed circuit board (PCB) or a component that is being tested in a test system.

Printed Circuit Board (PCB) A circuit board where the signal lines are printed onto the board and components are mounted onto the board.

Semiconductor A material whose conductivity is between a conductor and an insulator (in the range of 10^3 to 10^{-8} siemens per centimeter). Semiconductors are commonly used in integrated circuit component technology.

Small Charge Model (SCM) A test method whose source is a capacitor network with a 10 pF capacitor.

Socketed Device Model (SDM) A test method for evaluation of the charging of a packaged semiconductor chip, followed by a grounding a pin. The semiconductor chip is socketed during the test.

Static Electricity Electrical charge generated from charging processes that are sustained and accumulated on an object.

Surface Mount Device (SMD) A component that mounts onto a PCB, which mounts directly on the surface.

Surface Resistivity The resistance of a material on its surface (as opposed to a bulk resistivity).

System Efficient ESD Design (SEED) A system level design practice that integrates ESD and system co-design.

System Level IEC 61000-4-2 A system level test that applies a pulse to a system using an ESD gun.

System Level IEC 61000-4-5 A system level test to address a transient surge that applies a pulse to a system.

Transient Voltage Suppression (TVS) A component or device that addresses electrical overstress by voltage suppression.

Transmission Line Pulse (TLP) A test method that applies a rectangular pulse to a component (10 ns rise and fall time; 100 ns plateau)

Triboelectric Charging A method that generates charging through contact electrification. Contact electrification is when certain materials become electrically charged after they come into contact with another different material and are then separated (such as through rubbing). The polarity and strength of the charges produced differ according to the materials, surface roughness, temperature, strain, and other properties.

Triboelectric Series The ordering of materials according to their triboelectric behavior. Materials are often listed in order of the polarity of charge separation when they are touched with another object. A material the bottom of the series, when touched to a material near the top of the series, will attain a more negative charge, and vice versa. Tribo is from the Greek word for "rubbing," τρίβω (τριβή: friction).

Ultra-Fast Transmission Line Pulse (UF-TLP) A test method that applies a fast pulse to a component on the order of tens of picoseconds. This is presently not a standard.

Vertical Coupling Plane (VCP) A surface that is typically a metal where coupling is made through the EM field in a vertical position.

Very Fast Transmission Line Pulse (VF-TLP) A test method that applies a rectangular pulse to a component (1 ns rise and fall time; 10 ns plateau).

Appendix B

Standards

B.1 ESD Association

ANSI/ESD S1.1 – 2006 Wrist Straps
ESD DSTM2.1 Garments
ANSI/ESD STM3.1 – 2006 Ionization
ANSI/ESD SP3.3 – 2006 Periodic Verification of Air Ionizers
ANSI/ESD STM4.1 – 2006 Worksurfaces – Resistance Measurements
ANSI/ESD STM3.1 – 2006 ESD Protective Worksurfaces – Charge Dissipation Characteristics
ANSI/ESD STM5.1 – 2007 Electrostatic Discharge Sensitivity (ESDS) Testing – Human Body Model (HBM) Component Level
ANSI/ESD STM5.1.1 – 2006 Human Body Model (HBM) and Machine Model (MM) Alternative Test Method: Supply Pin Ganging – Component Level
ANSI/ESD STM5.1.2 – 2006 Human Body Model (HBM) and Machine Model (MM) Alternative Test Method: Split Signal Pin – Component Level
ANSI/ESD S5.2 – 2006 Electrostatic Discharge Sensitivity Testing – Machine Model (MM) Component Level
ANSI/ESD S5.3.1 – 2009 Charged Device Model (CDM) – Component Level
ANSI/ESD SP5.3.2 – 2008 Electrostatic Discharge Sensitivity Testing – Socketed Device Model (SDM) Component Level
ANSI/ESD STM5.5.1 – 2008 Electrostatic Discharge Sensitivity Testing – Transmission Line Pulse (TLP) Component Level
ANSI/ESD SP5.5.2 – 2007 Electrostatic Discharge Sensitivity Testing – Very Fast Transmission Line Pulse (VF-TLP) Component Level
ANSI/ESD SP6.1 – 2009 Grounding
ANSI/ESD S7.1 – 2005 Resistive Characterization of Materials – Floor Materials
ANSI/ESD S8.1 – 2007 Symbols – ESD Awareness
ANSI/ESD STM9.1 – 2006 Footwear – Resistive Characterization
ESD SP9.2 – 2003 Footwear – Foot Grounders Resistive Characterization
ANSI/ESD SP10.1 – 2007 Automatic Handling Equipment (AHE)

ESD Testing: From Components to Systems, First Edition. Steven H. Voldman.
© 2017 John Wiley & Sons, Ltd. Published 2017 by John Wiley & Sons, Ltd.

ANSI/ESD STM11.11 – 2006 Surface Resistance Measurement of Static Dissipative Planar Materials
ESD DSTM11.13 – 2009 Two Point Resistance Measurement
ANSI/ESD STM11.31 – 2006 Bags
ANSI/ESD STM12.1 – 2006 Seating – Resistive Measurements
ESD STM13.1 – 2000 Electrical Soldering/Desoldering Hand Tools
ANSI/ESD SP14.1 System Level Electrostatic Discharge (ESD) Simulator Verification
ESD SP14.3 – 2009 System Level Electrostatic Discharge (ESD) Measurement of Cable Discharge Current
ANSI/ESD SP15.1 – 2005 In Use Resistance Testing of Gloves and Finger Cots
ANSI/ESD S20.20 – 2007 Protection of Electrical and Electronic Parts, Assemblies, and Equipment
ANSI/ESD STM97.1 – 2006 Floor Materials and Footwear – Resistance Measurements in Combination with a Person

B.2 International Organization of Standards

ISO Standard 10605 Road Vehicles – Test Methods for Electrical Disturbances from Electrostatic Discharge – 2008

B.3 IEC

IEC Medical Electrical Equipment – Part 1–2: General Requirements for Basic Safety and Essential Performance – Collateral Standard: Electromagnetic Disturbances – Requirement and Tests, 2007

B.4 RTCA

RTCA DO-160 Section 25 Environmental Conditions and Test Procedures for Airborne Equipment, DO-160F, 2007

B.5 Department of Defense

DOD HDBK 263 Electrostatic Discharge Control Handbook for Protection of Electrical and Electronic Parts, Assemblies and Equipment
DOD-STD-1686 Electrostatic Discharge Control Program for Protection of Electrical and Electronic Parts, Assemblies and Equipment
DOD-STD-2000-2A Part and Component Mounting for High Quality/High Reliability Soldered Electrical and Electronic Assembly
DOD-STD-2000-3A Criteria for High Quality/High Reliability Soldering Technology
DOD-STD-2000-4A General Purpose Soldering Requirements for Electrical and Electronic Equipment
FED Test Method STD-101 – Method 4046 – Electrostatic Properties of Materials

B.6 Military Standards

MIL-STD-454 Standard General Requirements for Electronic Equipment

MIL-STD-785 Reliability Program for System and Equipment Development and Production
MIL-STD-883 – Method 3015-4 – Electrostatic Discharge Sensitivity Classification
MIL-STD-1686A Electrostatic Discharge Control Program for Protection of Electrical and Electronic Parts, Assemblies and Equipment
MIL-E-17555 Electronic and Electrical Equipment, Accessories, and Provisioned Items (Repair Parts: Packaging of)
MIL-M-38510 Microcircuits, General Specification for
MIL-D-81705 Barrier Materials, Flexible, Electrostatic Free, Heat Sealable
MIL-D-81997 Pouches, Cushioned, Flexible, Electrostatic Free, Reclosable, Transparent
MIL-D-82646 Plastic Film, Conductive, Heat Sealable, Flexible
MIL-D-82647 Bags, Pouches, Conductive, Plastic, Heat Sealable, Flexible
IEC 801-2 Electromagnetic Compatibility for Industrial Process Measurements and Control Equipment, Part 2: Electrostatic Discharge (ESD) Requirements
EIA-541 Packaging Material Standards for ESD Sensitive Materials
JEDEC 108 Distributor Requirements for Handling Electrostatic Discharge Sensitive (ESDS) Devices

B.7 Airborne Standards and Lightning

DO-160E Environmental Conditions and Test Procedures for Airborne Equipment
SAE ARP5412 Aircraft Lightning Environment and Related Test Waveforms
EUROCAE/ED-14E Environmental Conditions and Test Procedures for Airborne Equipment

Index

ESD Testing: From Components to Systems, First Edition. Steven H. Voldman.
© 2017 John Wiley & Sons, Ltd. Published 2017 by John Wiley & Sons, Ltd.